George Facer

A2 Chemistry

2nd Edition

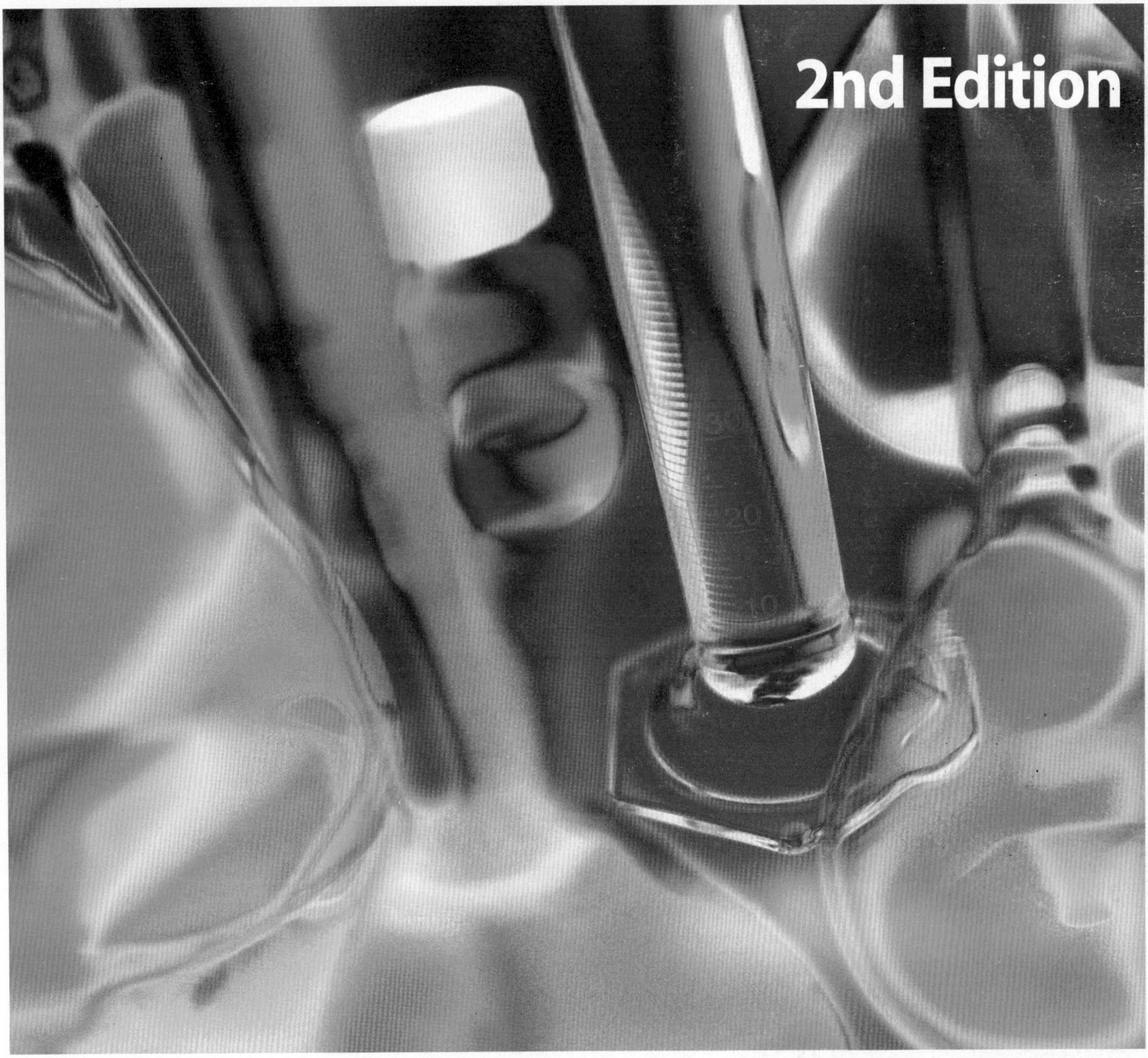

This book is dedicated to Judy, my wife of 43 years, who has supported me throughout the research and writing of this textbook.

Philip Allan Updates, an imprint of Hodder Education, an Hachette UK company, Market Place, Deddington, Oxfordshire OX15 0SE

Orders
Bookpoint Ltd, 130 Milton Park, Abingdon, Oxfordshire OX14 4SB
tel: 01235 827827 fax: 01235 400401
e-mail: education@bookpoint.co.uk
Lines are open 9.00 a.m.–5.00 p.m., Monday to Saturday, with a 24-hour message answering service. You can also order through the Philip Allan Updates website: www.philipallan.co.uk

ISBN 978-0-340-95761-5

First printed 2009
Impression number 7
Year 2013 2012

This textbook has been written specifically to support students studying Edexcel A2 Chemistry. The content has been neither approved nor endorsed by Edexcel and remains the sole responsibility of the author.

All efforts have been made to trace copyright on items used.
Front cover photograph reproduced by permission of Science Photo Library.

Design by Juha Sorsa
Printed in Dubai

Hachette UK's policy is to use papers that are natural, renewable and recyclable products and made from wood grown in sustainable forests. The logging and manufacturing processes are expected to conform to the environmental regulations of the country of origin.

Contents

Introduction v
- Onine resources v
- Required previous knowledge and skills v
- Scheme of assessment vi
- Examination technique vii

Unit 4 Rates, equilibria and further organic chemistry

Chapter 1 Rates: how fast? 2
- Introduction 2
- Experimental methods 4
- Rate equations 8
- Effect of temperature and a catalyst on the rate constant 15
- Mechanisms and the rate-determining step 18

Chapter 2 Entropy: how far? 30
- Introduction 30
- Entropy change of the system 31
- Entropy change of the surroundings 33
- Total entropy change 34
- Solubility 37
- Melting and boiling points 48

Chapter 3 Equilibrium 51
- Introduction 51
- The equilibrium constant, K_c 52
- The equilibrium constant in terms of partial pressure, K_p 60
- Calculations using K_c and K_p 64

Chapter 4 Application of rates and equilibrium 73
- Predicting the direction of change 73
- Altering equilibrium conditions 74

Chapter 5 Acid–base equilibria 88
- Introduction 88
- Acid–base conjugate pairs 88
- Strong and weak acids and bases 90
- Acid and base equilibrium constants 90
- Auto-ionisation of water and the pH scale 91
- The pH of acids, bases and salts 94
- Buffer solutions 104
- Acid–base indicators 109
- Titration curves 111
- Enthalpy of neutralisation of acids 115
- The structure of acids 116

Chapter 6 Isomerism 120
- Structural isomerism 120
- Stereoisomerism 121

Chapter 7 Carbonyl compounds 130
- Introduction 130
- Physical properties 132
- Laboratory preparation 133
- Reactions of aldehydes and ketones 134

Chapter 8 Carboxylic acids and their derivatives 144
- Introduction 144
- Carboxylic acids and their derivatives in nature 145
- Carboxylic acids 145
- Esters 151
- Acid chlorides 157

Chapter 9 Spectroscopy and chromatography 162
- Introduction 162
- Emission spectra 163
- Absorption spectra 164
- Infrared spectra 165
- Microwaves 168
- Radio waves (NMR) 168
- Mass spectroscopy 174
- Chromatography 175

Unit 5 Transition metals, arenes and organic nitrogen chemistry

Chapter 10 Electrochemistry and redox equilibria 180
Introduction 180
Redox as change in oxidation number 181
Redox as electron transfer 182
Electrode potentials 183
Evaluation of experimental results compared with calculated $E^{\ominus}_{cell}$ values 192
Practical aspects of electrochemistry 195
Oxidising and reducing agents 199
Uncertainty of measurement 204

Chapter 11 Transition metals and the *d*-block elements 208
Introduction 208
Common chemical properties of transition metals 212
Common physical properties of transition metal ions 218
Reactions of *d*-block ions with particular reference to chromium and copper 221
Uses of some *d*-block metals and their compounds 231

Chapter 12 Arenes and their derivatives 236
Introduction 236
Benzene, C_6H_6 241
Phenol, C_6H_5OH 249
Benzoic acid, C_6H_5COOH 254
2-methylnitrobenzene, $CH_3C_6H_4NO_2$ 255

Chapter 13 Organic nitrogen compounds 258
Amines 258
Phenylamine, $C_6H_5NH_2$ 262
Benzenediazonium compounds 264
Amides 266
Natural products containing the amine group 271
Amino acids 272
Proteins 276
Nitriles 277

Chapter 14 Organic analysis and synthesis 281
Introduction 281
Organic analysis 282
Methods of separation and purification 292
Organic synthesis 296
Yields 304
Stereochemistry of reactions 304
Safety issues 306
Combinatorial chemistry and pharmaceuticals 307

Index 314

Periodic table 326

Introduction

This textbook covers the Edexcel specification for A2 chemistry. Generally, the order follows that of the specification. To give the student a better understanding of, and feel for, some topics, the content of the book occasionally goes beyond the confines of the A2 specification.

The first nine chapters describe and explain the material of **Unit 4: Rates, equilibria and further organic chemistry**. The following eight chapters cover the material of **Unit 5: Transition metals and organic nitrogen chemistry**.

Margin comments are provided throughout the book. These comprise valuable reminders and snippets of information and include examiner's tips (indicated by an e symbol), which clarify what you need to know and common sources of confusion.

This book is not a guide to the practical chemistry that all A2 candidates will study. However, many of the reactions that will be met in the laboratory are detailed throughout the book. At the back of the book (p. 234), there is a periodic table. This should be referred to for atomic numbers, atomic masses and symbols of the elements. The table is similar to the one printed on the back of the examination papers and has relative atomic masses to one decimal place.

Online resources

To access your free online resources go to **www.hodderplus.co.uk/philipallan**. These resources comprise:

- end-of-chapter multiple-choice tests for students to print off and complete
- practice exam-style unit tests
- exam-style questions with graded student answers, supported by examiner's comments

Required previous knowledge and skills

It is assumed that all A2 chemistry students have successfully completed the AS course. All students should be:

- familiar with the use of a calculator
- able to change the subject of an algebraic equation
- able to draw straight-line and curved graphs from supplied data and to extrapolate graphs
- able to draw tangents to curves and to calculate the slope of the tangent and of straight-line graphs
- confident in the use of scientific (standard) notation, for example that the number 1234 can be written as 1.234×10^3 and that 1.234×10^{-3} is the same as 0.001234

Scheme of assessment

Assessment objectives

AO1: knowledge with understanding of science and How Science Works

Candidates should be able to remember specific chemical facts, such as reactions, equations and conditions. They should be able to use correct chemical terminology. This skill is primarily one of factual recall, which many students find difficult. It is a skill that needs much practice.

AO2: application of knowledge and understanding, analysis and of How Science Works

Candidates have to be able to:

- explain and interpret chemical phenomena
- select and use data presented in the form of continuous prose, tables or graphs
- carry out calculations
- apply chemical principles to compounds similar to those covered by the specification
- assess the validity and accuracy of chemical experiments and suggest improvements

AO3: How Science Works

There will be no questions that candidates will be able to recognise as being 'How Science Works'. In the theory papers, examples of this assessment objective include questions on the interpretation of data and planning a multi-step synthesis. AO3 is also examined in the internal assessment of practical work.

Details of this assessment objective can be found in Appendix 1 of the Edexcel specification.

The unit tests

Candidates take two theory papers. Each lasts 1 hour and 40 minutes and has a maximum of 90 marks available. They can be taken more than once, with the best scores of the AS and A2 papers counting towards the A-level grade.

Quality of written communication is tested in both Unit Tests 4 and 5. There are no separate marks for this, but candidates will only be awarded the 'chemistry' marks if their explanations have been communicated clearly. Questions testing this are marked with an asterisk, *.

The Edexcel data booklet is needed for Unit Tests 4 and 5. A clean copy must be available for each candidate.

In both Unit Tests 4 and 5, Section A consists of about 20 multiple-choice questions. In each question, four options, A to D, are offered as answers; only one of these options is correct.

In Unit Tests 4 and 5, Section B consists of a mixture of short-answer questions and some questions that require longer and well expressed answers.

Section C in Unit Test 4 is based on data. Candidates are expected to use data supplied in the question and data obtained from the data booklet.

Section C in Unit Test 5 starts with a passage that the candidate has to read; some of the questions in this section relate directly to the passage. Other parts of the question do not require an understanding of the passage but are loosely connected to it.

In Unit Test 4, students are expected to draw on their knowledge of relevant material from Units 1 and 2; in Unit Test 5 knowledge from Unit 4 is also required. In this way, both unit tests have an element of synopticity.

Sections B and C of both tests contain some challenging questions to differentiate between the able and the exceptionally able. This is what QCA has called 'stretch and challenge'.

The marks from both units (and from the practical assessments in Unit 6) are converted to uniform marks and then added to the uniform mark scale (UMS) marks from AS to determine the A-level grade.

The A-grade boundary for each paper will vary from year to year as the challenge of the paper varies. As a guide, a mark well into the seventies is likely to be awarded an A grade and then be scaled to 80% UMS.

An A* grade will be awarded to those candidates who do particularly well in Unit Tests 4 and 5. A total UMS mark of over 90% will be required.

Examination technique

Mark allocation

In both A2 papers the marks for each part of the question are given in brackets. This is a much better guide as to how much to write than the number of dotted lines provided for the answer. If there are 2 marks, two statements must be made. For example, if the question asks for the conditions for a particular reaction and there are 2 marks available, there must be two different conditions given, such as solvent, temperature or catalyst.

Alternative answers

Do *not* give alternative answers. If one of them is wrong, the examiner will not award any marks for this part of the question. If both answers are correct, you *would* score the mark. However, there is no point in risking one answer being wrong. Beware also of contradictions, such as giving the reagent as concentrated sulfuric acid and then writing $H_2SO_4(aq)$ in an equation.

Writing your answers

In Edexcel A2 chemistry exams, the answers are written in the spaces on the question paper. If part of your answer is written elsewhere on the page, alert the examiner by writing, for example, 'see below' or 'continued on page 5'. Exam papers will be marked online, so question papers and answers will be electronically scanned. For this reason, it is *essential not to write outside* the borders marked on the page.

Multiple-choice questions

You must put a cross in the box corresponding to your chosen answer. If you change your mind, put a horizontal line through the cross and then mark your new answer with a cross. Be careful about negative questions — e.g. 'Which is **not**...'.

In numerical questions, you should just work out the answer. Otherwise, read *all* the options. This will help you not to be misled by a half-correct response.

Writing a tick or cross after the end of each response could help to focus your mind. If you find that you have two ticks, then think about which is really the correct answer.

If you are having difficulty with a question, put a large ring around the question number, leave it and go on to the next question. If you have time, come back when you have finished Sections B and C. Remember that you should not spend more than 30 minutes on the multiple-choice section, i.e. one question every 90 seconds. Some questions will take less time than this and others will take longer, especially if a calculation is involved.

Correction fluid and red pens or pencils

Do not use any of these. Mistakes should be crossed out neatly before writing the new answer. Red ink and pencil will not be seen when the paper is scanned electronically.

Command words

It is important that you respond correctly to key words or phrases in the question.

- **Define** — definitions of important terms such as relative atomic mass or standard enthalpy of formation are frequently asked for. You must know these definitions. They are printed in red in this book.
- **Name** — give the full name of the substance, not its formula.
- **Identify** — give either the name or the formula.
- **Write the formula** — a molecular formula, such as C_2H_5Cl, will suffice, as long as it is unambiguous. It is no use writing C_2H_4O for the formula of ethanal, or C_3H_7Br for the formula for 2-bromopropane. This also applies to equations. For example, the equation $C_2H_4 + Br_2 \rightarrow C_2H_4Br_2$ would not score a mark, because the formula $C_2H_4Br_2$ is ambiguous.
- Draw or write the structural formula — this must clearly show any double bonds and the position of the functional group. For example, CH_3COOH is not acceptable as the structural formula of ethanoic acid. It must be written as:

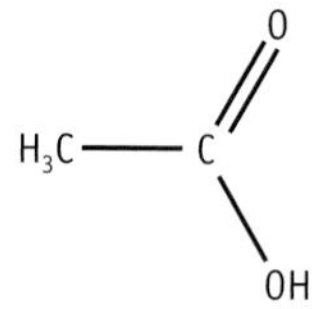

O
//
H_3C—C
\
OH

- An acceptable structural formula of but-1-ene is $H_2C{=}CHCH_2CH_3$.
- **Draw the full structural formula** — all the atoms and all the bonds in the molecule *must* be shown. For example, the full structural formula of ethanoic acid is:

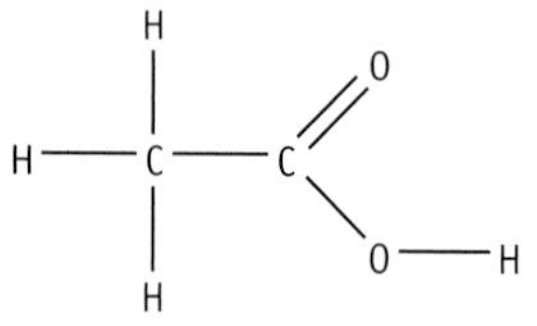

H
|
H—C—C=O
| \
H   O—H

- **State** — give the answer without any explanation. For example, if asked to state in which direction the position of equilibrium moves, the answer is simply 'to the left' or 'to the right'.
- **State, giving your reasons** — this is a difficult type of question. First, look at the mark allocation. Then state the answer (1 mark) and follow this with an explanation containing enough chemical points to score the remaining marks.
- **Explain** — look at the mark allocation and then give the same number of pieces of chemical explanation, or even one extra. For example, in answer to the question 'explain why but-2-ene has two geometric isomers (2)', the first point is that there is restricted rotation about the double bond and the second point is that there are two different groups on each double-bonded carbon atom.
- **Deduce** — the data supplied in the question, or an answer from a previous part of the question, are used to work out the answer. The data could be numerical or they could be the results of qualitative tests on an unknown substance. Alternatively, knowledge from another part of the specification may be needed to answer a question about a related topic or similar substance.
- **Suggest** — candidates are not expected to know the answer. Knowledge and understanding of similar compounds have to be used to work out (deduce) the answer. For example, the shape of SF_6 is covered in the specification, so students should be able to deduce the shape of the PCl_6^- ion. Alternatively, the question might ask candidates to suggest the identity of an organic compound because there are not sufficient data to decide between two possible isomers.
- **Compare or explain the difference between** — valid points must be made about both substances. For example, if the question asks for an explanation for the difference in the boiling points of hydrogen fluoride and hydrogen chloride, the different types and strengths of intermolecular forces in both substances must be described, together with an explanation of what causes these differences.
- **Calculate** — it is essential to show all working. Calculations at A2 are not structured in the same way as they are at AS. This means that you must make clear what you are doing at each stage. If you make a mistake early in the calculation, you could be awarded marks for the subsequent steps. However, this depends on the examiner being able to follow your working. An answer without working will score a maximum 1 mark. Always give your final answer to the number of significant figures justified by the number of significant figures in the data.
- **Identify the reagent** — give the full name or formula. Answers such as 'acidified dichromate' or 'OH^- ions' do not score full marks.
- **State the conditions** — do not automatically write down 'heat under reflux'. The answer might be the necessary solvent (e.g. ethanol for the elimination of HBr from bromoalkanes) or a specific catalyst (e.g. platinum or nickel in the addition of hydrogen to alkenes). If a concentrated acid is needed in a reaction, this must be stated. In the absence of any knowledge of the reaction, then try 'heat under reflux' — it might be correct!

Equations

- Equations must always be balanced. Word equations never score any marks.
- Ionic equations and half-equations must also balance for charge.

- State symbols must be included:
  - if the question asks for them
  - in all thermochemical equations
- The use of the symbols [O] and [H] in organic oxidation and reduction reactions, respectively, is acceptable. Equations using these symbols must still be properly balanced.
- Organic formulae used in equations must be written in such a way that their structures are unambiguous.

Stability

'A secondary carbocation is stable' has no meaning. 'Stability' must be used only when comparing two states or two sets of compounds. You have to know and understand the difference between thermodynamic stability (ΔS and E_{cell}) and kinetic stability (activation energy and rate of reaction).

Graphs

Normally, there is a mark for labelling the axes. When sketching a graph, make sure that any numbers are on a linear scale. The graph should start at the right place, have the correct shape and end at the right place. An example is the Maxwell–Boltzmann distribution, which starts at the origin, rises in a curve to a maximum and tails off as an asymptote to the x-axis.

Diagrams of apparatus

Make sure that a flask and condenser are not drawn as one continuous piece of glassware. The apparatus must work. There must be an outlet to the air somewhere in the apparatus. In distillation, the top should be closed and the outlet should be at the end of the condenser. For heating under reflux, the top of the condenser must be open. Never draw a Bunsen burner as the heater. It is always safer to draw an electrical heater, in case one of the reagents is flammable.

Safety

A statement that laboratory coats or safety glasses must be worn will not score a mark. Safety issues must be linked to particular dangers associated with the chemicals referred to in the question.

Read the question

Questions are often very similar to, but slightly different from, those previously asked. Make sure that you answer this year's question, not last year's! Look for the words 'using your answer to...' or 'hence...'. For example, if you have been asked to calculate oxidation numbers and are then asked to 'hence explain why this is a redox reaction', your answer must be in terms of changes in oxidation number and not in terms of loss or gain of electrons.

Quality of written communication

The most important thing is for you to convey the meaning clearly, accurately and in a logical order. Minor spelling and punctuation errors will not be penalised as long as they do not distort the meaning. Note the subtle difference between 'more successful collisions' (a total number) and 'more of the collisions are successful' (a proportion of the number).

George Facer

Unit 4

Rates, equilibria and further organic chemistry

Chapter 1

Rates: how fast?

Introduction

The rate of a chemical reaction is the rate of change of concentration of a reactant or product with time.

Its units are mol dm^{-3} s^{-1}.

The *average* rate of reaction is defined as:

$$\text{rate} = \frac{\Delta\text{concentration}}{\Delta\text{time}}$$

where Δconcentration is the change in concentration of a reactant or product and Δtime is the time over which this change takes place.

This is only a reasonable assumption if the concentration of a reactant has fallen by less than 10% during the time elapsed.

e The value of Δ[reactant] is negative because its concentration decreases with time. Therefore, a more correct definition of rate is:

$$\text{rate} = \frac{-\Delta[\text{reactant}]}{\Delta\text{time}}$$

or

$$\text{rate} = \frac{+\Delta[\text{product}]}{\Delta\text{time}}$$

This gives a positive value for the rate in both cases.

Required AS chemistry

Collision theory

For a reaction to take place, reactant molecules must collide:

- with kinetic energy greater than or equal to the activation energy of the reaction
- with the correct orientation

Maxwell–Boltzmann distribution of energy

The molecules in a gas or liquid and the molecules or ions in a solution move at different speeds. They possess different amounts of kinetic energy. This is shown by the blue line in Figure 1.1. The total number of molecules with energy equal to or greater than a particular energy value is given by the area under the graph to the *right* of that energy. Thus the blue area to the right of the **activation energy**, E_a, is the fraction of molecules that have sufficient energy (at temperature T_1) to react on collision, providing that the orientation of collision is correct.

Effect of temperature on rate

When the temperature is increased, the molecules or ions gain kinetic energy. They have a greater range of energies (greater entropy) and the average energy is increased. This means that the peak of the Maxwell–Boltzmann distribution is lowered and moved to the right. This is shown by the red line (at temperature T_2) in Figure 1.1. The red area to the right of the activation energy is greatly increased because a much greater *proportion* of the colliding molecules have energy greater than or equal to the activation energy. Therefore, a greater proportion of collisions will result in reaction.

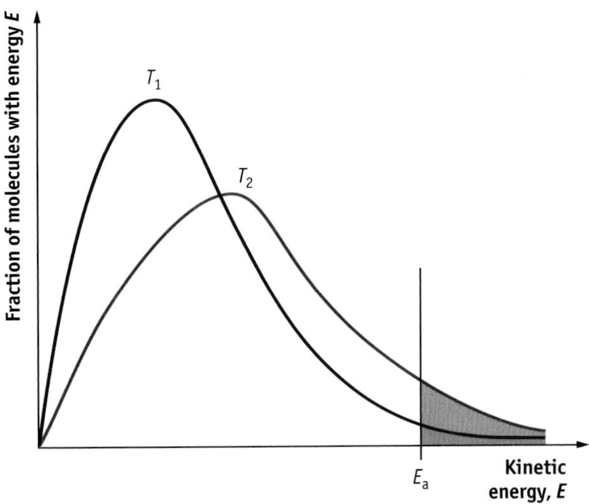

Figure 1.1 Maxwell–Boltzmann distribution of kinetic energy

A reaction that takes place fairly quickly at room temperature has an activation energy of about 60 kJ mol^{-1}. This means that less than one in a billion collisions will have the necessary energy for a reaction to take place.

An approximate guide is that the rate doubles for a 10 K increase in temperature. The magnitude of the effect of increasing temperature depends on the value of the activation energy. A rise from 298 K to 308 K will cause the rate to increase by a factor of:

$$\frac{e^{(-E_a/308R)}}{e^{(-E_a/298R)}}$$

where R is the gas constant (8.13 J K^{-1} mol^{-1}) and E_a is the activation energy for the reaction.

If E_a = 60 kJ mol^{-1}, the rate increases by a factor of 1.96 ≈ 2 or 100%.

An increase in temperature also increases the average speed of the molecules and so increases the collision *frequency*. For a 10 K rise from 298 K, this increases the rate by a factor of 1.02 (2%). This is negligible compared with the increase in rate caused by the increased proportion of collisions that result in reaction.

Effect of pressure on the rate of a gaseous reaction

If the pressure on a gaseous system is increased at constant temperature, the molecules become packed more closely together. There is no change in their speed or energy, but the collision *frequency* increases. The same proportion of the collisions results in reaction. However, because the frequency of collisions increases, the rate of reaction also goes up.

The situation is different if a gas is reacting with a solid, such as a catalyst. The surface area of the solid is usually the limiting factor, so the rate is independent of the pressure of the gas (p. 25).

Experimental methods

The rate of a reaction cannot be measured directly. It can only be determined from concentration and time data. There are a number of methods for 'following' a reaction which enable these data to be measured.

Titration

If the concentration of a reactant or product can be estimated by a titration, the reaction can be followed using this technique:

- Measure out samples of the reactants with known concentration.
- Mix them together, start a clock and stir the mixture thoroughly.
- At regular time intervals, withdraw samples using a pipette and quench (stop) the reaction. Quenching can usually be achieved either by adding the solution from the pipette to ice-cold water or to a solution that reacts with one of the reactants, to prevent further reaction from taking place. The time at which half the contents of the pipette have been added to the quenching solution is noted.
- The quenched solution is then titrated against a suitable standard solution.

The titre is proportional to the concentration of the reactant or product being titrated.

This method can be used when an acid, alkali or iodine is a reactant or product. Acids can be titrated with a standard alkali, alkalis with a standard acid and iodine with a standard solution of sodium thiosulfate.

e Care must be taken that the quenching reagent does not react with one of the reagents to give the same product that is being measured. For example, the acid hydrolysis of an ester cannot be quenched by adding alkali because this would react with the ester and increase the amount of product. It is always safer to quench with iced water. Sodium hydrogencarbonate will remove acid without making the solution alkaline.

Worked example

In the presence of an acid catalyst, aqueous solutions of iodine and propanone react according to the equation:

$$CH_3COCH_3 + I_2 \rightarrow CH_2ICOCH_3 + HI$$

Describe a method to find how the concentration of iodine varies with time in this reaction.

Answer

- Place 25 cm^3 of propanone solution in a beaker, followed by 25 cm^3 of dilute sulfuric acid.
- Place 25 cm^3 of iodine solution of known concentration in a second beaker.
- Simultaneously, mix the two solutions and start a clock.
- Stir the mixture thoroughly. After 5 minutes remove 10 cm^3 of the solution in a pipette and run it into a cold sodium hydrogencarbonate solution. Note the time when half the liquid in the pipette has run into the sodium hydrogencarbonate solution.
- Titrate the iodine present with standard sodium thiosulfate solution, adding starch when the iodine colour has faded to a straw colour.
- Stop when the blue-black colour of the starch–iodine complex has vanished. Read the burette volume.
- Repeat the process every 5 minutes until there is no solution left.

The concentration of iodine is proportional to the volume of sodium thiosulfate solution required to decolorise the iodine.

In the experiment described in the worked example above, in order to reduce the number of variables, it is usual to have the concentrations of the acid and the propanone about ten times greater than the concentration of iodine. Then, the only two variables are time and the concentration of iodine. The concentrations of the acid and the propanone remain approximately constant during the experiment.

Colorimetry

If a reactant or product is coloured, the concentration of the coloured species can be measured using a spectrophotometer. The amount of light of a particular frequency that is absorbed depends on the concentration of the coloured substance.

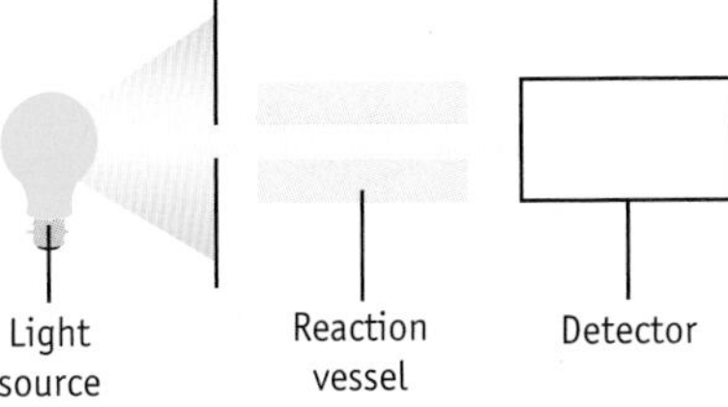

Figure 1.2
A spectrophotometer

The reactants are mixed and a clock started. The light absorbed is measured at set time intervals.

A suitable example is the reaction between bromine and methanoic acid:

$$Br_2(aq) + HCOOH(aq) \rightarrow 2Br^-(aq) + CO_2(g) + 2H^+(aq)$$

If this reaction is done in a beaker, the colour of bromine can be seen to fade gradually. A spectrophotometer is used to follow the absorption of light by bromine.

The fading colour of bromine as it reacts with methanoic acid

Infrared spectroscopy

Infrared spectroscopy can be used in a similar way to colorimetry. The spectrometer is set at a particular frequency and the amount of infrared radiation absorbed at that frequency is measured at regular time intervals. The oxidation of propan-2-ol to propanone by acidified potassium dichromate(VI) can be followed by setting the spectrometer at $1700\,cm^{-1}$ (the absorption frequency due to the stretching of the C=O bond) and measuring the increase in absorption as the CHOH group is oxidised to the C=O group.

Polarimetry

If a reactant is optically active and the product either has a different optical activity or is a racemic mixture, the reaction can be followed by measuring the extent to which the plane of polarisation of plane-polarised light is rotated.

The reaction mixture is placed in a cell in the polarimeter and the angle of rotation is measured at regular time intervals. The angle of rotation is proportional to the concentration of the optically active substance.

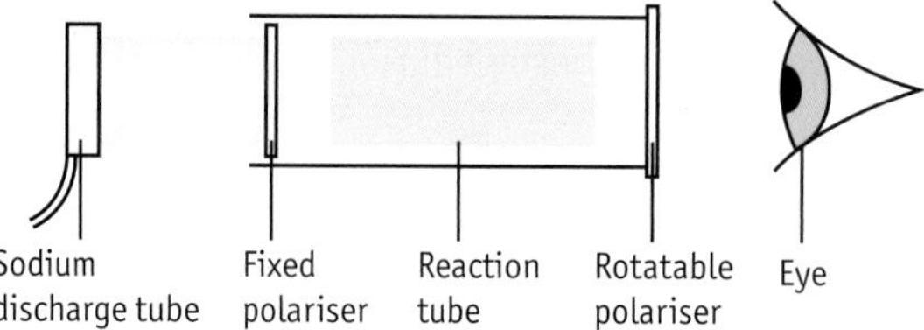

Figure 1.3 *A polarimeter*

If a sample of one of the optical isomers of 2-iodobutane is mixed with aqueous sodium hydroxide solution, hydrolysis occurs. The product is the racemic mixture of butan-2-ol. The angle of rotation of the plane of polarisation of the plane-polarised light gradually decreases as the single chiral isomer of 2-iodobutane is hydrolysed.

The rate of the hydrolysis of sucrose by the enzyme invertase can also be studied using a polarimeter. The products are fructose and glucose. Sucrose is dextrorotatory (rotates the plane clockwise) and the final mixture is laevorotatory (rotates the plane counter clockwise).

$$\underset{\text{sucrose}}{C_{12}H_{22}O_{11}} + H_2O \rightarrow \underset{\text{glucose}}{C_6H_{12}O_6} + \underset{\text{fructose}}{C_6H_{12}O_6}$$

Volume of gas evolved

If the reaction produces a gas, the volume of gas produced can be measured at regular time intervals. The volume of gas is proportional to the moles of gas and can, therefore, be used to measure the concentration of the product.

The rate of the reaction of an acid with a solid carbonate can be studied this way. The acid is added to the carbonate and the volume of carbon dioxide noted every 30 seconds.

$$CaCO_3(s) + 2H^+(aq) \rightarrow Ca^{2+}(aq) + H_2O(l) + CO_2(g)$$

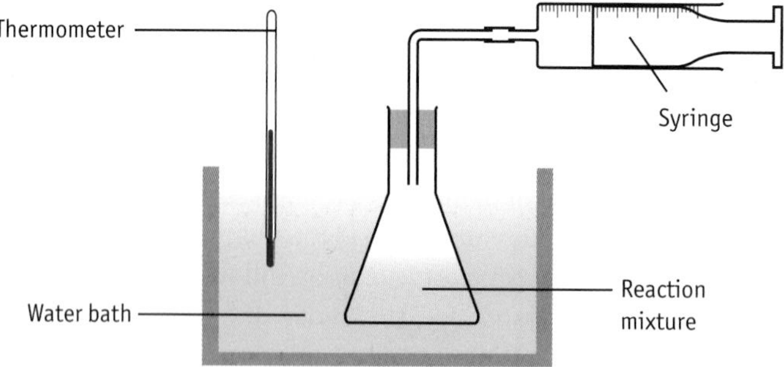

Figure 1.4 *Measuring the volume of carbon dioxide produced in a reaction*

This method can be modified by measuring the loss of mass as the gas produced escapes from the solution. The reactants are mixed and placed on a top-pan balance. The mass is measured at set times.

The problem with this method is that the changes in mass are very small. 50 cm^3 of carbon dioxide weighs less than 0.1 g at room temperature and pressure and 50 cm^3 of hydrogen weighs less than 0.005 g. Very sensitive balances are needed to measure the mass change accurately.

The assumption is often made that the rate of reaction is proportional to 1/time for a certain volume of gas to be produced. This is only an acceptable approximation if less than 10% of the acid is used up.

Time for the reaction to finish

An approximate method is to add the two reactants and time how long it takes for the reaction to stop. An example would be to add a strip of magnesium to an excess of dilute sulfuric acid and time how long it takes for the production of hydrogen bubbles to stop. The experiment is repeated with either a different concentration of the acid or at a different temperature. The assumption is then made that the rate is proportional to 1/time. However, this is only valid if the concentration of the acid has fallen by less than 10–15% and if the temperature did not change by more than 1°C during the measurement.

'Clock' reactions

In a 'clock' reaction, the reactants are mixed and the time taken to produce a fixed amount of product is measured. The experiment is then repeated several times using different starting concentrations.

The iodine 'clock'

The oxidation of iodide ions by hydrogen peroxide in acid solution can be followed as a 'clock' reaction:

$$H_2O_2(aq) + 2I^-(aq) + 2H^+(aq) \rightarrow I_2(s) + 2H_2O(l)$$

- 25 cm^3 of hydrogen peroxide solution is mixed in a beaker with 25 cm^3 of water and a few drops of starch solution are added.
- 25 cm^3 of potassium iodide solution and 5 cm^3 of a dilute solution of sodium thiosulfate are placed in a second beaker.
- The contents of the two beakers are mixed and the time taken for the solution to go blue is measured.
- The experiment is repeated with the same volumes of potassium iodide and sodium thiosulfate but with 20 cm^3 of hydrogen peroxide and 30 cm^3 of water, and then with other relative amounts of hydrogen peroxide and water, totalling 50 cm^3.

The reaction produces iodine, which reacts with the sodium thiosulfate. When all the sodium thiosulfate has been used up, the next iodine that is produced reacts with the starch to give an intense blue-black colour.

The amount of iodine produced in the measured time is proportional to the volume of sodium thiosulfate solution taken. Therefore, the average rate of reaction for each experiment is proportional to 1/time.

The sulfur 'clock'

Sodium thiosulfate is decomposed by acid, producing a precipitate of sulfur:

$$S_2O_3^{2-}(aq) + 2H^+(aq) \rightarrow S(s) + SO_2(aq) + H_2O(l)$$

- A large X is drawn on a white tile with a marker pen.
- 2 cm^3 of sodium thiosulfate solution is mixed with 25 cm^3 of water in a beaker.
- 25 cm^3 of dilute nitric acid is placed in a second beaker.
- The first beaker is placed on top of the X and the contents of the second one are added.
- The mixture is stirred and the time (t) taken for sufficient sulfur to be produced to hide the X when looking down through the beaker is measured.
- The experiment is repeated with different relative amounts of sodium thiosulfate and water, totalling 50 cm^3.

Instead of varying the concentration of sodium thiosulfate, the temperature could be altered. This would enable the activation energy to be calculated (see p. 16).

The number of moles of sulfur produced is the same in all experiments. Therefore, the average rate of reaction for each experiment is proportional to $1/t$.

pH measurements

If one of the reactants or one of the products in a reaction is an acid or an alkali and the reaction takes place in aqueous solution, the change in pH with time can be measured.

The problem with this technique is that pH is a logarithmic quantity. If a strong acid is a reactant and the starting concentration is 1.0 mol dm^{-3}, the pH only changes by 1 unit (from 0 to 1) when 90% of the acid has reacted. The pH rises to 2 when 99% of the acid has reacted. This method requires a very accurate, and hence expensive, pH meter to monitor the change in acid concentration. This makes the method unsuitable for school laboratory use.

Rate equations

The purpose of the experimental methods described above is to find the rate equation. Consider the reaction:

$$nA + mB \rightarrow \text{products}$$

The rate equation for this reaction is of the form:

$$\text{rate} = k[A]^p[B]^q$$

where n and m are the stoichiometries in the chemical equation and p and q are the powers of the concentrations of the substances in the rate equation.

The quantity k is called the **rate constant** and varies with the nature of the reaction, the temperature and the presence of any catalyst.

The *order* of reaction is $p + q$. The order with respect to substance A (also called the partial order) is p. The order with respect to substance B is q.

The order of a reaction is the *sum* of the powers to which the concentrations of the reactants are raised in the experimentally determined rate equation.

The partial order of one reactant is the power to which the concentration of that reactant is raised in the rate equation.

e The rate constant is always represented by a *lower-case k*. An upper-case *K* is the symbol for equilibrium constant.

e The general formula given for the rate equation is not always correct. Some reversible reactions have much more complex rate equations, often containing fractional partial orders. These are beyond the scope of A-level.

The values of p and q cannot be predicted from the chemical equation. They depend on both the stoichiometry *and* the mechanism of the reaction. Therefore, they have to be found by experiment.

Deduction of order of reaction

From initial rates

The initial rate of a reaction is the rate at the instant that the chemicals are mixed. This is normally found by measuring the time taken for the concentration of a reactant or product to change by a known amount, which must be less than 10% of the initial concentration of the reactant. The experiment is then repeated, changing the concentration of one of the reactants but keeping the concentration of all the others constant.

Consider a reaction:

$$A + B + C \rightarrow \text{products}$$

The experimental method is as follows:

- The initial rate is measured when all three reactants have the same concentration, for example, 1.0 mol dm^{-3}.
- The experiment is repeated with [A] = 2.0 mol dm^{-3} and [B] and [C] unchanged at 1.0 mol dm^{-3}.
  - If the rate does not alter, the reaction is zero order with respect to substance A.
  - If the rate doubles (increases by a factor of 2^1), it is first order in A.
  - If the rate increases by a factor of 4 (2^2) it is second order in A.
- A third experiment is performed with [A] and [C] equal to 1.0 mol dm^{-3} and [B] = 2.0 mol dm^{-3}. This enables the partial order with respect to B to be deduced.
- A fourth experiment is carried out in which [C] is altered but [A] and [B] are kept the same as in one of the previous experiments. This enables the order with respect to C to be deduced.

The overall order is the *sum* of all the partial orders.

The change in concentration can usually be measured by one of the procedures described earlier in this chapter.

Worked example

The reaction between nitrogen(II) oxide and hydrogen at 1000°C is:

$$2NO(g) + 2H_2(g) \rightarrow N_2(g) + 2H_2O(g)$$

Use the data below to deduce the order of reaction with respect to hydrogen and nitrogen(II) oxide. Calculate the overall order of the reaction. Write the rate equation and calculate the value of the rate constant.

Experiment	[NO]/mol dm^{-3}	[H_2]/mol dm^{-3}	Initial rate/mol dm^{-3} s^{-1}
1	4.0×10^{-3}	1.0×10^{-3}	1.2×10^{-5}
2	8.0×10^{-3}	1.0×10^{-3}	4.8×10^{-5}
3	8.0×10^{-3}	4.0×10^{-3}	1.92×10^{-4}

Answer

Consider experiments 1 and 2:

- [NO] is increased by a factor of 2. $[H_2]$ is unchanged.
- The rate increases by a factor of 2^2. Therefore, the reaction is second order in nitrogen(II) oxide.

Consider experiments 2 and 3:

- $[H_2]$ goes up by a factor of 4. [NO] is unchanged.
- The rate increases by a factor of 4^1. Therefore, the reaction is first order with respect to hydrogen (even though there are two hydrogen molecules in the chemical equation for the reaction).

overall order = 2 + 1 = 3

rate = $k\,[NO]^2[H_2]^1$

rate constant, $k = \dfrac{\text{rate}}{[NO]^2[H_2]^1}$

Using the data from experiment 1:

$$k = \frac{1.2 \times 10^{-5}}{(4.0 \times 10^{-3})^2 \times (1.0 \times 10^{-3})} = 750\,\text{mol}^{-2}\,\text{dm}^6\,\text{s}^{-1}$$

$$\text{units} = \frac{\text{concentration} \times \text{time}^{-1}}{\text{concentration}^2 \times \text{concentration}} = \text{concentration}^{-2} \times \text{time}^{-1}$$

Note that if the reaction were second order in hydrogen, when $[H_2]$ was increased by a factor of 4, the rate would increase by a factor of 4^2 or 16 times. This was not the case in this reaction.

You must either indicate the experiments that you are comparing or state which concentrations remain constant and how the other changes. Both are indicated in the worked example.

e The units of the rate constant, k, vary according to the total order of reaction (see p. 15).

e The symbol for half-life is $t_{\frac{1}{2}}$.

From half-lives

Half-life is the time taken for the concentration of a reactant to halve.

- For a *first*-order reaction, the half-life is constant at a fixed temperature.

This means, for example, that if it takes 25 s for the concentration of any reactant to fall from 8 units to 4 units, then it also takes 25 s for it to fall from 4 units to 2 units or from 6 units to 3 units.

- For a *second*-order reaction, the half-life increases in a regular geometric manner.

This means, for example, that if it takes 25 s for the concentration of any reactant to fall from 8 units to 4 units, then it will take 50 s for it to fall from 4 units to 2 units.

If a graph of the concentration of a reactant is plotted against time and the concentration falls during the experiment to less than 25% of its initial value, two consecutive half-lives can be measured. If the fall is less than this, two non-consecutive half-lives can still be measured, but the fall must be greater than 66% to be certain that the half-lives are accurate within experimental error.

Worked example

Consider the reaction:

A + B → products

Use the data here to plot a graph of [A] against time. Measure two consecutive half-lives and hence deduce the order of reaction.

$[A]/\text{mol dm}^{-3}$	Time/s
1.00	0
0.90	5
0.75	12
0.42	38
0.30	52
0.20	70
0.15	82

Answer

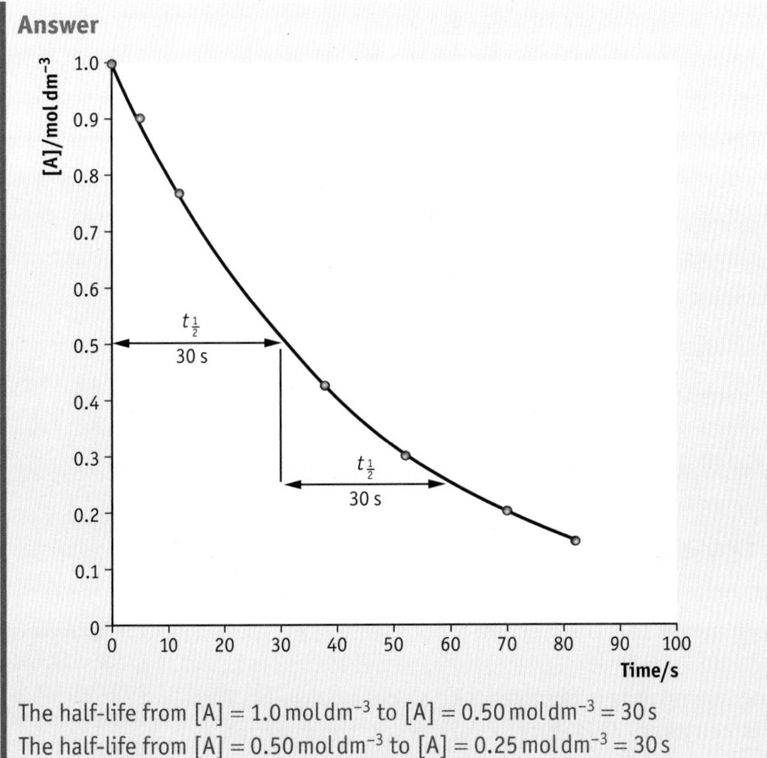

The half-life from [A] = 1.0 mol dm^{-3} to [A] = 0.50 mol dm^{-3} = 30 s
The half-life from [A] = 0.50 mol dm^{-3} to [A] = 0.25 mol dm^{-3} = 30 s
The half-life is constant, so the reaction is first order.

In the worked example above, note that the second half-life is not 60 s. This is the sum of the two consecutive half-lives and represents the concentration falling to one-quarter of the original value.

Note also that the two half-lives need not be consecutive. If the time for the concentration of A to fall from 0.40 mol dm^{-3} to 0.20 mol dm^{-3} had been measured from the graph, it would also have been 30 s. Likewise, the time for the concentration of A to fall from 0.60 mol dm^{-3} to 0.30 mol dm^{-3} would also be 30 s.

Extension for students studying A-level mathematics

For a reaction A + 2B → C + 3D, the expression:

$$\text{rate} = \frac{-\Delta[\text{A}]}{\Delta t}$$

is an approximation that measures the average rate during this period. The instantaneous rate is the differential of this:

$$\text{rate} = -\text{d}[\text{A}]/\text{d}t \ (\text{or} -\tfrac{1}{2}\text{d}[\text{B}]/\text{d}t \text{ or } + \text{d}[\text{C}]/\text{d}t \text{ or } + \tfrac{1}{3}\text{d}[\text{D}]/\text{d}t)$$

For a first-order reaction this becomes:

$$\text{rate} = -\text{d}[\text{A}]/\text{d}t = k[\text{A}]$$

which on integration becomes:

$$kt = \ln[\text{A}]_0 - \ln[\text{A}]_t = \ln\{[\text{A}]_0/[\text{A}]_t\}$$

where $[\text{A}]_0$ is the initial concentration of A and $[\text{A}]_t$ is its concentration at time t.

e All kinetic experiments should be carried out at constant temperature and for most reactions the laboratory itself acts as a constant-temperature medium. However, most reactions are exothermic, so the temperature may rise spontaneously. This can cause the half-life to be slightly shorter.

After one half-life the initial concentration of A has halved:

$[A]_{t_{\frac{1}{2}}} = \frac{1}{2}[A]_0$

Therefore:

$$kt_{\frac{1}{2}} = \ln\{[A]_0/\tfrac{1}{2}[A]_0\} = \ln 2$$

$$t_{\frac{1}{2}} = \frac{\ln 2}{k} \quad \text{or} \quad k = \frac{\ln 2}{t_{\frac{1}{2}}}$$

This proves that the half-life of a first-order reaction is constant at a given temperature and that its value can be used to calculate the rate constant, k. In the worked example above, the half-life is 30 s, so $k = \ln 2/30 = 0.023\,s^{-1}$.

From the slope of a concentration–time graph

Form of the graph

When a graph of [A] is plotted against time, the shape of the graph depends on the order of reaction.

The slope of the graph at any value of [A] is the rate of reaction at that concentration.

- If it is a horizontal line, it means that the reaction is not taking place (rate of reaction = 0).
- If a *straight* line sloping downwards is obtained, the slope is constant. This means that the rate is constant. This only occurs when the reaction is zero order.

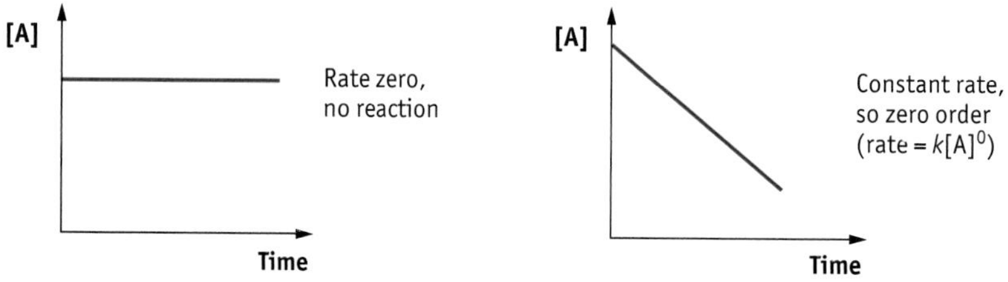

The half-lives on the graph in Figure 1.5(a) are constant. Therefore, this reaction is first order.
The half-lives on the graph in Figure 1.5(b) increase rapidly (they double as the concentration is halved). Therefore, this reaction is second order.

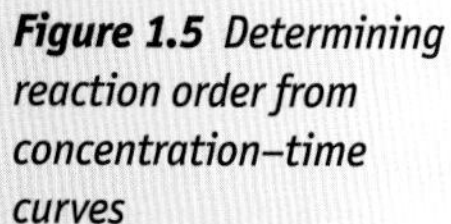

***Figure 1.5** Determining reaction order from concentration–time curves*

- If a downward *curve* is obtained, with decreasing slope, the rate is decreasing as [A] falls. Therefore, the reaction is first order or greater (Figure 1.5).

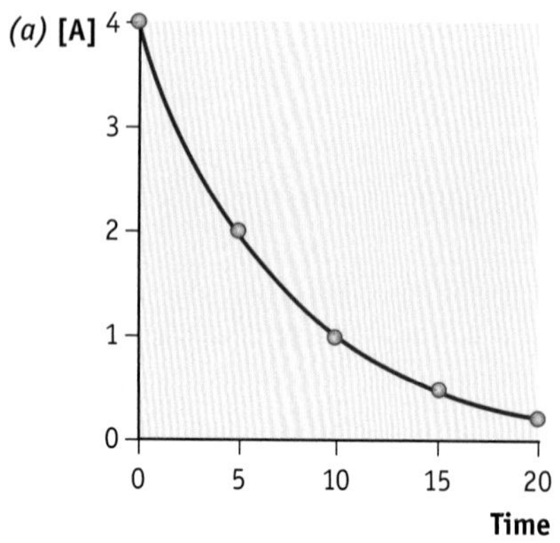

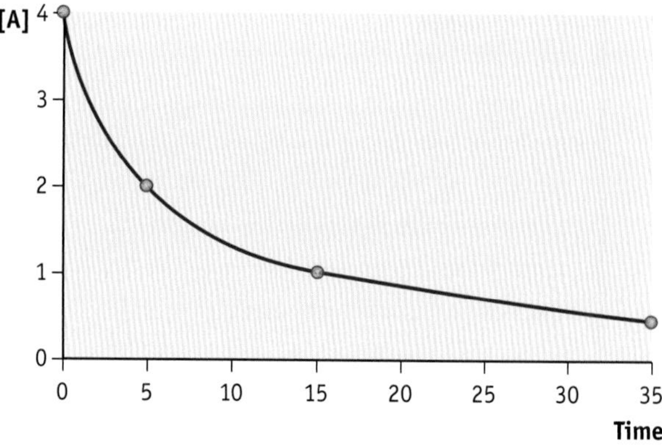

Drawing tangents

The rate at any particular concentration can be calculated by drawing a tangent to the curve at that point and measuring the slope of the tangent.

For a plot of concentration of reactant against time, the rate of reaction is equal to minus the slope of the graph.

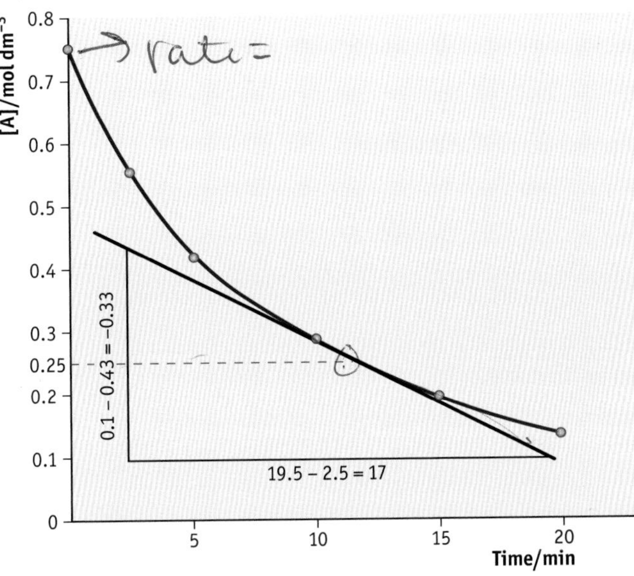

***Figure 1.6** Calculating the rate of a second-order reaction by drawing a tangent*

It is difficult to draw a tangent accurately. The tangential line must be long enough for the coordinates at the top and the bottom of the line to be read from the graph without a large error.

By drawing tangents at two points, or by comparing the initial rate with the rate measured by the tangent, the order of reaction can be deduced.

Worked example

The reaction $A + 2B \rightarrow C + D$ was studied and the graph of [A] as a function of time was drawn, as in Figure 1.6.

Draw the tangent at $[A] = 0.25\,mol\,dm^{-3}$ and measure its slope.

The initial rate, when [A] was $0.75\,mol\,dm^{-3}$, was $0.17\,mol\,dm^{-3}\,min^{-1}$.

Deduce the order of the reaction and suggest rate equations that fit the order.

Answer

$$\text{gradient of the slope} = \frac{(0.1 - 0.43)}{(19.5 - 2.5)} = \frac{-0.33}{17} = -0.019\,mol\,dm^{-3}\,min^{-1}$$

rate of reaction when $[A] = 0.25\,mol\,dm^{-3} = +0.019\,mol\,dm^{-3}\,s^{-1}$

initial rate when $[A] = 0.75\,mol\,dm^{-3} = +0.17\,mol\,dm^{-3}\,min^{-1}$

The concentration of [A] was decreased by a factor of 3 and the rate decreased by a factor of $0.17/0.019 = 8.9 \approx 9\ (3^2)$, so the reaction is second order.

Possible rate equations are:

rate = $k[A]^2$

rate = $k[B]^2$

rate = $k[A][B]$

e Following the concentration of one of the reactants gives the overall order of reaction, not the partial order with respect to the substance whose concentration is being measured. Thus, if [B] as a function of time were measured, the order would still be 2, regardless of which of the three possible rate equations is correct.

From the slope of rate–concentration graphs

If the rate of a reaction, or some quantity that is proportional to the rate, is plotted against the concentration of one reactant, the order with respect to that reactant can be found.

1/time for the reaction to proceed to a certain point is often used as a measure of the rate.

- If the graph of rate (or 1/time) against [reactant] is a straight line, the reaction is first order with respect to that reactant.
- If the graph of rate (or 1/time) against $[\text{reactant}]^2$ is a straight line, the reaction is second order with respect to that reactant.

Worked example

Colourless nitrogen monoxide reacts with oxygen to form the brown gas nitrogen dioxide:

$$2NO(g) + O_2(g) \rightarrow 2NO_2(g)$$

The time taken for a certain depth of brown colour to appear was measured. The data obtained are shown in the table:

Time/s	1/time (proportional to rate)/s^{-1}	[NO]/mol dm^{-3}	$[NO]^2$/$mol^2\,dm^{-6}$
60	0.017	0.045	2.0×10^{-3}
100	0.010	0.035	1.2×10^{-3}
200	0.005	0.024	0.60×10^{-3}

(a) Plot:

(i) a graph of 1/time against [NO]

(ii) a graph of 1/time against $[NO]^2$

Use these graphs to determine the order of reaction with respect to nitrogen monoxide.

(b) A new series of readings was obtained using twice the initial concentration of oxygen. The gradient of the graph of 1/time against $[NO]^2$ doubled. Determine the order of reaction with respect to oxygen.

Answer

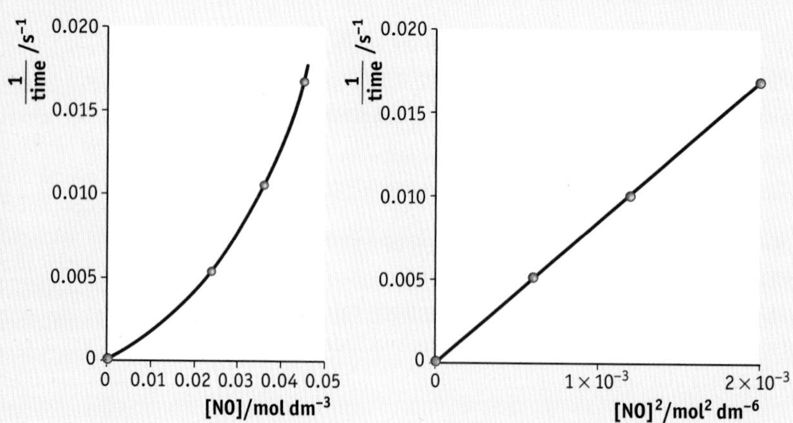

(a) The graph of 1/time against [NO] is a curve. So the reaction is *not* first order with respect to nitrogen monoxide, as the rate is *not* proportional to [NO].

The answer would have been the same had the data been about partial pressures of the two gases rather than about their concentrations.

e The origin must be plotted as one point. The rate is zero when [A] = 0.

The graph of 1/time against $[NO]^2$ is a straight line, so the rate is proportional to $[NO]^2$. Therefore, the reaction is second order with respect to nitrogen monoxide.

(b) The gradient of the line is proportional to $k[O_2]^x$. The gradient doubles as $[O_2]$ doubles, so the reaction is first order with respect to oxygen.

Units of rate constants

For a first-order reaction, the rate equation is rate = $k[A]^1$.

Therefore, k = rate/[A]. The units of the rate constant are:

$$\frac{\text{concentration} \times \text{time}^{-1}}{\text{concentration}} = \text{time}^{-1} \text{ (e.g. s}^{-1} \text{ or min}^{-1})$$

For a second-order reaction, rate = $k[A]^2$ or rate = $k[A][B]$

$k = \text{rate}/[A]^2$ or $k = \text{rate}/[A][B]$

The units of the rate constant are

$$\frac{\text{concentration} \times \text{time}^{-1}}{\text{concentration}^2} = \text{concentration}^{-1}\,\text{time}^{-1} \text{ (e.g. mol}^{-1}\,\text{dm}^3\,\text{s}^{-1})$$

The units of the rate constant for other orders can be worked out in a similar way.

Effect of temperature and a catalyst on the rate constant

The value of the rate constant depends on:

- the complexity of the geometry of the molecules. This is also called the orientation factor. If only 1 in 10 collisions occurs with the correct orientation, the orientation factor equals 0.1. This factor is a constant for a particular reaction.
- the activation energy of the reaction
- the temperature

The second and third factors cause the rate constant to be proportional to:

$$e^{(-E_a/RT)}$$

where E_a is the activation energy, R is the gas constant and T is the temperature in kelvin.

A large E_a results in a large negative exponent and, therefore, a small value for the rate constant. A catalyst effectively lowers the activation energy (by providing a different route for the reaction). Therefore, the exponent becomes less negative and k gets larger, and so the rate of reaction increases.

A rise in temperature increases the denominator of the exponential term and makes its value less negative, increasing the value of k. This means that the rate of reaction increases.

e Remember that reactions that have a high activation energy have a small value for the rate constant and are, therefore, slow at room temperature. An increase in temperature increases the value of the rate constant and so the reaction is faster.

The relationship between the rate constant and temperature is described by the Arrhenius equation:

$$\ln k = \ln A - E_a/RT$$

where A is a constant.

If the value of the rate constant is measured at different temperatures, a graph can be plotted of $\ln k$ against $1/T$. The graph is a straight line of slope $-E_a/R$, allowing the activation energies to be determined.

The rate of reaction at a particular initial concentration of reactant is proportional to the rate constant. This means that a graph of ln (1/time) against $1/T$ has the same slope as the graph of $\ln k$ against $1/T$. This is true providing that the approximation that 1/time is a measure of the rate of reaction is reasonable.

Worked example

The second-order rate constant for the reaction of 1-bromopropane with aqueous hydroxide ions was measured as a function of temperature.

$$CH_3CH_2CH_2Br(aq) + OH^-(aq) \rightarrow CH_3CH_2CH_2OH(aq) + Br^-(aq)$$

The results are shown in the table below.

Temperature/°C	Temperature/K	$1/T$/K^{-1}	k/mol^{-1} dm^3 s^{-1}	ln k
25	298	0.00336	1.4×10^{-4}	−8.9
35	308	0.00325	3.0×10^{-4}	−8.1
45	318	0.00314	6.8×10^{-4}	−7.3
55	328	0.00306	1.4×10^{-3}	−6.6

Plot a graph of $\ln k$ against $1/T$. Measure the slope (gradient) of the line and hence calculate the activation energy of the reaction.

Answer

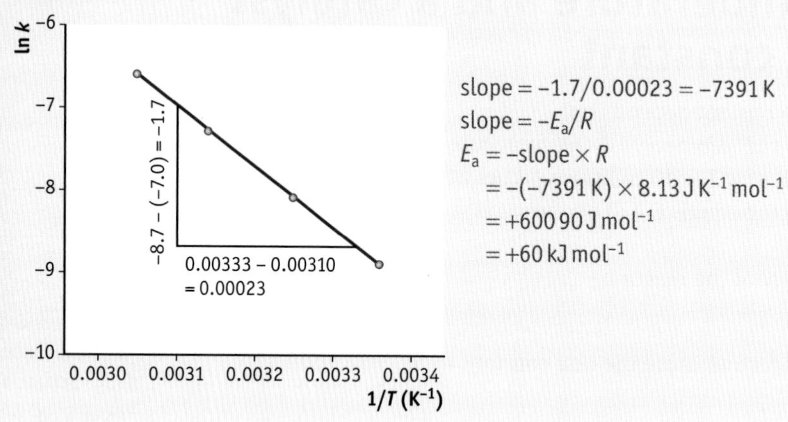

slope = $-1.7/0.00023 = -7391$ K

slope = $-E_a/R$

$E_a = -\text{slope} \times R$

$= -(-7391\text{ K}) \times 8.13\text{ J K}^{-1}\text{ mol}^{-1}$

$= +600\,90\text{ J mol}^{-1}$

$= +60\text{ kJ mol}^{-1}$

e The activation energy is always positive because it is the energy that the colliding molecules must have for a reaction to take place.

The activation energy can be found using a clock reaction and making the assumption that 1/time is proportional to the rate of reaction.

Worked example

The activation energy of the oxidation in acid solution of iodide ions by iodate(V) ions can be found by experiment:

- Place 25 cm^3 of a solution of potassium iodate(V) in a beaker and add 5 drops of starch solution.
- Place 25 cm^3 of a solution of potassium iodide and 5 cm^3 of a solution of sodium thiosulfate in another beaker.

- Mix the two solutions, start a stop clock immediately and stir with a thermometer. Read and record the temperature of the mixture.
- Record the time when the solution turns intense blue.
- Repeat the experiment with the same volumes of the solutions but at a higher temperature.
- Repeat the experiment at two more different temperatures.

The equation for the reaction is:

$$IO_3^-(aq) + 5I^-(aq) + 6H^+(aq) \rightarrow 3I_2(aq) + 3H_2O(l)$$

The iodine produced then reacts with the sodium thiosulfate:

$$I_2(aq) + 2S_2O_3^{2-}(aq) \rightarrow 2I^-(aq) + S_4O_6^{2-}(aq)$$

When all the sodium thiosulfate has been used up, the next iodine that is produced forms an intense blue colour with the starch.

The results are shown in the table below.

Time/s	1/time/s^{-1}	ln (1/time)	Temperature/°C	Temperature/K	1/temperature/K^{-1}
135	0.0074	−4.9	25	298	0.00336
46	0.0217	−3.8	40	313	0.00319
14	0.0714	−2.6	60	333	0.00300

Plot a graph of ln (1/time) on the *y*-axis against $1/T$ on the *x*-axis. Measure the gradient of the line and evaluate the activation energy.

You may assume that the gradient of this line is the same as that of $\ln k$ against $1/T$ and that its value is $-E_a/R$.

The temperature must be in kelvin. The conversion necessary is $x°C = (x + 273)\,K$.

Answer

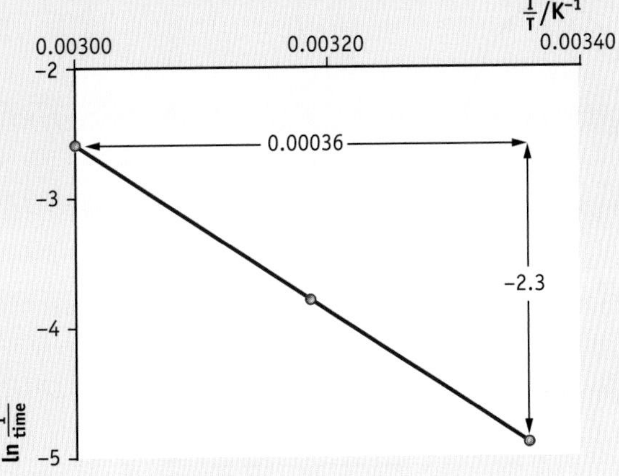

The gradient of the line = −2.3/0.00036 = −6389 K

As the Arrhenius equation is $\ln k = -E_a/RT$ + a constant, the gradient equals $-E_a/R$

$E_a = -\text{gradient} \times RT = -(-6389)\,K \times 8.31\,J\,K^{-1}\,mol^{-1} = +53\,090\,J\,mol^{-1} = +53.1\,kJ\,mol^{-1}$

e Note that the units of the gas constant, *R*, are $J\,K^{-1}\,mol^{-1}$ and so the activation energy will be calculated in $J\,mol^{-1}$.

Mechanisms and the rate-determining step

Single-step reactions

Some reactions take place in a single step. For example, the reaction between aqueous hydroxide ions and a primary halogenoalkane is a one-step reaction that is thought to involve a collision between the two species.

During the collision, the C–halogen bond begins to break and a new O–C bond forms. At this halfway point, the system is said to have reached a position of maximum potential energy. This is the **transition state** between the reactants and the products — for example, with CH_3I:

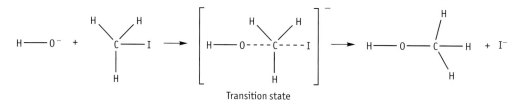

$$H{-}O^{-} + CH_3{-}I \longrightarrow \left[H{-}O\cdots CH_3 \cdots I\right]^{-} \longrightarrow H{-}O{-}CH_3 + I^{-}$$

Transition state

The reaction energy profile for a single-step reaction is shown in Figure 1.7.

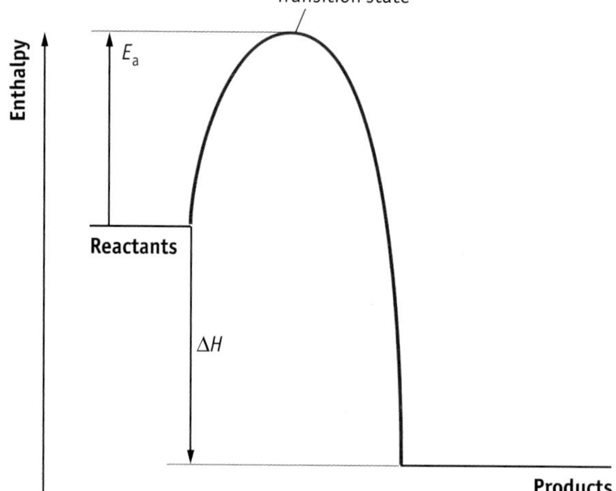

***Figure 1.7** Reaction profile for a single-step reaction*

As this is a single-step reaction between two species, the reaction is second order and the rate equation is:

rate = k[A][B]

In this example,

rate = $k[OH^-][CH_3I]$

The powers to which the concentrations are raised in the rate equation for a *single*-step reaction are the same as the stoichiometry.

Multi-step reactions

Many reactions take place in more than one step, via intermediate compounds, ions or radicals.

The hydrolysis of a tertiary halogenoalkane, for example 2-chloro-2-methylpropane, with aqueous hydroxide ions is an example of a two-step reaction.

Step 1: the C–Cl bond breaks heterolytically, forming a carbocation and a chloride anion:

$$(CH_3)_3CCl \rightarrow (CH_3)_3C^+ + Cl^-$$

The rate equation for step 1 is:

$$\text{rate} = k_1[(CH_3)_3CCl]$$

Step 2: the lone pair of electrons on the oxygen of the OH^- ion forms a new bond with the positive carbon atom:

$$(CH_3)_3C^+ + OH^- \rightarrow (CH_3)_3COH$$

The rate equation for step 2 is:

$$\text{rate} = k_2[(CH_3)_3C^+][OH^-]$$

The overall rate is controlled by the rate of the slowest step in the mechanism. This step is called the **rate-determining step**.

The rate-determining step is the slowest step in a multi-step mechanism.

For the hydrolysis of 2-chloro-2-methylpropane, the first step is the slower step and is therefore rate determining. Hence, the rate equation for this reaction is:

$$\text{rate} = k_1[(CH_3)_3CCl]$$

Evidence for a mechanism

The main evidence for a mechanism is the order of reaction with respect to each reactant.

The hydrolysis of halogenoalkanes can take place by two mechanisms. Both are nucleophilic substitutions, but in one the rate of reaction depends only on the concentration of the halogenoalkane and not on the nucleophile. This is called a S_N1 reaction (S = substitution, N = nucleophilic and 1 = first order). As there are two reactants and the reaction is zero order with respect to the nucleophile, the reaction must take place in at least two steps with the nucleophile entering the reaction *after* the rate-determining step.

In the other type of mechanism, the rate equation is:

$$\text{rate} = k\,[\text{halogenoalkane}][\text{nucleophile}]$$

This is called a S_N2 reaction and is thought to take place in a single step, going via a transition state.

Halogenoalkanes react with nucleophiles such as OH^-, H_2O, NH_3 and CN^- in substitution reactions. This is because the carbon atom joined to the halogen is slightly δ^+ and is, therefore, attacked by nucleophiles.

Which mechanism is followed depends on whether the halogenoalkane is primary, secondary or tertiary.

Primary halogenoalkanes

These react in an **S_N2 reaction.** The mechanism is a single step that goes through a **transition state.**

e A transition state is not a species that can be isolated. It changes immediately into the product.

An example of an S_N2 mechanism involving a transition state is the reaction between hydroxide ions and bromoethane:

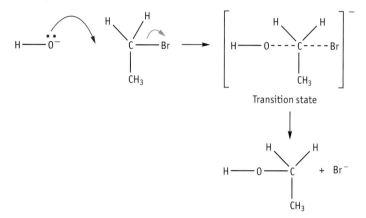

$$H{-}O^- + H_2C(CH_3){-}Br \longrightarrow \left[H{-}O\cdots C(H_2)(CH_3)\cdots Br\right]^- \longrightarrow H{-}O{-}CH_2CH_3 + Br^-$$

Transition state

Do not forget to include the negative charge on the transition state.

The red curly arrow shows the movement of a lone pair of electrons from the oxygen to the carbon as a covalent bond forms. The green arrow represents the electrons in the C–Br σ-bond moving to the bromine atom as the bond breaks.

A curly arrow represents the movement of a pair of electrons.

The transition state occurs when the new O–C bond is half-formed and the C–Br bond is half-broken.

The **reaction profile** diagram for this type of reaction is shown in Figure 1.8.

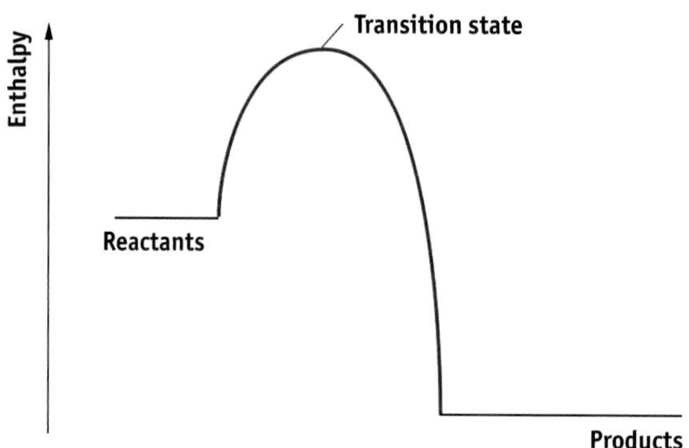

***Figure 1.8** Reaction profile for an S_N2 reaction involving a transition state*

The energy released in the formation of the O–C bond is enough to provide the energy to break the C–halogen bond. The weaker the C–halogen bond, the faster is the rate of the reaction. Therefore, since the C–Cl bond is the strongest and the C–I bond is the weakest, the rate order is:

C–I > C–Br > C–Cl

Further evidence comes from the optical activity of the product compared with that of the reactant (see p. 128).

Tertiary halogenoalkanes

Tertiary halogenoalkanes react by a $\mathbf{S_N1}$ mechanism. This type of reaction takes place in two steps.

Step 1: the carbon–halogen bond breaks, an intermediate carbocation is formed and a halide ion is released. This is the slow rate-determining step.

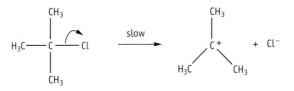

$$(CH_3)_3C{-}Cl \xrightarrow{\text{slow}} (CH_3)_3C^{+} + Cl^{-}$$

A carbocation is an ion in which there is a positive charge on a carbon atom.

Step 2: the carbocation is attacked by the nucleophile in a fast reaction.

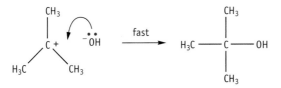

$$(CH_3)_3C^{+} + {}^{-}\ddot{O}H \xrightarrow{\text{fast}} (CH_3)_3C{-}OH$$

As with the S_N2 mechanism, some evidence for the S_N1 mechanism comes from optical activity.

Secondary halogenoalkanes

- The rate of reactions with S_N2 mechanisms decreases in the order primary > secondary > tertiary halogenoalkane. This is because of the increasing steric hindrance by the alkyl groups on the attacking nucleophile.
- The rate of reactions with S_N1 mechanisms increases in the order primary < secondary < tertiary halogenoalkane. This is because the intermediate carbocation is increasingly stabilised by the electron pushing effect of the alkyl groups.

These relationships are shown graphically in Figure 1.9.

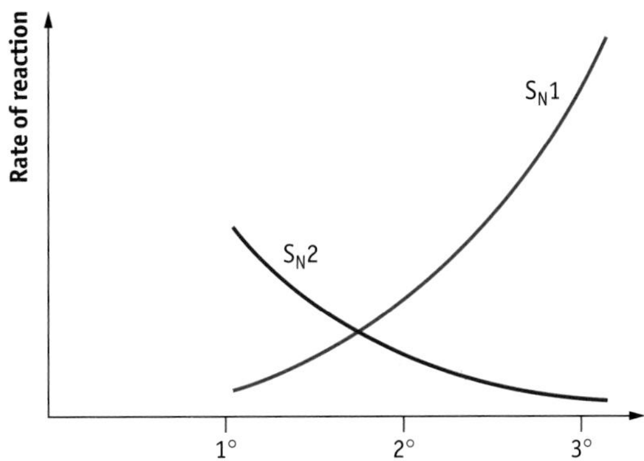

***Figure 1.9** Different rates of S_N1 and S_N2 reactions*

Figure 1.9 shows that primary halogenoalkanes react almost entirely by an S_N2 mechanism and that tertiary halogenoalkanes react by an S_N1 mechanism.

Secondary halogenoalkanes react by both mechanisms.

The overall rate is fastest with a tertiary halogenoalkane and slowest with a primary. For example, 2-chloro-2-methylpropane, $(CH_3)_3CCl$, produces an instant precipitate of silver chloride with aqueous silver nitrate, whereas 1-chloropropane gives a precipitate only after heating for a long period.

First step is rate determining

If the rate-determining step is the first step, then the rate equation for the overall reaction is the same as that for the rate-determining step.

For the alkaline hydrolysis of 2-chloro-2-methylpropane, the rate equation is:

$$\text{rate} = k[(CH_3)_3CCl]$$

This means that if a substance enters the mechanism *after* the rate-determining step, its partial order is zero. In this example the partial order of the alkali is zero.

The iodination of propanone in alkaline solution takes place in several steps. The overall equation is:

$$CH_3COCH_3 + I_2 + OH^- \rightarrow CH_2ICOCH_3 + H_2O + I^-$$

The reaction is found to be second order overall and zero order with respect to iodine.

The rate equation is:

$$\text{rate} = k[CH_3COCH_3][OH^-]$$

The most likely explanation is that iodine enters the mechanism after the rate-determining step. A suggested mechanism is:

Step 1: the base removes an H^+ from a $-CH_3$ group — this is the rate-determining (slower) step:

$$CH_3COCH_3 + OH^- \rightarrow CH_3COCH_2^- + H_2O$$

Step 2: the lone pair of electrons on the C^- of the carbanion forms a bond with an iodine atom in I_2 and the I_2 bond breaks:

$$CH_3COCH_2^- + I_2 \rightarrow CH_3COCH_2I + I^-$$

As the reaction is first order in propanone and OH^- ions and zero order in I_2, step 1 must be the rate-determining step.

Further evidence for this is that the rate is exactly the same whether iodine, bromine or chlorine is the reactant.

Using curly arrows, the mechanism is shown below:

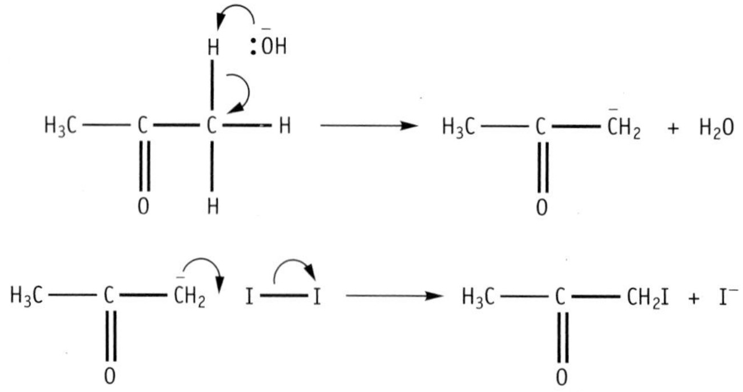

$$H_3C-C(=O)-CH_2-H \;\; :\overline{O}H \longrightarrow H_3C-C(=O)-\overline{C}H_2 + H_2O$$

$$H_3C-C(=O)-\overline{C}H_2 \;\; I-I \longrightarrow H_3C-C(=O)-CH_2I + I^-$$

The mechanism for the acid-catalysed reaction of propanone and iodine is much more complex, but the iodine also enters the mechanism after the rate-determining step.

Step 1: protonation of the carbonyl oxygen atom — this is the rate-determining step:

$$H_3C{-}C({=}O){-}CH_3 \xrightarrow{H^+} H_3C{-}\overset{+}{C}(OH){-}CH_3$$

Step 2: loss of H^+:

$$H_3C{-}\overset{+}{C}(OH){-}CH_3 \longrightarrow H_3C{-}C(OH){=}CH_2 + H^+$$

Step 3: addition of iodine followed by loss of H^+:

$$H_3C{-}C(OH){=}CH_2 + I{-}I \longrightarrow H_3C{-}C({=}O){-}CH_2I + H^+ + I^-$$

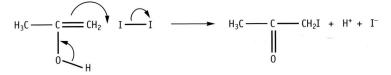

Second or subsequent step is rate determining

If the second (or a subsequent) step is rate determining, the derivation of the overall rate equation is more complex.

Consider the reaction:

$$A + 2B + C \rightarrow D + E$$

Step 1 is a rapid step that is reversible:

$$A + B \rightleftharpoons Int$$

where Int is an intermediate.

Step 2 is the slowest step and hence is rate determining:

$$Int + B \rightarrow X$$

Step 3 is faster than step 2:

$$C + X \rightarrow D + E$$

The overall rate is determined by the rate of the slowest step 2:

overall rate = rate of step 2 = $k_2[B][Int]$

The value of [Int] can be found by treating step 1 as an equilibrium reaction:

$$K_1 = \frac{[Int]}{[A][B]}$$

$$[Int] = K_1 \times [A][B]$$

Substituting [Int] into the rate equation for step 2:

overall rate = rate of step 2 = $k_2K_1[A][B]^2$

The reaction is third order and $k_2 K_1$ equals the rate constant.

e The powers to which the concentrations are raised in the rate equation for an individual step in a multi-step reaction are the same as the stoichiometry for that step.

Reaction profile diagrams for multi-step reactions

In a multi-step reaction, the reactants go via a transition state to an intermediate, which then reacts via a further transition state to form the products. The intermediate is usually at a lower energy level than the reactants.

e If the partial order of a reactant is zero, that reactant enters the mechanism *after* the rate-determining step.

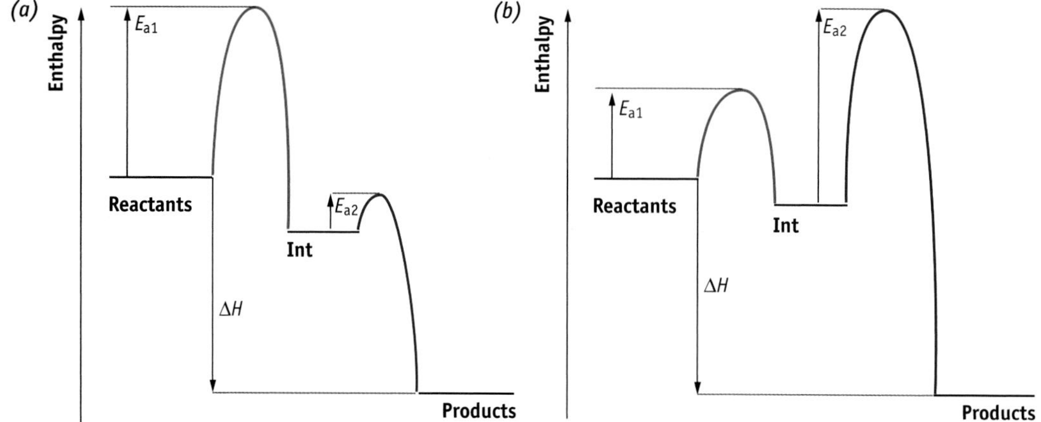

Figure 1.10 *Energy profile diagrams for two different reactions*

In Figure 1.10(a), the activation energy for the production of the intermediate is greater than the activation energy for the intermediate going to the products. This means that the first step is the rate-determining step.

In Figure 1.10(b), the second activation energy is greater, so the second step is the rate-determining step.

Pseudo-zero-partial-order reactions

The classic examples of pseudo-zero-partial-order reactions are the enzyme-catalysed reactions in living organisms.

The rate of reaction depends on the concentration of the enzyme and not on the concentration of the substrate.

A simplified explanation of enzyme activity is that the enzyme rapidly adsorbs the reactant and slowly converts it to a product that is then rapidly released by the enzyme. As long as there is enough reactant (the substrate) to saturate the enzyme, the reaction rate is not increased by increasing the concentration of reactant. Therefore, the reaction has an apparent zero partial order, even though the substrate enters the mechanism before the rate-determining step.

If the concentration of the substrate drops too far, the enzyme ceases to be saturated and the reaction becomes dependent on the concentration of the substrate.

Catalysis

Heterogeneous catalysts

These are catalysts that are in a different phase from the reactants. Many industrial processes use this type of catalyst; they have the advantage that the catalyst can be separated from the equilibrium mixture by simple physical means.

For example, the solid iron catalyst in the Haber process is in a different phase to the nitrogen and hydrogen, and stays in the reaction chamber as the product gases pass through.

A similar mechanism occurs in almost all metal-catalysed gaseous reactions. The catalyst has active sites on its surface that rapidly become saturated by reactants, which are then slowly converted into products. These leave the metal surface, thus allowing more of the reactant to be adsorbed. This means that the rate of a reaction is not altered by an increase in pressure of the gaseous reactants. This is true of:

- the Haber process, in which the reaction between nitrogen and hydrogen is catalysed by iron
- the manufacture of hydrogen from steam and methane, which is catalysed by nickel
- the oxidation of ammonia by air, which is catalysed by platinum

These reactions are all zero order with respect to both reactant gases, unless the pressure is very low.

Homogeneous catalysts

The catalyst and the reactants are in the same phase, usually the gas phase or in solution.

The rate of reaction depends on the concentrations of the catalyst and one of the reactants. An example is the oxidation of iodide ions by persulfate ions.

$$2I^-(aq) + S_2O_8^{2-}(aq) \rightarrow I_2(aq) + 2SO_4^{2-}(aq)$$

This is a slow reaction because it requires two negative ions to collide. The reaction is catalysed by iron(III) ions. The catalysed, route is oxidation of iodide ions by iron(III) ions, followed by reduction of persulfate ions by iron(II) ions and the regeneration of the catalyst of iron(III) ions:

$$2Fe^{3+}(aq) + 2I^-(aq) \rightarrow I_2(aq) + 2Fe^{2+}(aq)$$
$$2Fe^{2+}(aq) + S_2O_8^{2-}(aq) \rightarrow 2Fe^{3+}(aq) + 2SO_4^{2-}(aq)$$

Altering the order by the method of excess reagent

Consider a reaction:

$$A + 2B \rightarrow C + D$$

The rate equation is of the form:

$$\text{rate} = k[A]^p[B]^q$$

However, if the initial concentration of B is made at least ten times that of A, the change in [B] during the reaction will be negligible. This means that [B] is constant within experimental error, and so the rate equation becomes:

$$\text{rate} = \text{constant} \times [A]^p$$

where the constant $= k \times$ the approximately constant value of $[B]^q$.

The value of p (the order with respect to substance A) can be found by the usual methods of initial rate or half-life.

If the experiment is repeated with the same initial concentration of A but 20 times as much B (doubling the concentration of B from the first experiment) the way in which the initial rate alters will depend on the order with respect to B. If the rate doubles, the reaction is first order with respect to B. If the rate quadruples, the reaction is second order in B.

Questions

1 Hydrogen peroxide slowly oxidises ethanol in acid solution. In an experiment, after 45 s the amount of hydrogen peroxide in 50 cm^3 of solution had fallen from 1.46×10^{-3} moles to 1.32×10^{-3} moles. Calculate the rate of the reaction.

2 a Draw the distribution of energies of a mixture of hydrogen and iodine at a temperature T_1 and at a lower temperature, T_2.

b Explain, in terms of energy and frequency of collisions, why the reaction between hydrogen and iodine is slower at the lower temperature.

c Which of energy and frequency of collisions is more important in causing the rate to decrease?

d What effect would an increase in pressure have on the frequency and energy of collisions and hence on the rate of the reaction?

3 Potassium manganate(VII) reacts slowly with a solution of ethanedioic acid in dilute sulfuric acid at room temperature. The equation is:

$$2MnO_4^-(aq) + 6H^+(aq) + 5(COOH)_2(aq) \rightarrow 2Mn^{2+}(aq) + 8H_2O(l) + 10CO_2(g)$$

a Describe a *chemical* method by which the progress of this reaction could be followed.

b Describe a *physical* method by which the progress of this reaction could be followed.

c If this reaction is carried out at a constant higher temperature, the rate at first increases and then slows down. Suggest an explanation for this and suggest an experiment to confirm your hypothesis for the initial increase in rate.

4 Consider the reaction:

$$A + 2B + 3C \rightarrow \text{products}$$

It was found to be first order in A and B and second order in C. Write the rate equation for this reaction.

5 Consider the reaction:

$$2A + B \rightarrow \text{products}$$

It was found to be second order. Write two rate equations that fit these data.

6 The kinetics of the reaction:

$$C_2H_5I + KOH \rightarrow C_2H_5OH + KI$$

were studied at a temperature *T*. The following initial rate data were obtained:

Experiment	$[C_2H_5I]$/ $mol\,dm^{-3}$	$[KOH]$/ $mol\,dm^{-3}$	Initial rate/ $mol\,dm^{-3}\,s^{-1}$
1	0.20	0.10	2.2×10^{-5}
2	0.40	0.10	4.4×10^{-5}
3	0.20	0.20	4.4×10^{-5}

a Deduce the partial orders of reaction with respect to iodoethane (C_2H_5I), and potassium hydroxide.

b Write the rate equation for the reaction.

c Calculate the value of the rate constant at this temperature and give its units.

7 The reaction between three reactants, A, B and C, was studied. The initial rates of reaction at different concentrations of the three reactants were measured and are given in the table.

Experiment	[A]/ $mol\,dm^{-3}$	[B]/ $mol\,dm^{-3}$	[C]/ $mol\,dm^{-3}$	Initial rate/ $mol\,dm^{-3}\,s^{-1}$
1	1.0	1.0	1.0	2.3×10^{-3}
2	1.0	3.0	1.0	6.9×10^{-3}
3	2.0	3.0	1.0	1.4×10^{-2}
4	2.0	1.0	2.0	4.6×10^{-3}

a Deduce the partial orders of reaction of A, B and C.

b State the overall order of this reaction.

c Write the rate equation.

d Calculate the value of the rate constant, giving its units.

8 The reaction between persulfate ions, $S_2O_8^{2-}$, and iodide ions was studied by an iodine clock method.

$S_2O_8^{2-}(aq) + 2I^-(aq) \rightarrow 2SO_4^{2-}(aq) + I_2(aq)$

a Describe the iodine clock method.

b The initial rates of reaction were measured at different concentrations. The results are shown below.

Experiment	$[S_2O_8^{2-}]$/ mol dm^{-3}	$[I^-]$/ mol dm^{-3}	Initial rate/ mol dm^{-3} s^{-1}
1	0.038	0.050	1.2×10^{-5}
2	0.076	0.050	2.4×10^{-5}
3	0.152	0.100	9.6×10^{-5}

Deduce the overall order of reaction.

c Write the rate equation and calculate the value of the rate constant.

9 The reaction between 2-bromopropane and an aqueous solution of sodium hydroxide is as follows:

$CH_3CHBrCH_3 + NaOH \rightarrow CH_3CH(OH)CH_3 + NaBr$

In an experiment, the initial concentration of both reagents was 0.10 mol dm^{-3}. The concentration of hydroxide ions was measured at set time intervals, the graph of $[OH^-]$ against time was plotted and the rate of reaction found by drawing tangents to the graph at two different values of $[OH^-]$:

- $[OH^-] = 0.10$ mol dm^{-3}; slope of tangent $= 1.6 \times 10^{-3}$
- $[OH^-] = 0.05$ mol dm^{-3}; slope of tangent $= 0.8 \times 10^{-3}$

a What is the order of reaction?

b In a separate experiment, [2-bromopropane] was measured at intervals of time. The starting concentration of each reactant was 0.10 mol dm^{-3}. What would be the relative values of the slopes of tangents drawn to a graph of [2-bromopropane] against time at the points where [2-bromopropane] were 0.10 mol dm^{-3} and 0.050 mol dm^{-3}?

c How would you modify the experiment to find out the partial orders of the two reagents?

10 Cyclopropane can be converted into its isomer, propene by heating to 500 °C:

CH_2 / H_2C—CH_2 (cyclopropane) → $CH_3CH{=}CH_2$

[Cyclopropane]/ mol dm^{-3}	Time/min
0.080	0
0.062	5
0.048	10
0.038	15
0.023	25
0.014	35
0.0065	50

a Use the data in the table to plot a graph of concentration (y-axis) against time (x-axis).

b Measure three consecutive half-lives and hence deduce the order of the reaction.

c Calculate the value of the rate constant, stating its units.

11 The half-lives of two different reactions were measured. The results are shown in the table.

Reaction I [Reactant]/ mol dm^{-3}	Half-life/ min	Reaction II [Reactant]/ mol dm^{-3}	Half-life/ min
0.8	20.0	0.8	2.2
0.4	19.7	0.4	4.3
0.2	19.9	0.2	8.5
0.1	20.4	0.1	17.0

Deduce the order of both reactions.

12 The data in the table refer to a reaction between A and B.

[A]/mol dm^{-3}	Time/s
0.20	0
0.16	10
0.13	20
0.11	30
0.07	50
0.04	80

a Plot a graph of [A] against time.

b Draw tangents at $[A] = 0.16\,mol\,dm^{-3}$ and at $[A] = 0.08\,mol\,dm^{-3}$. Measure the slope of both tangents.

c Use your answers from **b** to estimate the order of the reaction.

13 Consider a reaction that has a two-step mechanism in which both steps are exothermic. Step 2 is the rate-determining step.

Step 1: A + B → Intermediate + C

Step 2: A + Intermediate → D

a Write the overall equation for the reaction.

b Draw the reaction profile diagram for the overall reaction.

c Predict the rate equation for the reaction.

d How would the rate equation differ if step 1 were the rate-determining step?

14 The Arrhenius equation is:

$\ln k = \ln A - E_a/RT$

where *k* is the rate constant, *A* (the Arrhenius constant) is specific to the reaction, E_a is the activation energy, *R* is the universal gas constant and *T* is the temperature in kelvin.

The acid hydrolysis of sucrose has an activation energy of $1.1 \times 10^5\,J\,mol^{-1}$.
The Arrhenius constant, *A*, for this reaction is $1.2 \times 10^{15}\,mol^{-1}\,dm^3\,s^{-1}$. (The gas constant, $R, = 8.31\,J\,K^{-1}\,mol^{-1}$)

Calculate the value of the rate constant at a temperature of 303 K.

15 The values of the rate constant, *k*, at different temperatures for the cracking of ethane into ethene and hydrogen are given in the table.

Temperature/K	660	680	720	760
Rate constant/s^{-1}	0.00037	0.0011	0.0082	0.055

The Arrhenius equation is:

$\ln k = \ln A - E_a/RT$

a Draw a graph of ln *k* against 1/*T*.

b Measure the slope of the line and hence calculate the value of the activation energy of this reaction.

c Use the graph to calculate the value of the rate constant at a temperature of 700 K.

16 The reaction between small equal-sized lumps of calcium carbonate and dilute hydrochloric acid was studied at different temperatures. The acid was in excess and the time taken for the production of bubbles to stop was determined.

- $50\,cm^3$ of $0.50\,mol\,dm^{-3}$ hydrochloric acid were added to one lump of calcium carbonate of mass 0.20 g. The time for bubble production to stop was measured.
- The experiment was repeated at different temperatures. The results are shown in the table.

Time/s	$\frac{1}{time}/s^{-1}$	$\ln\left(\frac{1}{time}\right)$	Temperature/°C	Temperature/K	$\frac{1}{temperature}/K^{-1}$
120			17		
69			25		
36			35		
14			50		

a Complete the table.

b Calculate the percentage by moles of the acid used up in the experiment. Hence, suggest whether 1/time is a reasonable measure of the rate of reaction.

c Plot a graph of ln (1/time) against 1/kelvin temperature and measure its gradient.

d Assuming that the gradient of the line is the same as that of the graph of ln *k* against 1/*T*, calculate the activation energy for the reaction. The gradient of the graph of ln *k* against $1/T = -E_a/R$.

(The gas constant $R = 8.31\,J\,K^{-1}\,mol^{-1}$)

e The value of ΔH for this reaction is $-351\,kJ\,mol^{-1}$. Calculate the heat energy released when 0.20 g of calcium carbonate reacts with dilute hydrochloric acid. Hence, calculate the temperature change during the reaction. The solution has a specific heat capacity of $4.2\,J\,g^{-1}\,°C^{-1}$.

17 The alkaline hydrolysis of 2-chlorobutane, $CH_3CHClC_2H_5$, was studied. The results are shown in the table. The equation for the hydrolysis is:

$$CH_3CHClC_2H_5 + OH^- \rightarrow CH_3CH(OH)C_2H_5 + Cl^-$$

Experiment	$[CH_3CHClC_2H_5]$/ $mol\,dm^{-3}$	$[OH^-]$/ $mol\,dm^{-3}$	Initial rate/ $mol\,dm^{-3}\,s^{-1}$
1	0.10	0.10	1.4×10^{-4}
2	0.20	0.20	2.9×10^{-4}
3	0.30	0.10	4.1×10^{-4}

a Deduce the order of reaction with respect to:
(i) 2-chlorobutane
(ii) hydroxide ions

b Write the rate equations for this reaction.

c Calculate the rate constant, based on the data from experiment 3. Give your answer to two significant figures and include a unit.

d Draw the mechanism for this reaction that is consistent with your answer to **b**.

e If a single optical isomer of 2-chlorobutane had been used, what would be the effect of the product on plane-polarised light? Justify your answer in terms of the mechanism.

Entropy: how far?

Chapter 2

Introduction

It is often assumed that exothermic reactions will take place and endothermic reactions will not. This is an oversimplification, as can be seen by studying the solubilities and the enthalpies of solution of many salts. For example:

- ΔH_{soln} of ammonium nitrate is endothermic, yet it is very soluble in water.
- ΔH_{soln} of calcium carbonate is exothermic, yet it is insoluble in water.

During an exothermic process, the energy of the chemicals decreases (ΔH is negative). However, the energy of the surroundings increases by exactly the same amount. In an endothermic reaction, the chemicals gain energy (ΔH is positive) and the surroundings lose an equal quantity of energy. So what is the driving force of spontaneous change?

The answer lies in the simple concept that **energy and matter tend to spread out or disperse**.

When a highly ordered crystalline solid dissolves in water, the solid becomes dispersed throughout the liquid. When the denser gas carbon dioxide is added to air, it does not form a lower layer, but spreads throughout the air. The same happens with energy. If a hot piece of iron is placed in a beaker of water, the heat from the iron disperses into the water until both the iron and the water are at the same temperature. You cannot boil a kettle of water by putting it on a block of ice and expect the ice to become colder as the water heats up. Such a change would not break the first law of thermodynamics (the conservation of energy), but experience tells us that it never happens. Heat spontaneously flows from a hotter body to a colder body.

The spreading out of a solute into water, the spontaneous mixing of carbon dioxide and air, the heat transfer from the hot iron to the colder water are all examples of an increase in disorder. The scientific term for disorder is **entropy**.

Left: Ordered, therefore low entropy

Right: Disordered, therefore high entropy

The second law of thermodynamics states that spontaneous changes result in an increase in disorder or entropy.

The second law of thermodynamics determines:

- whether a physical or chemical change is likely to happen at a particular temperature
- whether redox reactions will take place
- the position of equilibrium

It can be said that the second law of thermodynamics explains all of chemistry.

Care must be taken to include not only the entropy change of the chemicals (ΔS_{system}) but also the entropy change of the surroundings (ΔS_{surr}). For example, when solid sodium hydroxide is added to water, the mixture of the two chemicals is called the system. The test tube and the air in the room are regarded as the surroundings. For a change to happen spontaneously, ΔS_{total} must be positive:

$$\Delta S_{total} = \Delta S_{system} + \Delta S_{surr}$$

This is another way of expressing the second law of thermodynamics.

In any spontaneous change, ΔS_{total} will be positive.

Entropy change of the system

A solid is much more ordered (or less disordered) than a liquid, which in turn is more ordered than a gas. So gaseous water is more disordered and has a larger entropy than liquid water, which has a greater entropy than ice. In general, this can be expressed as:

$$S_{solid} < S_{liquid} < S_{gas}$$

Table 2.1 shows the entropy changes for melting and boiling some substances.

***Table 2.1** Some entropy changes*

	$\Delta S_{melting}$/ $J\,K^{-1}\,mol^{-1}$	Melting temperature/K	$\Delta S_{boiling}$/ $J\,K^{-1}\,mol^{-1}$	Boiling temperature/K
O_2	8.2	54	76	90
H_2O	22	273	109	373
NH_3	29	195	97	246

In the combustion of phosphorus, the reaction goes from a solid plus a gas to a solid:

$$P_4(s) + 5O_2(g) \rightarrow P_4O_{10}(s)$$

The disorder of a gas is replaced by the order of a solid. Therefore, the extent of disorder decreases and ΔS_{system} is negative.

When dilute hydrochloric acid is added to solid calcium carbonate, carbon dioxide gas is produced:

$$CaCO_3(s) + 2HCl(aq) \rightarrow CaCl_2(aq) + H_2O(l) + CO_2(g)$$
$$\text{solid} + \text{solution} \rightarrow \text{solution} + \text{gas}$$

The disorder increases, so ΔS_{system} is positive.

The entropy change of the system can be calculated from the formula:

$$\Delta S^{\ominus}_{system} = \Sigma nS^{\ominus}(\text{products}) - \Sigma nS^{\ominus}(\text{reactants})$$

In this equation, n represents the stoichiometric numbers in the chemical equation.

e Note the similarity between this expression and the one used to find ΔH_r from enthalpy of formation data:

$\Delta H_r = \Sigma n\Delta H_f(\text{products}) - \Sigma n\Delta H_f(\text{reactants})$

Table 2.2 shows the standard entropy values of some substances. A more complete list can be found in the Edexcel data booklet that is used in the A2 exam.

Gas	Entropy $S^\ominus$/ $J K^{-1} mol^{-1}$	Liquid	Entropy $S^\ominus$/ $J K^{-1} mol$	Solid	Entropy $S^\ominus$/ $J K^{-1} mol^{-1}$
H_2	131	C_2H_5OH	161	P_4	164
O_2	205	CCl_4	216	P_4O_{10}	229
N_2	192	C_6H_6	174	C	5.7
$H_2O(g)$	189	$H_2O(l)$	70	$H_2O(s)$	43
CO_2	214			CaO	40
NH_3	192			$CaCO_3$	93
CH_4	186				
C_2H_6	230				

***Table 2.2** Standard entropies of some elements and compounds at 25°C*

Table 2.2 shows a second trend, which is that entropy increases as the complexity of a substance increases. For example, the entropy of ethane is greater than that of methane; the entropy of calcium carbonate is greater than that of calcium oxide. Note the difference between this table and one of enthalpies of formation. The standard enthalpy of formation of an element in its standard state is defined as zero. The same is not true about the standard entropy values of an element: $\Delta H^\ominus_f$ of $O_2(g) = 0\,kJ\,mol^{-1}$; $S^\ominus$ of $O_2(g) = -205\,J\,K^{-1}\,mol^{-1}$.

Effect of temperature on entropy

The third law of thermodynamics states that the entropy of a perfect crystalline substance at absolute zero (0 K or –273°C) is zero. As the crystalline substance is heated, it gains in entropy until its melting temperature is reached. On melting there is a large jump in entropy, followed by a steady increase as the liquid is heated to its boiling temperature. There is another large jump in entropy as its physical state changes, followed by a gradual increase as the gas is heated. This is shown in Figure 2.1.

The values in Table 2.2 are standard entropy values, which means that they are the values at a stated temperature, usually 298 K (25°C), and 1 atm pressure. The entropy of liquid water at 100°C is $80\,J\,K^{-1}\,mol^{-1}$, which is $10\,J\,K^{-1}\,mol^{-1}$ more than its value at 25°C.

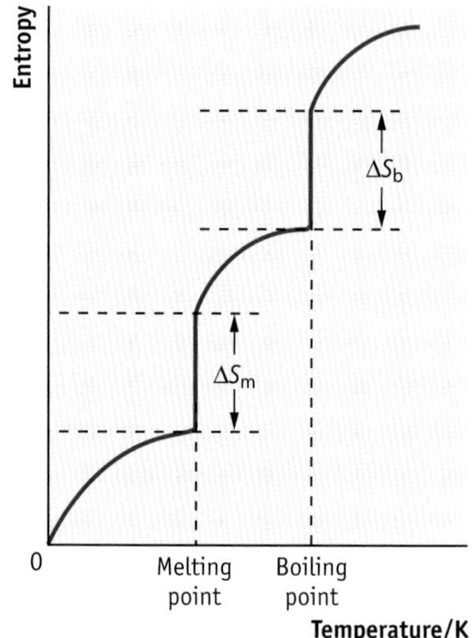

Figure 2.1

> ***Worked example***
>
> Use data from Table 2.2 to calculate the standard entropy change of the system for the reaction between phosphorus and oxygen:
>
> $P_4(s) + 5O_2(g) \rightarrow P_4O_{10}(s)$
>
> **Answer**
>
> $\Delta S^\ominus_{system} = \sum nS^\ominus(\text{products}) - \sum nS^\ominus(\text{reactants})$
>
> $= 229 - (+164 + 5 \times 205) = -960\,J\,K^{-1}\,mol^{-1}$

e Note that S has joules in the units, whereas ΔH has kilojoules.

At first sight you might think that the reaction in the worked example should not take place because the entropy decreases. However, this is only the entropy change of the *system*. Both this reaction (negative ΔS value) and the reaction of acid with calcium carbonate (positive ΔS value) take place spontaneously. What has not been taken into account is the entropy change of the surroundings (the reactions are exothermic).

Entropy change of the surroundings

When an exothermic reaction takes place, heat energy is transferred to the surrounding air, causing an increase in disorder of the air molecules. This can be seen from the Maxwell–Boltzmann distribution of energies at two temperatures.

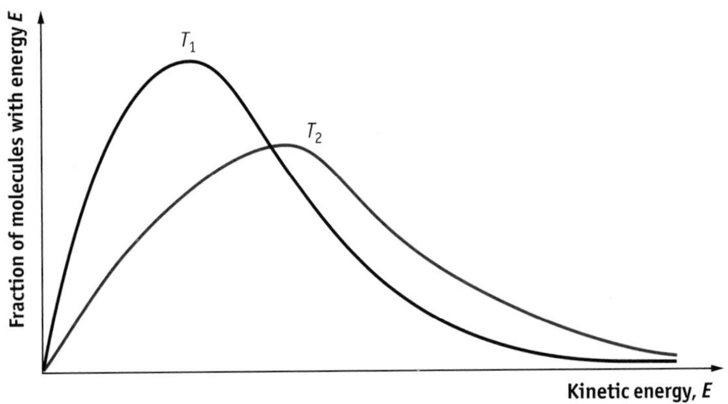

Figure 2.2 *Maxwell–Boltzmann distribution of molecular energies at two temperatures ($T_2 > T_1$)*

At the higher temperature (T_2), the molecules have a much greater range of energy and so are more random or disordered. This leads to the important conclusions:

- $\Delta S^\ominus_{surr}$ is positive for all exothermic reactions.
- $\Delta S^\ominus_{surr}$ is negative for all endothermic reactions.

This is shown pictorially in Figure 2.3.

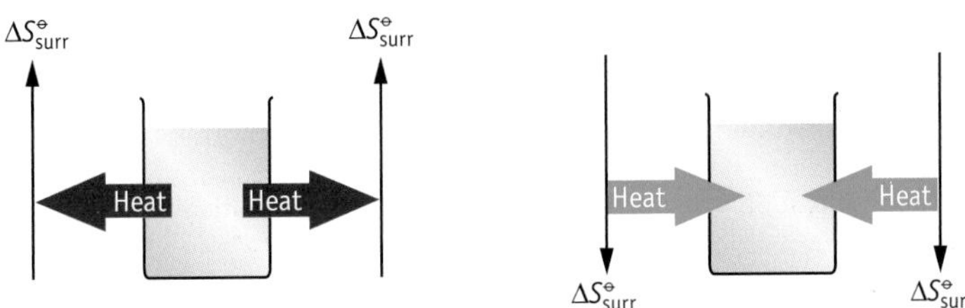

Figure 2.3 $\Delta S^\ominus_{surr}$ *and exothermic and endothermic reactions*

If the surroundings are hot, the entropy increase is small because the molecules have high entropy and are already in chaotic motion. Conversely, if the surroundings are cold, the entropy change is much greater. The entropy change in the surroundings, caused by transfer of heat, depends on the value of the heat change and is also inversely proportional to the temperature of the surroundings. The heat change of the surroundings is the negative of the enthalpy change of the system:

$$\Delta S^\ominus_{surr} = \frac{-\Delta H^\ominus}{T}$$

e If the system is exothermic, it loses enthalpy which is transferred as heat to the surroundings.

The temperature is given in kelvin (K = °C + 273).

Worked example

Calculate the entropy change of the surroundings when 1 mol of phosphorus, P_4, burns in air.

$P_4(s) + 5O_2(g) \rightarrow P_4O_{10}(s)$ $\Delta H_c = -2984\,kJ\,mol^{-1}$ at 25°C (298 K)

Answer

$$\Delta S^{\ominus}_{surr} = \frac{-\Delta H^{\ominus}}{T}$$

$$= \frac{-(-2984)}{298} = +10.0\,kJ\,K^{-1}\,mol^{-1} = +10\,000\,J\,K^{-1}\,mol^{-1}$$

As with ΔH calculations, you should always include a sign in the answer for ΔS.

Total entropy change

The total entropy change (sometimes called the entropy change of the universe) is the sum of the entropy changes of the system and the surroundings:

$$\Delta S^{\ominus}_{total} = \Delta S^{\ominus}_{system} + \Delta S^{\ominus}_{surr}$$

Worked example

Calculate the total entropy change for the combustion of phosphorus. The entropy change of the *system* is $-960\,J\,K^{-1}\,mol^{-1}$ and the entropy change of the *surroundings* is $+10\,000\,J\,K^{-1}\,mol^{-1}$.

Answer

$$\Delta S^{\ominus}_{total} = \Delta S^{\ominus}_{system} + \Delta S^{\ominus}_{surr} = -960 + (+10\,000)$$

$$= +9040\,J\,K^{-1}\,mol^{-1}$$

Note that entropy data are in joules (per kelvin per mole) and enthalpy data are in kilojoules (per mole). You must either divide the entropy value by 1000 or multiply the enthalpy value by 1000.

Changes are thermodynamically feasible if the *total* entropy change is positive.

This means that an unfavourable (negative) entropy change of the system can be compensated for by a favourable (positive) entropy change of the surroundings.

$\Delta S^{\ominus}_{system}$	$\Delta S^{\ominus}_{surr}$	Feasible
Positive	Positive (exothermic reaction)	Always
Negative	Negative (endothermic reaction)	Never
Negative	Positive (exothermic reaction)	If the value of $\Delta H/T > \Delta S^{\ominus}_{system}$ (more likely at low temperatures)
Positive	Negative (endothermic reaction)	If the value $\Delta S^{\ominus}_{system} > \Delta H/T$ (more likely at high temperatures)

Table 2.3 *Entropy changes and feasibility*

Direction of change

If ΔS_{total} is positive the change is said to be thermodynamically spontaneous. Thus the change will occur, providing that the kinetics of the change are favourable. If ΔS_{total} is negative, the reverse reaction is thermodynamically spontaneous. The value of ΔS_{total} can be altered by altering the temperature. For an exothermic reaction an increase in temperature will cause ΔS_{surr} to become less positive, which in turn will make ΔS_{total} less positive.

Worked example

Calculate the value of ΔS_{surr} at 25°C and at 100°C for a reaction with $\Delta H = -123\,kJ\,mol^{-1}$.

Answer

$\Delta S_{surr} = -\Delta H/T = -(-123\,000\,J\,mol^{-1})/298\,K = +413\,J\,K^{-1}\,mol^{-1}$ at 25°C

$\Delta S_{surr} = -(-123\,000\,J\,mol^{-1})/373\,K = +330\,J\,K^{-1}\,mol^{-1}$ at 100°C

Note that the value of ΔH was converted from kJ to J and the temperature from °C to K.

The value of ΔS_{total} also determines the extent of the reaction. The more positive its value, the more the position of equilibrium will lie to the right. This is explained in more detail in Chapter 3.

Worked example 1

Comment on the feasibility of the following reaction occurring at a temperature of 298 K:

$C_2H_5OH(l) + PCl_5(s) \rightarrow C_2H_5Cl(g) + HCl(g) + POCl_3(l)$

$\Delta H = -107\,kJ\,mol^{-1}$; $\Delta S_{system} = +368\,J\,K^{-1}\,mol^{-1}$

ΔS_{system} is positive because the reaction involves liquid + solid → two gases + liquid (getting more random).

Answer

As ΔH is negative, ΔS_{surr} will be positive. Both ΔS_{surr} and ΔS_{system} are favourable (positive), so the reaction is thermodynamically feasible (spontaneous) at all temperatures.

e As both factors are favourable, there is no need to work out the value of ΔS_{total}. However, the reaction may be too slow (reactants kinetically stable) at low temperatures.

Worked example 2

Comment on the feasibility of the following reaction occurring at a temperature of 298 K:

$2C(s) + 2H_2(g) \rightarrow CH_2{=}CH_2(g)$

$\Delta H = +52.2\,kJ\,mol^{-1}$; $\Delta S_{system} = -184\,J\,K^{-1}\,mol^{-1}$

Answer

Both ΔH and ΔS_{system} are unfavourable, so the reaction will not take place. Carbon and hydrogen are thermodynamically stable relative to ethene at all temperatures.

e As both factors are unfavourable, there is no need to work out the value of ΔS_{total}.

Worked example 3

Comment on the feasibility of the following reaction occurring at a temperature of 298 K:

$H_2(g) + \frac{1}{2}O_2(g) \rightarrow H_2O(l)$

$\Delta H = -286\,kJ\,mol^{-1}$; $\Delta S_{system} = -45\,J\,K^{-1}\,mol^{-1}$

This reaction mixture is kinetically stable at room temperature. Either a catalyst or a spark is needed for reaction to occur. Water decomposes into its elements only at very high temperatures.

Answer

ΔH is favourable (exothermic) but ΔS_{system} is unfavourable (negative). The reaction will take place only if $\Delta H/T$ is greater than ΔS_{system}.

$\Delta S_{total} = \Delta S_{system} + \Delta S_{surr} = \Delta S_{system} - \Delta H/T$

$= -45\,J\,K\,mol^{-1} - (-286\,000\,J\,K^{-1}\,mol^{-1})/298\,K = +915\,J\,K^{-1}\,mol^{-1}$

This is a positive value and so the reaction is thermodynamically feasible at 298 K. At very high temperatures the value of $-\Delta H/T$ will become too small to overcome the negative value of ΔS_{system} and the reaction will not be feasible.

Worked example 4

Comment on the feasibility of the following reaction occurring at a temperature of 298 K:

$$CaCO_3(s) \rightarrow CaO(s) + CO_2(g)$$

$$\Delta H = +178\,\text{kJ mol}^{-1};\ \Delta S_{system} = +164\,\text{J K}^{-1}\,\text{mol}^{-1}$$

Answer

ΔH is unfavourable (endothermic), but ΔS_{system} is favourable (positive). Therefore, the reaction will only be thermodynamically feasible when ΔS_{system} is greater than $\Delta H/T$.

$$\Delta S_{total} = \Delta S_{system} + \Delta S_{surr} = \Delta S_{system} - \Delta H/T$$

$$= +164\,\text{J K mol}^{-1} - (+178\,000\,\text{J K}^{-1}\,\text{mol}^{-1})/298\,\text{K} = -433\,\text{J K}^{-1}\,\text{mol}^{-1}$$

This is a negative value and so the reaction is not feasible at 298 K.

This reaction occurs at high temperatures because the value of ΔS_{surr} becomes less negative and eventually smaller than 164. Thus, calcium carbonate is both thermodynamically and kinetically stable at room temperature, but will decompose on strong heating.

e Calcium carbonate is kinetically stable as the activation energy for its decomposition is high.

e At A-level, it is normally assumed that an increase in temperature results in negligible change in the value of ΔS_{system} because the entropies of the reactants and products change by similar amounts.

At a higher temperature, the magnitude of ΔS_{surr} always gets smaller. Thus a negative ΔS_{surr} becomes less negative and a positive ΔS_{surr} becomes less positive.

e Thermodynamics gives no information about reaction rate. A thermodynamically feasible reaction might have such high activation energy that it does not proceed at room temperature.

Spontaneous endothermic reactions

When solid ammonium carbonate is added to pure ethanoic acid, bubbles of gas are rapidly produced. This appears to be a violent reaction, but if a thermometer is placed in the acid before the ammonium carbonate is added, it will be observed that the temperature falls considerably as the reaction takes place.

$$2CH_3COOH(l) + (NH_4)_2CO_3(s) \rightarrow 2CH_3COONH_4(s) + H_2O(l) + CO_2(g)$$

Even though the reaction is endothermic, there is a considerable increase in entropy of the system because a gas is produced. This makes ΔS_{total} positive and the reaction thermodynamically spontaneous.

Hydrated barium hydroxide reacts with solid ammonium chloride in a rapid endothermic reaction at room temperature:

$$Ba(OH)_2.8H_2O(s) + 2NH_4Cl(s) \rightarrow BaCl_2(s) + 10H_2O(l) + 2NH_3(g)$$

As with the previous example, the driving force of the reaction is ΔS_{system}, which overcomes the endothermic nature of the reaction. The reactants are solids and the products are a solid, a liquid and a gas, and three substances make 13 substances, so there is a considerable gain in disorder.

Another easily observed endothermic change is the dissolving of ammonium nitrate in water.

$$NH_4NO_3(s) + aq \rightarrow NH_4^+(aq) + NO_3^-(aq)$$

As the reaction is spontaneous, the negative value of ΔS_{surr} must be outweighed by the positive value of ΔS_{system}.

All three of the above examples are thermodynamically spontaneous. The activation energies are low and so all three reactions take place rapidly at room temperature. They are examples of reactants that are thermodynamically and kinetically unstable relative to their products.

- A positive ΔS_{total} means that the reactants are thermodynamically unstable relative to the products.
- A negative ΔS_{total} means that the reactants are thermodynamically stable relative to the products.
- A small activation energy means that the reactants are kinetically unstable relative to the products.
- A large activation energy means that the reactants are kinetically stable relative to the products.

Solubility

Dissolving a gas always results in a negative ΔS_{system} because the system becomes more ordered. Therefore, for a gas to be soluble it must always dissolve exothermically (the surroundings become more disordered). This means that the equilibrium:

$$X(g) \rightleftharpoons X(aq)$$

is driven to the left by an increase in temperature. Gases, such as carbon dioxide, are less soluble in hot water than in cold water.

$$CO_2(g) \rightleftharpoons CO_2(aq) \quad \Delta H \text{ negative}$$

An interesting point follows about global warming. The accepted theory is that, historically, periods of high levels of atmospheric carbon dioxide produced high temperatures. This is questionable for a number of reasons. First, there is some evidence from ice cores that implies that the warming *preceded* the rise in carbon dioxide levels, rather than following it. This can be explained easily. As the sea temperature rises, due to changes in activity of the sun, the equilibrium shown above is driven to the left (the endothermic direction), so more carbon dioxide enters the atmosphere. The second flaw in the argument that carbon dioxide caused higher temperatures before the industrial revolution is that there is no reason why the atmospheric carbon dioxide level should have varied before man started to burn fossil fuels in large quantities.

Contrary to the expected approach to entropy, dissolving solids does not always result in a positive $\Delta S^{\ominus}_{system}$ (the system becoming more disordered). The solute becomes more disordered as it goes from highly ordered solid to a more random solution, but the solvent can become more ordered due to the forces of attraction between solute and solvent. This is particularly the case when compounds containing ions of high charge density dissolve in water (see p. 44). $\Delta H^{\ominus}_{soln}$ can be either negative (hence the surroundings also become more disordered) or slightly positive (hence the surroundings become slightly less disordered). The equilibrium:

$$X(s) \rightleftharpoons X(aq)$$

is driven to the left (less soluble) if $\Delta H^{\ominus}_{soln}$ is exothermic and to the right (more soluble) if $\Delta H^{\ominus}_{soln}$ is endothermic.

Solubility of ionic compounds

When an ionic solid dissolves in water, the lattice breaks down and the ions are separated. This is very endothermic, so you might expect that ionic solids would not dissolve in water. To explain this apparent paradox, you must think about what happens to the ions as the solid dissolves.

Cations become surrounded by water molecules. Strong ion–dipole forces act between the positive cations and the δ^- oxygen atoms in the water. Similarly, the anions become surrounded by water molecules with the δ^+ hydrogen atoms of the water molecules being strongly attracted to the negative anions. This process is called **hydration**. It is the highly exothermic nature of hydration that compensates for the endothermic break-up of the lattice.

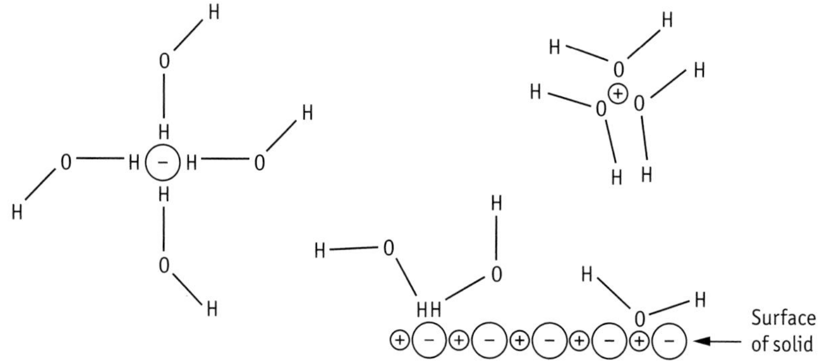

***Figure 2.4** An ionic solid dissolving*

The enthalpy of solution of a solid, ΔH_{soln}, is the enthalpy change when 1 mol of the solid is dissolved in sufficient solvent to give an infinitely dilute solution.

The hydration enthalpy of an ion, ΔH_{hyd}, is the enthalpy change when 1 mol of **gaseous** ions is dissolved in sufficient solvent to give an infinitely dilute solution.

e An infinitely dilute solution can be thought of as one in which further dilution does not cause a heat change.

Step 1:

$$NaCl(s) \rightarrow Na^+(g) + Cl^-(g) \quad \Delta H = -\Delta H_{latt}$$

Step 2:

$$Na^+(g) \xrightarrow{H_2O} Na^+(aq) \quad \Delta H = \Delta H_{hyd}(Na^+)$$

$$Cl^-(g) \xrightarrow{H_2O} Cl^-(aq) \quad \Delta H = \Delta H_{hyd}(Cl^-)$$

On addition, the gaseous ions cancel, giving:

$$NaCl(s) \xrightarrow{H_2O} Na^+(aq) + Cl^-(aq) \quad \Delta H_{soln} = -\Delta H_{latt} + \Delta H_{hyd}(Na^+) + \Delta H_{hyd}(Cl^-)$$

This can be shown as a Hess's law cycle:

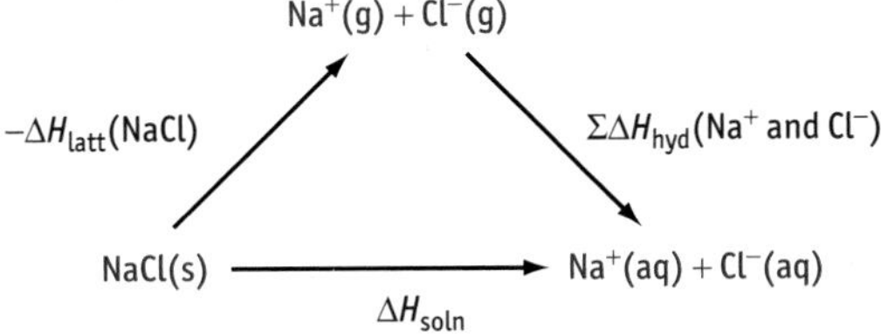

$$\Delta H_{soln} = -\Delta H_{latt} + \Delta H_{hyd}(Na^+) + \Delta H_{hyd}(Cl^-)$$

From this equation, it can be seen that:

- the more exothermic the lattice energy, the more endothermic the enthalpy of solution
- the more exothermic either of the hydration enthalpies, the more exothermic the enthalpy of solution

The sign of the enthalpy of solution is determined by the difference in magnitude between the lattice energy and the sum of the hydration energies.

Remember that the lattice energy is defined in the exothermic direction. It is the energy change when 1 mol of an ionic solid is formed from its constituent gaseous ions infinitely far apart.

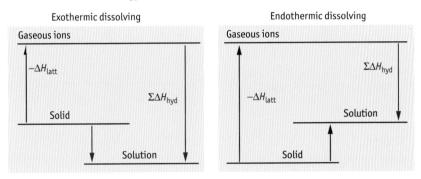

Exothermic dissolving

Gaseous ions

$-\Delta H_{latt}$

$\Sigma\Delta H_{hyd}$

Solid

Solution

Endothermic dissolving

Gaseous ions

$\Sigma\Delta H_{hyd}$

$-\Delta H_{latt}$

Solution

Solid

If the magnitude of the lattice enthalpy (blue arrow) is *less* than the sum of the hydration enthalpies of the two ions (red arrow), dissolving will be exothermic.

If the magnitude of the lattice enthalpy is *greater* than the sum of the hydration enthalpies of the two ions, dissolving will be endothermic.

A general relationship can be deduced from the energy-level diagrams above:

$$\Delta H_{soln} = -\text{lattice energy} + \text{the sum of the hydration energies of all the ions}$$

Factors that affect lattice energy

The magnitude of the lattice energy depends on the strength of the forces acting on the ions. In a lattice, each ion is surrounded by a number of ions of opposite charge, resulting in strong forces of attraction (red) and some forces of repulsion (blue). This is illustrated in Figure 2.5.

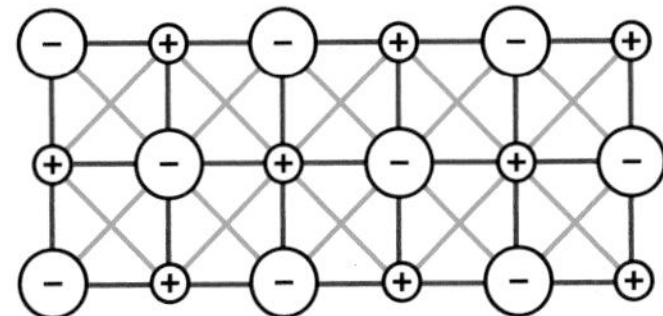

***Figure 2.5** The forces acting between the ions in a planar slice through a crystal of sodium chloride*

The strength of these forces depends mainly on:

- the product of the charges on the ions — the larger the product, the greater is the magnitude of the lattice energy
- the sum of the radii of the cation and the anion — the larger the sum, the smaller the magnitude of the lattice energy
- the extent of covalency (a small effect) — the greater the extent of covalency, the greater is the magnitude of the lattice energy

Worked example 1

Explain why the lattice energy of sodium fluoride is more exothermic than the lattice energy of potassium chloride.

Answer

It is because Na^+ has a smaller ionic radius than K^+, and F^- is smaller than Cl^-. Therefore, the forces between sodium ions and fluoride ions are stronger than those between potassium ions and chloride ions.

Worked example 2

Explain why the lattice energy of calcium oxide is approximately four times larger than that of potassium fluoride.

Answer

This is because in calcium oxide the product of the ionic charges is 4, whereas in potassium fluoride it is 1. The sums of the ionic radii of the two compounds are not very different. However, the value is smaller in CaO than in KF, so the lattice energy ratio is further increased from 4.

Table 2.4 shows lattice energies for some ionic solids.

Table 2.4 Some lattice energies

	Lattice energy/kJ mol^{-1}				
Halides	LiF −1022	NaF −902	KF −801	RbF −767	CsF −716
	LiCl −846	NaCl −771	KCl −701	RbCl −675	CsCl −645
	LiBr −800	NaBr −733	KBr −670	RbBr −647	CsBr −619
	LiI −744	NaI −684	KI −629	RbI −609	CsI −585
	$BeCl_2$ −3006	$MgCl_2$ −2500	$CaCl_2$ −2237	$SrCl_2$ −2112	$BaCl_2$ −2018
Oxides	Li_2O −2814	Na_2O −2478	K_2O −2232	Rb_2O −2161	Cs_2O −2063
	BeO −4444	MgO −3890	CaO −3513	SrO −3310	BaO −3152
Sulfides	Li_2S −2500	Na_2S −2200	K_2S −2052	Rb_2S −1944	Cs_2S −1850
	BeS −3832	MgS −3300	CaS −3013	SrS −2850	BaS −2725
Hydroxides		$Mg(OH)_2$ −2842	$Ca(OH)_2$ −2553	$Sr(OH)_2$ −2354	$Ba(OH)_2$ −2228

Table 2.4 shows that:

- the magnitude of the lattice energy steadily decreases down a group of the periodic table as the size of the cation increases
- the magnitude of the lattice energy steadily decreases down the group as the size of the anion increases
- the magnitude of the lattice energy increases as the charge on either or both the cation and the anion increases

Factors that affect hydration enthalpy

The magnitude of the hydration enthalpy of an ion depends on the strength of the force between the ion and the water molecules surrounding it. Positive ions are attracted to the δ^- oxygen atoms of the water and negative ions to the δ^+ hydrogen atoms.

The strength of the forces depends on:

- the magnitude of the charge on the ion — the greater the charge, the greater is the force
- the radius of the ion — the smaller the radius, the greater is the force

Hydration enthalpies of some gaseous ions are shown in Table 2.5.

***Table 2.5** Hydration enthalpies of ions*

Cation	ΔH_{hyd}/kJ mol^{-1}	Anion	ΔH_{hyd}/kJ mol^{-1}
Li^+	−519	F^-	−506
Na^+	−406	Cl^-	−364
K^+	−322	Br^-	−335
Mg^{2+}	−1920	I^-	−293
Ca^{2+}	−1650	OH^-	−460
Sr^{2+}	−1480		
Ba^{2+}	−1360		

Table 2.5 shows that

- the hydration energies become less exothermic as the radius of the ions in a group increases
- the magnitude of the hydration energy increases as the charge on the cation increases

Calculation of enthalpy of solution

Worked example

Use data from Tables 2.4 and 2.5 to predict the enthalpy of solution of sodium chloride.

Answer

$$\Delta H_{soln} = -\Delta H_{latt} + \Delta H_{hyd}(Na^+) + \Delta H_{hyd}(Cl^-)$$
$$= -(-771) + (-406) + (-364) = +1\,\text{kJ mol}^{-1}$$

Using lattice energies and hydration energies to predict the enthalpy of solution may be inaccurate, because slight errors in any of the quantities could result in an answer with the wrong sign. For example, if the data in the above worked example had been taken from a different source, the calculation could have been:

$$\Delta H_{soln} = -\Delta H_{latt} + \Delta H_{hyd}(Na^+) + \Delta H_{hyd}(Cl^-)$$
$$= -(-780) + (-444) + (-340) = -4\,kJ\,mol^{-1}$$

The true value of ΔH_{soln} of NaCl(s) is +3.9 kJ mol^{-1}.

Extra care must be taken with ionic compounds of formula MX_2. This can be illustrated using a Hess's law cycle:

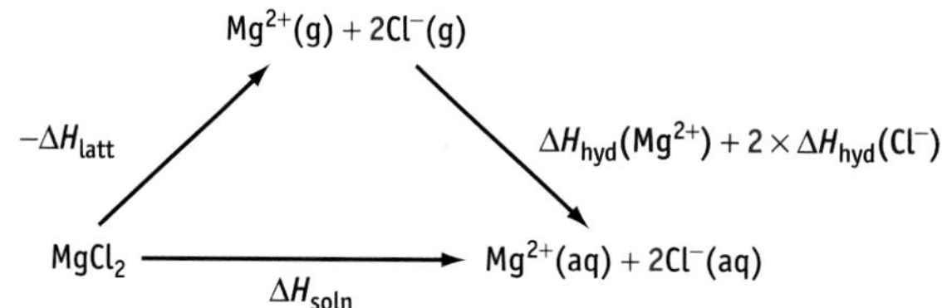

$$\Delta H_{soln} = -\Delta H_{latt}(MgCl_2) + \Delta H_{hyd}(Mg^{2+}) + 2 \times \Delta H_{hyd}(Cl^-)$$
$$= -(-2500) + (-1920) + 2(-364) = -148\,kJ\,mol^{-1}$$

The hydration enthalpy of the chloride ion must be multiplied by two, because there are two Cl^- ions in the equation.

As can be seen from Table 2.6, many ionic solids have endothermic enthalpies of solution and are still soluble. Others have exothermic enthalpies of solution and are insoluble. The concept that exothermic changes will take place spontaneously and endothermic changes will not is an oversimplification. The criteria for spontaneity include the **entropy** change of the system (p. 31).

***Table 2.6** Enthalpies (kJ mol^{-1}) of solutions of anhydrous compound at 25°C*

Cation	Anion							
	F^-	Cl^-	Br^-	I^-	OH^-	CO_3^{2-}	NO_3^-	SO_4^{2-}
Li^+	+4.9**	−37	−49	−63	−24	−18	−3	−30
Na^+	+1.9	+3.9	−0.6	−7.5	−45	−27	+21	−2.4
K^+	−18	+17	+20	+20	−57	−31	+35	+24
NH_4^+	−1.2	+15	+16	+14	–	–	+26	+6.6
Ag^+	−23	+66*	+84*	+112*	–	+42*	+23	+18
Mg^{2+}	−18*	−160	−186	−213	+2.3*	−25.3*	−91	−91
Ca^{2+}	+12*	−81	−103	−120	−16.7**	−13*	−19	−18**
Al^{3+}	−27**	−330	−370	−390	*	–	–	−350

* insoluble
** slightly soluble

Extent of solubility

Solubility of a solid is determined by the total entropy change for 1 mol of that solid.

$$\Delta S^{\ominus}_{total} = \Delta S^{\ominus}_{system} + \Delta S^{\ominus}_{surr} = \Delta S^{\ominus}_{system} + (-\Delta H^{\ominus}_{soln}/T)$$

For an ionic solid to dissolve significantly, $\Delta S^{\ominus}_{total}$ must be positive. Its value depends on both the entropy change of the system and the enthalpy change.

Entropy changes on dissolving an ionic solid

When ammonium chloride dissolves in water, the enthalpy change is +15 kJ mol^{-1}.

This is endothermic and the system moves spontaneously to a state of higher enthalpy. The entropy of the surroundings *decreases* by 15 000/298 = 50 J K^{-1} mol^{-1} (ΔS°_{surr} = −50 J K^{-1} mol^{-1}). However, this is balanced by an increase in the entropy of the system, as the ordered ammonium chloride lattice is broken down and the ions dispersed into the solvent. The value of $\Delta S^{\circ}_{system}$ is +167 J K^{-1} mol^{-1}. The total entropy change is:

$$\Delta S^{\circ}_{total} = \Delta S^{\circ}_{system} + \Delta S^{\circ}_{surr} = +167 + (-50) = +117\,\mathrm{J\,K^{-1}\,mol^{-1}}$$

This is a positive number and so the dissolving of ammonium chloride is thermodynamically spontaneous.

Entropy of the surroundings

Because $\Delta S^{\circ}_{surr} = -\Delta H^{\circ}_{soln}/T$, the sign of the entropy change depends on the sign of the enthalpy change:

- For endothermic enthalpies of solution, ΔS°_{surr} is negative

Some ionic solids have an endothermic enthalpy of solution (see Table 2.6). This means that ΔS°_{surr} is negative. If these compounds are to be soluble, $\Delta S^{\circ}_{system}$ must be positive and must outweigh the negative ΔS°_{surr}. Examples of soluble substances with endothermic ΔH°_{soln} are sodium chloride (ΔH°_{soln} = +3.9 kJ mol^{-1}), potassium sulfate (ΔH°_{soln} = +23.8 kJ mol^{-1}) and ammonium nitrate (ΔH°_{soln} = +25.8 kJ mol^{-1}).

Many insoluble substances have positive enthalpies of solution and hence negative ΔS°_{surr}. Because they are insoluble, ΔS°_{total} is negative. An example is silver chloride: ΔH°_{soln} = +66 kJ mol^{-1}; ΔS°_{surr} = −66 000/298 = −220 J K^{-1} mol^{-1}. This means that $\Delta S^{\circ}_{system}$ must either have a smaller positive value than 220 J K^{-1} mol^{-1} or be negative. (In fact, the value of $\Delta S^{\circ}_{system}$ for silver chloride is +33 J K^{-1} mol^{-1}.)

Values of enthalpy and entropy changes in dissolving are shown in Table 2.7.

- For exothermic enthalpies of solution, ΔS°_{surr} is positive.

Some ionic solids have an exothermic enthalpy of solution and so ΔS°_{surr} is positive. At first sight, it might be expected that these solids would be soluble. This is the case if $\Delta S^{\circ}_{system}$ is positive. An example is sodium hydroxide, ΔH°_{soln} = −32.7 kJ mol^{-1}. It is also the case if $\Delta S^{\circ}_{system}$ is negative but its magnitude is less than ΔS°_{surr}. An example is barium chloride, ΔH°_{soln} = −13.2 kJ mol^{-1}.

If the solid is *insoluble*, ΔS°_{total} is negative. This means that $\Delta S^{\circ}_{system}$ must be negative and its magnitude must outweigh the positive ΔS°_{surr}. An example is calcium carbonate, ΔH°_{soln} = −12.3 kJ mol^{-1} and ΔS°_{surr} = +41 J K^{-1}mol^{-1}. For ΔS°_{total} to be negative, $\Delta S^{\circ}_{system}$ must be more negative than −41 J K^{-1} mol^{-1}. (Its value is −202 J K^{-1} mol^{-1}.)

Entropy of the system

This is made up of the entropy change of the solute and the entropy change of the solvent.

$$\Delta S^{\circ}_{system} = \Delta S^{\circ}_{solute} + \Delta S^{\circ}_{solvent}$$

When any solute is dissolved, its entropy increases as the particles go from being arranged in a regular pattern in the solid to being distributed randomly in the solution. This applies to both covalent substances, such as glucose, and to ionic substances such as sodium chloride.

$\Delta S^{\ominus}_{solute} > 0$ ($\Delta S^{\ominus}_{solute}$ **positive)**

In the case of many solutions, there is also an increase in the entropy of the solvent as it becomes mixed with solute particles. However, when anhydrous ionic substances are dissolved in water, there is also a certain amount of ordering of water molecules. The positive ions become surrounded by water molecules, as the δ^- oxygen atoms in the water bond with the positive cations. This makes the δ^+ hydrogen atoms more positive causing them to bind a second sphere of water molecules. The extent to which this happens depends on the charge density of the cation. The charge density of the ammonium ion is small because the single positive charge is delocalised over the whole ion and so the entropy of the water is not significantly altered. The lithium ion, Li^+, decreases the entropy of the solvent water to a greater extent than the other group 1 ions, as its ionic radius is much smaller. The group 2 cations have a much larger charge density than group 1 ions. They are doubly charged and, for each, the ionic radius is much less than that of the group 1 ion in the same period. The extent to which the $\Delta S^{\ominus}_{water}$ is negative (more ordered) decreases as the group is descended. Thus barium ions, Ba^{2+}, order the water less than magnesium ions, Mg^{2+}.

This can be demonstrated by filling a burette with water and then adding some anhydrous aluminium chloride or iron(III) chloride. The volume of the water shrinks noticeably as water molecules bond to the ions and become less randomly arranged. The high charge density of Al^{3+} ions means that when aluminium chloride is dissolved, ΔS_{system} is negative.

Negative ions also cause a slight ordering of water molecules, as the negative ions are surrounded by δ^+ hydrogen atoms in water molecules.

For the process:

ionic solid + water → aqueous ions

fhe following assumptions can be made:

- Dissolving group 1 compounds, ammonium compounds and silver compounds causes only a small change in the entropy of the water, so $\Delta S^{\ominus}_{system}$ is always positive.
- Dissolving compounds containing doubly charged cations, such as those of the group 2 metals, causes a large decrease in the entropy of the water, so $\Delta S^{\ominus}_{system}$ is negative.

It is possible to calculate the value of $\Delta S^{\ominus}_{system}$. It has been estimated that $\Delta S^{\ominus}_{system}$ for dissolving ammonium chloride in water is $+167\,J\,K^{-1}\,mol^{-1}$; the theoretical value for barium sulfate is $-104\,J\,K^{-1}\,mol^{-1}$.

Some enthalpy and entropy changes for ionic solids dissolving in water, i.e. for the change

$MX(s) + aq \rightarrow M^{n+}(aq) + X^{n-}(aq)$

are shown in Table 2.7.

It can be seen from the table that all the compounds containing singly charged cations have a positive $\Delta S^{\ominus}_{system}$. This is the main reason why group 1 compounds are water soluble, the exception being lithium fluoride. This is caused by the fluoride ion forming strong hydrogen bonds with water molecules and by the

Substance	$\Delta H^\ominus_{soln}$/ kJ mol^{-1}	$\Delta S^\ominus_{surr}$/ J K^{-1} mol^{-1}	$\Delta S^\ominus_{system}$/ J K^{-1} mol^{-1}	$\Delta S^\ominus_{total}$/ J K^{-1} mol^{-1}	Solubility
LiCl(s)	−37	+124	+11	+135	Soluble
NaCl(s)	*+4*	−13	+43	+30	Soluble
KCl(s)	*+17*	−57	+77	+20	Soluble
LiF(s)	+5	−17	−37	−54	Insoluble
NH_4Cl(s)	*+15*	−51	+167	+116	Soluble
AgCl(s)	*+66*	−221	+33	−188	Insoluble
$MgSO_4$(s)	−91	+305	−213	−92	Soluble
$CaSO_4$(s)	−18	+60	−145	−85	Insoluble
$BaSO_4$(s)	*+19*	−65	−104	−169	Insoluble
$CuSO_4$(s)	−73	+245	−192	+53	Soluble

Table 2.7 ***Enthalpy and entropy changes for ionic solids dissolving in water***

The substances with an exothermic $\Delta H^\ominus_{soln}$ are shown in red; those that with an endothermic $\Delta H^\ominus_{soln}$ are in blue.

e Exothermic indicates a positive $\Delta S^\ominus_{surr}$; endothermic indicates a negative $\Delta S^\ominus_{surr}$

high charge density of the small Li^+ ion. Both these factors cause a more ordered arrangement of the water molecules. Silver chloride is insoluble in spite of the favourable $\Delta S^\ominus_{system}$, because its $\Delta H^\ominus_{soln}$ is highly endothermic.

The reverse is true for compounds containing doubly positive cations. All have a negative $\Delta S^\ominus_{system}$. This is why many of their compounds (carbonates, phosphates and hydroxides) are insoluble. They are only soluble if the enthalpy of solution is sufficiently exothermic. This is the case with magnesium sulfate and copper sulfate, but not with calcium sulfate. For barium sulfate, $\Delta H^\ominus_{soln}$ is endothermic *and* $\Delta S^\ominus_{system}$ is negative, so it is extremely insoluble.

The relationship between enthalpy of solution, solubility and the entropy of the system is summarised in Table 2.8.

$\Delta H^\ominus_{soln}$	Solubility	$\Delta S^\ominus_{system}$
Endothermic (+)	Soluble	Must be more positive than $\Delta H^\ominus_{soln}/T$
Endothermic (+)	Insoluble	Negative or less positive than $\Delta H^\ominus_{soln}/T$
Exothermic (−)	Soluble	Positive or less negative than $\Delta H^\ominus_{soln}/T$
Exothermic (−)	Insoluble	Must be more negative than $\Delta H^\ominus_{soln}/T$

Table 2.8 ***Enthalpy, solubility and entropy***

Solubility trends in a group

The enthalpy of solution of an ionic solid can be calculated by means of a Hess's law cycle (p. 42). The reason for the trend in the value of $\Delta H^\ominus_{soln}$ is found in the way that the lattice energies and hydration energies change down a group.

The enthalpy of solution is a balance between lattice energy and the sum of the hydration energies of the ions:

$$\Delta H_{soln} = -\text{lattice energy} + \text{sum of the hydration energies of the ions}$$

In any group of the periodic table:

- The lattice energies become less exothermic as the group is descended.
- The hydration energies of the cations become less exothermic down the group.

Therefore, the change in ΔH_{soln} down the group is determined by which quantity shows the *greater decrease*.

If the lattice energy decreases more than the hydration energy, the process of dissolving is more exothermic (or less endothermic).

The lattice energy and hydration enthalpy values for group 2 hydroxides and sulfates are shown in Table 2.9.

Substance	Lattice energy/ kJ mol^{-1}	Hydration enthalpy of cation/kJ mol^{-1}
$Mg(OH)_2(s)$	−2842	−1920
$Ca(OH)_2(s)$	−2553	−1650
$Sr(OH)_2(s)$	−2354	−1480
$Ba(OH)_2(s)$	−2228	−1360
Change down the group	614	560
$MgSO_4(s)$	−2874	−1920
$CaSO_4(s)$	−2677	−1650
$SrSO_4(s)$	−2516	−1480
$BaSO_4(s)$	−2424	−1360
Change down the group	450	560

***Table 2.9** Lattice energy and hydration enthalpy values for group 2 hydroxides and sulfates.*

Table 2.9 shows that, for the hydroxides of group 2, on descending the group there is a greater change in lattice energy than there is in hydration enthalpy. This results in the enthalpy of solution becoming steadily more exothermic.

The lattice energy changes more than the hydration enthalpy because of the way in which the two factors depend on the ionic radius:

- The hydration energy of a cation depends upon its charge density — the charge divided by the radius.
- The lattice energy depends upon the charges of the two ions multiplied together divided by the sum of the two ionic radii — $\{r(+) + (r(-)\}$.
- The OH^- ion is a small anion, similar in size to the group 2 cations. Therefore, the value of $\{r(+) + r(-)\}$ increases considerably as the value of the radius of the cation, r(+), increases.

The opposite is true for the group 2 sulfates. The sulfate ion is much larger than any of the group 2 cations. Therefore, as $r(-) >> r(+)$, the value of $\{r(+) + r(-)\}$ changes by only a small amount. This means that the decrease in the value of the lattice energy is more than the decrease of the hydration enthalpy of the ions. This makes the enthalpy of solution increasingly less exothermic as the group is descended.

The values of $\Delta H^{\ominus}_{soln}$ are shown in Table 2.10.

The value of $\Delta H^{\ominus}_{soln}$ is not the only factor determining solubility. The other factor is the change in entropy of the system. For many ionic solids, this is difficult to determine accurately. A guide is the relative entropy values of the ions, which are also shown in Table 2.10. Note how the entropy of the ion increases as the ionic radius increases.

Solubility of silver halides

Substance	$\Delta H^\ominus_{soln}$ /kJ mol^{-1}	$\Delta S^\ominus_{surr} = -\Delta H^\ominus_{soln}/T$ /J K^{-1} mol^{-1}	Hydrated ion	Relative* entropy value of hydrated ion/J K^{-1} mol^{-1}
AgF	−20	+67	F^-(aq)	−14
AgCl	+66	−221	Cl^-(aq)	+57
AgBr	+85	−285	Br^-(aq)	+82
AgI	+113	−379	I^-(aq)	+111
Change down the group from F to I	More endothermic	−446, so less likely to dissolve	Change down the group from F^-(aq) to I^-(aq)	−125, so more likely to dissolve

***Table 2.10** Enthalpy and entropy changes involved in dissolving ionic solids*

* The values of the entropy of hydrated ions are relative to the value for H^+(aq).

Note that the change in $\Delta S^\ominus_{surr}$ down a group is more than the change in $\Delta S^\ominus_{system}$. This means that the change in enthalpy of solution is the main factor determining the change in solubility down a group.

The change in value of $\Delta H^\ominus_{soln}$ and hence of $\Delta S^\ominus_{surr}$ indicates that the solubility should decrease from silver fluoride, AgF, to silver iodide, AgI. The change in entropy of the halide ion indicates the opposite, because the values become more favourable. However, the change in the entropy of the hydrated ions (125 J K^{-1} mol^{-1}) is significantly less than the change in $\Delta S^\ominus_{surr}$ (446 J K^{-1} mol^{-1}). This explains the fall in solubility of silver fluoride to silver iodide.

Solubility of group 2 hydroxides

Substance	$\Delta H^\ominus_{soln}$ /kJ mol^{-1}	$\Delta S^\ominus_{surr} = -\Delta H^\ominus_{soln}/T$ /J K^{-1} mol^{-1}	Hydrated ion	Relative* entropy value of hydrated ion/J K^{-1} mol^{-1}
$Mg(OH)_2$	+3	−10	Mg^{2+}(aq)	−138
$Ca(OH)_2$	−16	+54	Ca^{2+}(aq)	−53
$Sr(OH)_2$	−46	+154	Sr^{2+}(aq)	−33
$Ba(OH)_2$	−52	+174	Ba^{2+}(aq)	+10
Change down the group from Mg to Ba	More exothermic	−184, so more likely to dissolve	Change down the group from Mg^{2-}(aq) to Ba^{2-}	−148, so more likely to dissolve

The enthalpy of solution becomes more negative down the group. This means that $\Delta S^\ominus_{surr}$ becomes more positive and so favours solubility. The change in the entropy of the cation gets less negative and this also favours solubility. As both factors favour an increase in solubility, barium hydroxide is more soluble than magnesium hydroxide.

Solubility of group 2 sulfates

Substance	$\Delta H^\ominus_{soln}$ /kJ mol^{-1}	$\Delta S^\ominus_{surr} = -\Delta H^\ominus_{soln}/T$ /J K^{-1} mol^{-1}	Hydrated ion	Relative* entropy value of hydrated ion/J K^{-1} mol^{-1}
$MgSO_4$	−91	+305	Mg^{2+}(aq)	−138
$CaSO_4$	−18	+60	Ca^{2+}(aq)	−53
$SrSO_4$	−9	+30	Sr^{2+}(aq)	−33
$BaSO_4$	+19	−63	Ba^{2+}(aq)	+10
Change down the group from Mg to Ba	Less exothermic	−368, so less likely to dissolve	Change down the group from Mg^{2+}(aq) to Ba^{2+}(aq)	+148, so more likely to dissolve

The enthalpy of solution becomes *less* negative down the group. This means that $\Delta S^\ominus_{surr}$ becomes less positive and, therefore, favours insolubility. The change in the entropy of the cation gets less negative, which favours solubility, but the change in entropy of the hydrated cation (148 J K^{-1} mol^{-1}) is much less than the change in entropy of the surroundings (368 J K^{-1} mol^{-1}). This results in the solubility of the sulfates decreasing down the group.

Melting and boiling points

At equilibrium, the value of ΔS_{total} is zero. At 0°C, there is equilibrium between ice and water. The ice does not melt, nor does the water freeze, unless heat is added to or taken from the system. Neither direction is thermodynamically feasible, so the two forms of water remain in equilibrium:

$$\Delta S_{total} = \Delta S_{system} + \Delta S_{surr}$$

$$= \Delta S_{system} - \frac{\Delta H}{T} = 0$$

$$\Delta S_{system} = \frac{\Delta H}{T} \quad \text{or} \quad T = \frac{\Delta H}{\Delta S_{system}}$$

ΔH for ice melting = +6012 J mol^{-1}

$\Delta S_{system} = S(\text{water}) - S(\text{ice}) = +22\ \text{J K}^{-1}\ \text{mol}^{-1}$

$$\text{melting temperature of ice} = \frac{\Delta H}{\Delta S_{system}} = \frac{6012}{22} = 273\ \text{K} = 0°\text{C}$$

ΔH for water boiling = +40 700 J mol^{-1}

$\Delta S_{system} = S(\text{steam}) - S(\text{water}) = +109\ \text{J K}^{-1}\ \text{mol}^{-1}$

$$\text{boiling temperature of water} = \frac{\Delta H}{\Delta S_{system}} = \frac{40\,700}{109} = 373\ \text{K} = 100°\text{C}$$

It can be seen from the expression $T = \Delta H/\Delta S$, that the melting or boiling temperature depends upon the amount of energy required for the change of state. This explains why boiling and melting temperatures depend on the strength of the forces between the particles:

strong force = large amount of energy needed to separate the particles
= high melting or boiling temperature

Summary

- The reactants are thermodynamically unstable relative to the products if ΔS_{total} for the change is positive. This means that the reaction is thermodynamically feasible.
- Endothermic reactions can happen only if the entropy change of the system is positive.
- Endothermic reactions are more likely to take place at higher temperatures.
- Exothermic reactions are always thermodynamically favourable if the entropy change of the system is positive.
- Exothermic reactions are thermodynamically favourable even when the entropy of the system is negative, if the entropy change of the surroundings outweighs the entropy change of the system.

Questions

1 Give an example of an endothermic reaction that takes place rapidly at room temperature.

2 What is the entropy of a perfect crystal of helium at absolute zero? Would a perfect crystal of sodium chloride have a different entropy value at absolute zero? Explain your answer.

3 State and explain which of the following would have the higher entropy:

a an aqueous solution of glucose, $C_6H_{12}O_6$, or an aqueous solution of carbon dioxide at 25°C

b a solution of sodium chloride at 50°C or a solution of sodium chloride at 25°C?

4 Explain why the copper sulfate in a solution of copper sulfate does not spontaneously sink to the bottom even though it is denser than water.

5 Use the data in Table 2.2 to calculate the thermodynamic feasibility of the following reactions at 298 K.

a $N_2(g) + 3H_2(g) \rightarrow 2NH_3(g) \quad \Delta H = -92\ kJ\ mol^{-1}$

b $CaCO_3(s) \rightarrow CaO(s) + CO_2(g) \quad \Delta H = +178\ kJ\ mol^{-1}$

6 Hydrated barium hydroxide reacts with solid ammonium chloride according to the equation:

$$Ba(OH)_2.8H_2O(s) + 2NH_4Cl(s) \rightarrow BaCl_2(s) + 10H_2O(l) + 2NH_3(g)$$

The total standard entropy change at 298K = +150 J K^{-1} mol^{-1}

Some enthalpy and entropy data are given in the table.

Substance	Standard enthalpy of formation /kJ mol^{-1}	Standard entropy/ J K^{-1} mol^{-1}
$Ba(OH)_2.8H_2O(s)$	−3245	
$NH_4Cl(s)$	−315	92
$BaCl_2(s)$	−860	130
$H_2O(l)$	−286	70
$NH_3(g)$	−46	193

Use the data to:

a Calculate $\Delta H^\ominus$ for this reaction

b Calculate ΔS_{system} and hence calculate the standard entropy of hydrated barium hydroxide.

7 The reaction between hydrogen and oxygen does not take place at a temperature of 298 K.

$$H_2(g) - \tfrac{1}{2}O_2(g) \rightarrow H_2O(l) \quad \Delta H = -286\ kJ\ mol^{-1}$$

Use the data above and that in Table 2.2 to explain the concepts of thermodynamic stability and kinetic inertness.

8 Draw a Hess's law diagram, and use it together with the data here, to calculate ΔH_{soln} of lithium fluoride, LiF.

	Enthalpy change/ kJ mol^{-1}
$\Delta H_{hyd}(Li^+(g))$	−519
$\Delta H_{hyd}(F^-(g))$	−506
$\Delta H_{latt}(LiF(s))$	−1022

Comment on the likely solubility of lithium fluoride in water.

9 Draw a Hess's law diagram for dissolving calcium chloride. Use it and the data here to calculate ΔH_{hyd} of a chloride ion, Cl^-.

	Enthalpy change/ kJ mol^{-1}
$\Delta H_{hyd}(Ca^{2+}(g))$	−1650
$\Delta H_{latt}(CaCl_2(s))$	−2237
$\Delta H_{soln}(CaCl_2(s))$	−83

10 Given the ionic radii of the ions M^{2+} and Q^{3+}, explain which ion would have a higher exothermic hydration enthalpy.

Ion	Ionic radius/nm
M^{2+}	0.031
Q^{3+}	0.095

11 Study the data in the table.

Substance	$\Delta H^\ominus_{soln}$/ kJ mol^{-1}	Ion	$S^\ominus$/ J K^{-1} mol^{-1}
$MgF_2(s)$	−18	$Mg^{2+}(aq)$	−138
$CaF_2(s)$	+13	$Ca^{2+}(aq)$	−55
$AgBr(s)$	+85	$Br^-(aq)$	+57
$AgI(s)$	+112	$I^-(aq)$	+111

a Use the data to suggest and explain the relative solubility of the group 2 fluorides, magnesium fluoride, MgF_2, and calcium fluoride, CaF_2.

b Use the data to suggest and explain the relative solubility of the silver halides, silver bromide, AgBr, and silver iodide, AgI.

12 State and explain:

a whether $H_2O(l)$ at 25 °C has a higher or lower entropy than $H_2O(l)$ at 35 °C

b whether $H_2O(l)$ at 100 °C has a higher or lower entropy than $H_2O(g)$ at 100 °C

13 Use the following equations to state and explain whether the reactions result in an increase or decrease in the entropy of the systems:

a $SO_2(g) + \frac{1}{2}O_2(g) \rightarrow SO_3(g)$

b $NH_4Cl(s) + OH^-(aq) \rightarrow NH_3(g) + Cl^-(aq) + H_2O(l)$

c $CaCO_3(s) \rightarrow CaO(s) + CO_2(g)$

d $NH_4NO_3(s) + aq \rightarrow NH_4^+(aq) + NO_3^-(aq)$

e $Ca(NO_3)_2(s) + aq \rightarrow Ca^{2+}(aq) + 2NO_3^-(aq)$

14 a Calculate ΔS_{system} for the following reaction at 85 °C:

$$N_2O_4(g) \rightarrow 2NO_2(g) \quad \Delta H = +57.4 \text{ kJ mol}^{-1}$$

Compound	Entropy, S, at 85 °C/J K^{-1} mol^{-1}
$N_2O_4(g)$	325
$NO_2(g)$	256

b Calculate ΔS_{total} for the reaction at 85°C.

c Comment on the feasibility of this reaction.

d Explain the term **thermodynamic stability** with reference to this reaction.

15 a Predict which of the following changes, W to Z, will take place at a temperature of 298 K.

Change	ΔH/kJ mol^{-1}	ΔS_{system}/J K^{-1} mol^{-1}
W	−170	+500
X	−170	−500
Y	+170	−500
Z	+170	+500

b Which of the changes W to Z will become more favourable when the temperature is increased?

16 Dinitrogen tetroxide decomposes spontaneously at 50°C:

$$N_2O_4(g) \rightarrow 2NO_2(g) \quad \Delta H = +58 \text{ kJ mol}^{-1}$$

State and explain the sign of ΔS_{system} for this reaction.

17 Ethanoic acid can be prepared from carbon monoxide and methanol

$$CO(g) + CH_3OH(l) \rightarrow CH_3COOH(l) \quad \Delta H^\ominus = -137 \text{ kJ mol}^{-1}$$

The standard entropy values, at 298 K, of these substances are given in the table.

Substance	$S^\ominus$/J K^{-1} mol^{-1}
$CO(g)$	198
$CH_3OH(l)$	127
$CH_3COOH(l)$	160

a Suggest why the standard entropy of ethanoic acid is greater then that of methanol.

b Calculate the entropy change of the system for this reaction.

c Calculate the entropy change of the surroundings.

d Show, by calculation, whether this reaction is thermodynamically spontaneous at 298 K.

e This reaction does not take place at room temperature unless a catalyst is present. Use this fact and your answer from part **d** to explain the terms **thermodynamic stability** and **kinetic inertness.**

18 Ludwig Boltzmann could be described as the 'father of entropy'. Refer to the web and write a few lines about him. What was written on his tombstone?

Equilibrium

Chapter 3

Introduction

Some reactions go to completion, and others do not. The latter type of reaction is called a **reversible** reaction. Physical changes, such as evaporation, are also reversible, for example:

$$Br_2(l) \rightleftharpoons Br_2(g)$$

When liquid bromine is mixed with air in a sealed container, a dynamic equilibrium between the liquid and gaseous bromine is reached. Gaseous bromine molecules condense into the liquid at exactly the same rate as bromine molecules evaporate from the surface of the liquid.

ZAHOOR UL-HAQ

Liquid bromine in equilibrium with gaseous bromine

Consider the gaseous reaction:

$$H_2(g) + I_2(g) \rightleftharpoons 2HI(g)$$

When hydrogen and iodine are mixed in a sealed container at a temperature, T, and left, **dynamic equilibrium** is eventually reached. At this point, there is no further change in the concentrations of the reactants and products but the reactions have not stopped. The forward and backward reactions are continuing at the same rate.

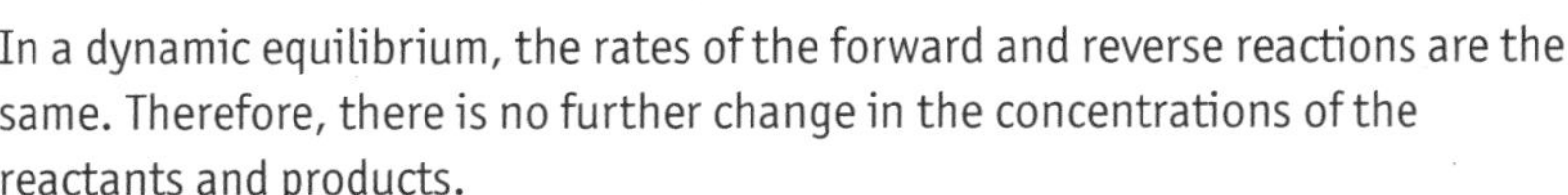

In a dynamic equilibrium, the rates of the forward and reverse reactions are the same. Therefore, there is no further change in the concentrations of the reactants and products.

Rate of forward reaction = rate of backward reaction

The reaction between hydrogen and iodine was studied by Guldberg and Waage in 1864. They mixed different amounts of hydrogen and iodine and allowed the mixtures to reach equilibrium at 480°C.

They then measured the concentration of the three substances at equilibrium and tried to find a mathematical relationship between these concentrations.

Typical results for this are shown in Table 3.1.

Square brackets around the symbol of a species mean the concentration, in mol dm^{-3}, of that substance.

Initial $[H_2]$	Initial $[I_2]$	$[H_2]$ at equilibrium	$[I_2]$ at equilibrium	[HI] at equilibrium	$\frac{[HI]}{[H_2][I_2]}$	$\frac{[HI]^2}{[H_2][I_2]}$
0.040	0.040	0.0089	0.0089	0.062	783	49
0.080	0.040	0.0426	0.0026	0.0748	675	51
0.080	0.080	0.018	0.018	0.124	383	47
0.020	0.080	0.0005	0.0605	0.0389	1286	50

***Table 3.1** Reaction between hydrogen and iodine*

The values in the final column are constant to within experimental error, so from these results, it appears that:

$$\frac{[HI]^2_{eq}}{[H_2]_{eq}[I_2]_{eq}} = \text{a constant}$$

The equilibrium constant, K_c

Law of mass action and K_c

The results given in Table 3.1, and those of other equilibrium reactions, enabled Guldberg and Waage to formulate the **law of mass action**. This states that when reactions reach equilibrium, the equilibrium concentrations of the products multiplied together and divided by the equilibrium concentrations of the reactants also multiplied together, with the concentration of each substance raised to the power appropriate to the reaction stoichiometry, are a constant at a given temperature.

For example, for the reaction:

$$N_2(g) + 3H_2(g) \rightleftharpoons 2NH_3(g)$$

$$\frac{[NH_3]^2_{eq}}{[N_2]_{eq}[H_2]^3_{eq}} = \text{a constant}$$

where $[NH_3]_{eq}$ is the concentration, in mol dm^{-3}, of ammonia at *equilibrium*.

The constant is called the **equilibrium constant** (measured in terms of concentrations) and has the symbol, K_c.

The value of the equilibrium constant depends on ΔS°_{total}:

$$\Delta S^{\circ}_{total} = R \ln K, \text{ where } R \text{ is the gas constant, which equals } 8.31\,\text{J K}^{-1}\,\text{mol}^{-1}$$

The value of ΔS°_{total} depends on the entropy change of the system and the entropy change of the surroundings (see Chapter 2):

$$\Delta S^{\circ}_{total} = \Delta S^{\circ}_{system} + \Delta S^{\circ}_{surroundings} = \Delta S^{\circ}_{system} - \Delta H/T$$

Thus ΔS°_{total} and hence K depend on:

- the nature of the reaction, which determines ΔH° and $\Delta S^{\circ}_{system}$
- the temperature at equilibrium

The value of K_c does *not* depend on the pressure or the presence of a catalyst.

In general, for a reaction:

$$xA + yB \rightleftharpoons nC + mD$$

where x, y, n and m are the stoichiometric amounts in the equation:

$$K_c = \frac{[C]^n_{eq}[D]^m_{eq}}{[A]^x_{eq}[B]^y_{eq}}$$

The right-hand side of this expression is called the 'concentration term' or the 'reaction quotient' and given the symbol Q.

It is important to realise that the equilibrium constant only equals the concentration term, Q, when the system it at equilibrium. The equilibrium constant is

e Remember, that in an equilibrium constant expression, the products are on top and the reactants are on the bottom.

a thermodynamic quantity, the value of which depends on the total entropy change of the reaction.

- $Q = K$: the system is in equilibrium and there will be no further change in concentration of the reactants and products.
- $Q > K$: the system is not in equilibrium and will react to reduce the value of the numerator. Thus products will be converted into reactants (the position of equilibrium will shift to the left).
- $Q < K$: the system is not in equilibrium and will react to increase the value of the numerator. Thus, reactants will be converted into products (the position of equilibrium will shift to the right).

e The numerator is the term on the top of the fraction

The chemical equation and the expression for K

Equilibrium constant in terms of concentration, K_c

This is defined in terms of the equilibrium concentrations of the reactants and products of the reversible reaction.

An equilibrium constant has no meaning unless it is linked to a chemical equation. Consider the equilibrium reaction of sulfur dioxide and oxygen reacting reversibly to form sulfur trioxide. This reaction can be represented by two equations and hence by two expressions for the equilibrium constant, K_c. The values given below are at 727°C (1000 K).

$$2SO_2(g) + O_2(g) \rightleftharpoons 2SO_3(g)$$

$$\frac{[SO_3]^2_{eq}}{[SO_2]^2_{eq}[O_2]_{eq}} = 2800\,mol^{-1}\,dm^3$$

or

$$SO_2(g) + \tfrac{1}{2}O_2(g) \rightleftharpoons SO_3(g)$$

$$K_c' = \frac{[SO_3]_{eq}}{[SO_2]_{eq}[O_2]^{\frac{1}{2}}_{eq}}$$

$$= 52.9\,mol^{-1/2}\,dm^{3/2}$$

Note that $52.9 = \sqrt{2800}$, so $K_c' = \sqrt{K_c}$

The reaction can also be written in the other direction, giving a third expression for K:

$$2SO_3(g) \rightleftharpoons 2SO_2(g) + O_2(g)$$

$$K_c'' = \frac{[SO_3]^2_{eq}[O_2]_{eq}}{[SO_3]^2_{eq}}$$

$$= 3.57 \times 10^{-4}\,mol\,dm^{-3}$$

Note that $\frac{1}{2800} = 3.57 \times 10^{-4}$

The three equilibrium constants are connected by the expression:

$$K_c = (K_c')^2 = \frac{1}{K_c''}$$

The reactions above are all examples of homogeneous reactions.

A homogeneous reaction is one in which all the reactants and products are in the same phase.

Gases always mix, so reactions involving only gases are homogeneous. Reversible reactions in solution are also examples of homogeneous equilibria. In such reactions, the concentration terms of all the reactants and products appear in the expression for the equilibrium constant. For example, for the homogeneous reaction:

$$CH_3COOH(l) + C_2H_5OH(l) \rightleftharpoons CH_3COOC_2H_5(l) + H_2O(l)$$

$$K_c = \frac{[CH_3COOC_2H_5]_{eq}[H_2O]_{eq}}{[CH_3COOH]_{eq}[C_2H_5OH]_{eq}}$$

$[H_2O]$ in equilibrium expressions

- When water is a reactant but not the solvent, the term $[H_2O]$ must always appear in the expression for the equilibrium constant.
- When water is in the gaseous state, $[H_2O]$ or $p(H_2O)$ (p. 60) must appear in equilibrium constant expressions.
- When water is the solvent, even if it is also a reactant or product, $[H_2O]$ does *not* appear in the expression for the equilibrium constant. This is because its concentration remains constant and its value is incorporated into K_c.

In aqueous solution or in pure water, $[H_2O]$ is the number of moles of water divided by the volume. For example:

- 1 dm^3 of water has a mass of 1000 g.
- It contains $\frac{1000\,g}{18.0\,g\,mol^{-1}} = 55.6\,mol$, so $[H_2O] = 55.6\,mol\,dm^{-3}$

Worked example 1

Write the expression for the equilibrium constant, K_c, for the reaction:

$$CH_3COOCH_3(l) + H_2O(l) \rightleftharpoons CH_3COOH(l) + CH_3OH(l)$$

Answer

$$K_c = \frac{[CH_3COOH]_{eq}[CH_3OH]_{eq}}{[CH_3COOCH_3]_{eq}[H_2O]_{eq}}$$

Water is a reactant but not the solvent, so $[H_2O]_{eq}$ appears in the equilibrium expression.

Worked example 2

Write the expression for the equilibrium constant, K_c, for the reaction:

$$Cr_2O_7^{2-}(aq) + H_2O(l) \rightleftharpoons 2CrO_4^{2-}(aq) + 2H^+(aq)$$

Answer

$$K_c = \frac{[CrO_4^{2-}]^2_{eq}[H^+]^2_{eq}}{[Cr_2O_7^{2-}]_{eq}}$$

Water is a reactant and the solvent, so $[H_2O]_{eq}$ is omitted from the equilibrium expression.

Calculation of K_c from experimental data

A typical question would give the starting amounts of the reactants, the total volume and the percentage that reacted and ask you to calculate the equilibrium constant.

The calculation requires the use of a table, as in the worked example below.

- Write the chemical equation.
- Construct a suitable table and write in the following:
  - the initial amounts (in moles) of the reactants and of the products if their initial amounts were not zero

- the amounts by which the reactants and the products change in reaching equilibrium — use the stoichiometry of the equation
- the amount, in moles, of each substance at equilibrium
- the equilibrium concentration in $mol\,dm^{-3}$ — divide the equilibrium number of moles by the total volume

- Below the table, write the expression for the equilibrium constant.
- Substitute the equilibrium concentrations into the expression and calculate its value. At the same time, determine the units of K_c and include them in your answer.

Worked example

When 0.0200 mol of sulfur trioxide is placed in a flask of volume 1.50 dm^3 and allowed to reach equilibrium at 600°C, 29% of it decomposes into sulfur dioxide and oxygen. Calculate the value of the equilibrium constant, K_c.

Answer

The equation is:

$$2SO_3(g) \rightleftharpoons 2SO_2(g) + O_2(g)$$

	SO_3	SO_2	O_2
Initial moles	0.0200	0	0
Change	-0.29×0.0200 $= -0.0058$	+0.0058	$+\frac{1}{2} \times 0.0058$ $= +0.0029$
Moles at equilibrium	$0.0200 - 0.0058$ $= 0.0142$	$0 + 0.0058$ $= 0.0058$	$0 + 0.0029$ $= 0.0029$
Concentration at equilibrium/$mol\,dm^{-3}$	0.0142/1.50 $= 0.00947$	0.0058/1.50 $= 0.00387$	0.0029/1.50 $= 0.00193$

$$K_c = \frac{[SO_2]^2_{eq}[O_2]_{eq}}{[SO_3]^2_{eq}} = \frac{(0.00387\ mol\,dm^{-3})^2 \times (0.00193\ mol\,dm^{-3})}{(0.00947\ mol\,dm^{-3})^2}$$

$$= 3.2 \times 10^{-4}\ mol\,dm^{-3}$$

Note that as 0.0058 mol of sulphur trioxide reacts, 0.0058 mol of sulphur dioxide and $\frac{1}{2}$ of $0.0058 = 0.0029$ mol of oxygen are produced. This is because the ratio of the three substances in the chemical equation is 2:2:1 or $1:1:\frac{1}{2}$.

The same method is used when some product, as well as the reactants, is initially present. In this type of question the equilibrium moles of reactants will be less than the initial amounts, but the equilibrium moles of the product will be more than the initial amount.

The units are determined as in worked examples 1 and 2 on p. 57.

Worked example 1

A vessel of volume 2.0 dm^3 was filled with 0.060 mol of methane, CH_4, 0.070 mol of steam and 0.010 mol of hydrogen and allowed to reach equilibrium at a temperature of T°C. 80% of the methane reacted. Calculate the value of the equilibrium constant, K_c, at this temperature.

Answer

The equation is

$$CH_4(g) + H_2O(g) \rightleftharpoons CO(g) + 3H_2(g)$$

	CH_4	H_2O	CO	H_2
Initial moles	0.060	0.070	0	0.010
Change	-0.80×0.060 $= -0.048$	-0.048	$+0.048$	$+3 \times 0.048$ $= +0.144$
Moles at equilibrium	$0.060 - 0.048$ $= 0.012$	$0.070 - 0.048$ $= 0.022$	$0 + 0.048$ $= 0.048$	$0.144 + 0.010$ $= 0.154$
Concentration at equilibrium/mol dm^{-3}	0.012/2.0 = 0.0060	0.022/2.0 = 0.011	0.048/2.0 = 0.024	0.154/2 = 0.077

$$K_c = \frac{[CO]_{eq}[H_2]^3_{eq}}{[CH_4]_{eq}[H_2O]_{eq}}$$

$$= \frac{0.024 \times (0.077)^3}{0.0060 \times 0.011} = 0.17\ mol^2\,dm^{-6}$$

Note that if 0.048 mol of methane reacts, then 0.048 mol of steam also reacts and 0.048 mol of carbon monoxide and 3×0.048 mol of hydrogen are produced. This is because the reaction stoichiometry is 1:1:1:3.

Worked example 2

Iron(II) sulfate, $FeSO_4$, and silver nitrate, $AgNO_3$, react according to the equation:

$$Ag^+(aq) + Fe^{3+}(aq) \rightleftharpoons Fe^{3+}(aq) + Ag(s)$$

To find the equilibrium constant K_c for this reaction, 25.0 cm^3 of 0.100 mol dm^{-3} solutions of iron(II) sulfate, $FeSO_4$, and silver nitrate, $AgNO_3$, were mixed and allowed to reach equilibrium.

The unreacted silver ions were then titrated using a 0.0600 mol dm^{-3} potassium thiocyanate solution. The titre was 21.00 cm^3.

In the titration, the Ag^+ ions react in a 1:1 ratio with the thiocyanate ions to form a precipitate of silver thiocyanate. The titration is self indicating — an intense red colour forms with one drop of excess potassium thiocyante and the iron(III) ions present in the equilibrium mixture.

Calculate:

a the initial amount (moles) of Ag^+ and Fe^{2+}

b the final amount (moles) of Ag^+ and hence the equilibrium amounts of Ag^+, Fe^{2+} and Fe^{3+}

c the final concentrations of all three ions and hence K_c

Answer

a initial amount (moles) of $Ag^+ = 0.100\ mol\,dm^{-3} \times 0.0250\ dm^3 = 0.00250\ mol$ = initial amount (moles) of Fe^{2+}

b amount (moles) of thiocyanate in titre $= 0.060\ mol\,dm^{-3} \times 0.02100\ dm^3 = 0.00126\ mol$ = amount of Ag^+ at equilibrium = amount of Fe^{2+} at equilibrium
amount Ag^+ reacted $= 0.0025 - 0.00126 = 0.00124$ = amount Fe^{3+} formed

c volume of solution at equilibrium $= 0.050\ dm^3$
so, $[Ag^+] = [Fe^{2+}] = 0.00126/0.050 = 0.0252\ mol\,dm^{-3}$
$[Fe^{3+}] = 0.00124/0.050 = 0.0248\ mol\,dm^{-3}$

$$K_c = \frac{[Fe^{3+}(aq)]}{[Ag^+(aq)][Fe^{2+}(aq)]}$$

$$= \frac{0.0248}{0.0252 \times 0.0252} = 39.1\ mol^{-1}\,dm^3$$

This method assumes that the position of equilibrium does not move to the left as the silver ions are removed in the titration. One way to tell would be to repeat the experiment, but pause for 5 minutes halfway through the titration and see if the titre (and hence the value of K_c) alters.

e $[Ag(s)]$ does not appear in the expression for K_c because it is a solid in a heterogeneous reaction (see p. 66).

Units of K_c

Care must be taken when evaluating the units of K_c. The simplest way is to look at the equilibrium constant expression and work out the resultant power (dimension) of the concentration, the unit of which is mol dm^{-3}. For example, for the expression:

$$K_c = \frac{[SO_2]^2_{eq}[O_2]_{eq}}{[SO_3]^2_{eq}}$$

the dimension of the top line is (concentration)^3 and that of the bottom line is (concentration)^2. Therefore, the resultant dimension is (concentration)^1 which has units of mol dm^{-3}. Therefore, K_c has units of mol dm^{-3}.

Worked example 1

Calculate the units of the equilibrium constant, K_c, for the following equilibrium reaction:

$$N_2(g) + 3H_2(g) \rightleftharpoons 2NH_3(g)$$

Answer

$$K_c = \frac{[NH_3]^2_{eq}}{[N_2]_{eq}[H_2]^3_{eq}}$$

$$\text{dimensions} = \frac{(\text{concentration})^2}{(\text{concentration})^4} = (\text{concentration})^{-2}$$

units of K_c = (mol dm^{-3})$^{-2}$ or mol^{-2} dm^6

Worked example 2

Calculate the units of the equilibrium constant, K_c, for the following equilibrium reaction:

$$CH_4(g) + H_2O(g) \rightleftharpoons CO(g) + 3H_2(g)$$

Answer

$$K_c = \frac{[CO]_{eq}[H_2]^3_{eq}}{[CH_4]_{eq}[H_2O]_{eq}}$$

$$\text{dimensions} = \frac{(\text{concentration})^4}{(\text{concentration})^2} = (\text{concentration})^2$$

units of K_c = (mol dm^{-3})^2 or mol^2 dm^{-6}

e The marks awarded in questions that ask for the value of K_c to be calculated are for:

- calculating the moles of each substance at equilibrium
- dividing these values by the volume to find the equilibrium concentrations
- correctly stating the expression for the equilibrium constant
- correctly substituting equilibrium concentrations into the expression and calculating the value of K_c
- working out the units (if there are no units, you must state this)

For some reactions the equilibrium constant has no units. This happens when there are an equal number of moles on each side of the equation. In examples like this, you may not be told the total volume, as it will cancel when the value of K_c is calculated.

> **Worked example**
>
> 1.00 mol of ethanol and 2.00 mol of ethanoic acid were mixed in a sealed flask at 35°C and left to reach equilibrium.
>
> $C_2H_5OH(l) + CH_3COOH(l) \rightleftharpoons CH_3COOC_2H_5(l) + H_2O(l)$
>
> The equilibrium mixture contained 1.15 mol of ethanoic acid. Calculate the value of the equilibrium constant, K_c, at 35°C.
>
> **Answer**
>
> initial amount (moles) of ethanoic acid = 2.00 mol
> number of moles at equilibrium = 1.15 mol
> number of moles that reacted = (2.00 – 1.15) = 0.85 mol

If the volume is not given, divide the moles by V, which will later cancel out when the concentration values are substituted into the expression for K_c.

	C_2H_5OH	CH_3COOH	$CH_3COOC_2H_5$	H_2O
Initial moles	1.00	2.00	0	0
Change	–0.85	–0.85	+0.85	+0.85
Moles at equilibrium	0.15	0.15	1.15	0.85
Concentration at equilibrium	$0.15/V$	$0.15/V$	$1.15/V$	$0.85/V$

where V = the total volume

$$K_c = \frac{[CH_3COOC_2H_5]\,[H_2O]}{[C_2H_5OH]\,[CH_3COOH]}$$

$$= \frac{0.85/V \times 0.85/V}{0.15/V \times 1.15/V} = 4.2$$

K_c has no units, as the volume, V, cancels.

Distribution of a solute between two immiscible solvents

Iodine is soluble in water and in some organic solvents, for example hexane. If iodine is added to a mixture of hexane and water, an iodine equilibrium is set up between the two layers:

$$I_2(aq) \rightleftharpoons I_2(hexane)$$

The expression for the equilibrium constant K_c is:

$$K_c = \frac{[I_2(hexane)]}{[I_2(aq)]}$$

This particular equilibrium constant is also called a **partition coefficient**.

> **Worked example**
>
> Iodine was shaken with a mixture containing 10.0 cm^3 of hexane and 100 cm^3 of water and allowed to reach equilibrium. The layers separated and each was titrated against 0.0200 mol dm^{-3} sodium thiosulfate solution
>
> $I_2 + 2Na_2S_2O_3 \rightarrow 2NaI + Na_2S_4O_6$

The aqueous layer required 8.9 cm^3 of sodium thiosulfate solution; the hexane layer required 76.0 cm^3. Calculate the value of the partition coefficient for iodine between water and hexane.

Answer

In the aqueous solution:

amount (moles) of sodium thiosulfate $= 0.0200\,mol\,dm^{-3} \times 8.9/1000\,dm^3$
$= 1.78 \times 10^{-4}\,mol$

amount (moles) of iodine $= \frac{1}{2} \times 1.78 \times 10^{-4} = 8.9 \times 10^{-5}\,mol$

$[I_2(aq)] = 8.9 \times 10^{-5}\,mol/0.100\,dm^3 = 8.9 \times 10^{-4}\,mol\,dm^{-3}$

Hexane solution:

amount (moles) of sodium thiosulfate $= 0.0200\,mol\,dm^{-3} \times 76.0/1000\,dm^3$
$= 1.52 \times 10^{-3}\,mol$

amount (moles) of iodine $= \frac{1}{2} \times 1.52 \times 10^{-3} = 7.6 \times 10^{-4}\,mol$

$[I_2(\text{in hexane})] = 7.6 \times 10^{-4}mol/0.0100\,dm^3 = 0.076\,mol\,dm^{-3}$

$$K_c = \frac{[I_2(\text{in hexane})]}{[I_2(aq)]} = \frac{0.076\,mol\,dm^{-3}}{8.9 \times 10^{-4}\,mol\,dm^{-3}} = 85$$

Partition can be used to determine the value of the equilibrium constant of a reaction.

Iodine in aqueous solution is in equilibrium with iodide ions:

$$I_2(aq) + I^-(aq) \rightleftharpoons I_3^-(aq)$$

The value of $[I_2(aq)]$cannot be found by titration against sodium thiosulfate, because the equilibrium shifts to the left as the iodine reacts. Instead, the equilibrium mixture is shaken with hexane. The hexane layer is removed and titrated against sodium thiosulfate solution. If the amount of iodine and iodide ions initially added is known, the equilibrium constant can be calculated.

Worked example

0.0118 mol of iodine was added to 100 cm^3 of 0.238 $mol\,dm^{-3}$ potassium iodide solution. This was shaken with 20.0 cm^3 of hexane. The hexane solution was found to have $[I_2]$ of 0.102 $mol\,dm^{-3}$. The partition coefficient for iodine is 85.0. Calculate the equilibrium constant for the reaction between iodine and iodide ions.

Answer

$[I_2(\text{hexane})] = 0.102\,mol\,dm^{-3}$

amount (moles) of iodine in hexane layer $= 0.102\,mol\,dm^{-3} \times 0.020\,dm^3$
$= 0.00204\,mol$

partition coefficient $= 85 = [I_2(\text{hexane})]/[I_2(aq)]$

$[I_2(aq)]$ after shaking with hexane $= [I_2(\text{hexane})]/85 = 0.102/85$
$= 0.00120\,mol\,dm^{-3}$

amount (moles) of iodine in aqueous layer $= 0.00120\,mol\,dm^{-3} \times 0.100\,dm^3$
$= 0.000120\,mol$

amount (moles) of iodine that reacted with I^- ions = initial amount –
(amount in hexane layer + amount in aqueous layer)
$= 0.0118 - (0.00204 + 0.000120) = 0.00964\,mol$ = amount of $I_3^-(aq)$ formed

$[I_3^-(aq)] = 0.00964\ mol/0.100\ dm^3 = 0.0964\ mol\ dm^{-3}$
amount (moles) of I^- ions added = $0.238\ mol\ dm^{-3} \times 0.100\ dm^3 = 0.0238\ mol$
amount (moles) of I^- at equilibrium = amount added – amount of iodine reacted
$= 0.0238 - 0.00964 = 0.01416\ mol$

$$[I^-(aq)] \text{ at equilibrium} = \frac{0.01416\ mol}{0.100\ dm^3} = 0.1416\ mol\ dm^{-3}$$

$$K_c = \frac{[I_3^-(aq)]}{[I_2(aq)]\,[I^-(aq)]}$$

$$= \frac{0.0964\ mol\ dm^{-3}}{0.00120\ mol\ dm^{-3} \times 0.1416\ mol\ dm^{-3}} = 567\ mol^{-1}\ dm^3$$

The equilibrium constant in terms of partial pressure, K_p

The molecules of a gas are in constant and random motion. The pressure of a gas is caused by the frequency and momentum of the collisions of its molecules with the container walls.

In a mixture of gases, every gas molecule contributes to the overall pressure. The sum of the individual contributions equals the total pressure. The contribution of one gas to the total pressure is called the **partial pressure** of that substance.

The partial pressure of a gas A in a mixture of gases is the pressure that the gas A would exert if it were alone in the container at that particular temperature.

This definition is better expressed as:

The partial pressure of a gas A, p(A), is equal to the mole fraction of gas A multiplied by the total pressure.

e mole fraction = $\frac{\text{number of moles of a gas}}{\text{total number of moles of gas}}$

$$p(A) = \frac{\text{moles of A}}{\text{total number of moles}} \times P$$

The sum of the partial pressures of the gases in a mixture equals the total pressure. For a mixture of three gases A, B and C:

$$p(A) + p(B) + p(C) = P$$

The symbol for partial pressure is a lower case p with the identity of the gas in brackets, i.e. $p(A)$, or as a subscript, i.e. p_A. The total pressure is an upper case P.

Dry air is a mixture of 78% nitrogen, 21% oxygen and 1% argon (plus small amounts of CO_2 and other gases). The partial pressure of nitrogen when the total air pressure is 1.0 atm is:

$$\frac{78}{100} \times 1.0\ \text{atm} = 0.78\ \text{atm}$$

When a diver descends to a depth of 10 m, the pressure doubles. The partial pressure of nitrogen is now:

$$\frac{78}{100} \times 2.0\ \text{atm} = 1.56\ \text{atm}$$

At this higher partial pressure, nitrogen dissolves in the blood. When the diver returns to the surface, the nitrogen comes out of solution. This causes pain and could even result in the death of the diver. This condition is called the 'bends'. To minimise the problem of the bends, experienced sports divers use an air mixture called 'nitrox', which has a lower mole fraction of nitrogen, equal to two-thirds instead of the four-fifths as in ordinary air.

The partial pressure of a gas is a measure of its concentration in the mixture. Therefore, partial pressures can be used to calculate equilibrium constants. The units are those of pressure and not mol dm^{-3}, so the value of the equilibrium constant will be different. The equilibrium constant in terms of pressures is given the symbol K_p.

The relation between the equilibrium constant, K_p, and partial pressures is similar to that for K_c and concentrations. For a reaction:

$$xA(g) + yB(g) \rightleftharpoons mR(g) + nS(g)$$

$$K_p = \frac{p(R)^m p(S)^n}{p(A)^x p(B)^y}$$

The units of K_p are $(\text{atm})^{m+n-x-y}$.

In all equilibrium constant calculations, the total pressure, and hence the partial pressures, must be measured in atmospheres (atm). The reason for this is beyond A-level.

The units for K_p can be worked out easily because the dimensions of K_p are obtained from the equilibrium expression. For the reaction $N_2(g) + 3H_2(g) \rightleftharpoons 2NH_3(g)$,

$$K_p = \frac{p(NH_3)^2}{p(N_2)p(H_2)^3}$$, and so the dimensions are:

$$\frac{(\text{pressure})^2}{(\text{pressure}) \times (\text{pressure})^3} = \frac{1}{(\text{pressure})^2} = (\text{pressure})^{-2}$$

Hence, the unit is atm^{-2}.

For the reaction:

$$H_2(g) + I_2(g) \rightleftharpoons 2HI(g)$$

$$K_p = \frac{p(HI)^2}{p(H_2)p(I_2)}$$

K_p has no units in this reaction, as the top line is in atm^2 and the bottom line atm × atm. The units cancel, leaving K_p as a dimensionless number.

The partial pressure term equals the value of K_p only when the system is in equilibrium, so equilibrium partial pressures must always be used in K_p calculations. If K_p does not equal the partial pressure term, the system is not at equilibrium — reaction will take place until equilibrium is established.

To summarise:

- If K_p equals the partial pressure term, the system is at equilibrium. There is no change to the relative amounts of the reactants and products.
- If the partial pressure term is smaller than K_p, the system is not at equilibrium. The system will react to form more products until the partial pressure term equals K_p. This means that the position of equilibrium moves to the right.
- If the partial pressure term is greater than K_p, the system is not at equilibrium. The system will react to form more reactants until the concentration term equals K_p. This means that the position of equilibrium moves to the left.

Calculation of K_p from experimental data

The calculation of K_p from experimental data is carried out in a similar way to the calculation of K_c. However, there is an extra step, which is the calculation of the total number of moles at equilibrium.

e Never use square brackets in K_p expressions. Square brackets around a formula mean the concentration, in mol dm^{-3}, of that substance and so must be used only in K_c expressions.

The calculation requires the use of a table, as in the worked example below.

- Write the chemical equation.
- Construct a suitable table and write in the following:
  - the initial amounts (in moles) of the reactants and of the products if their initial amounts were not zero
  - the amounts by which the reactants and the products change in reaching equilibrium — use the stoichiometry of the equation
  - the amount (in moles) of each substance at equilibrium; then add these values to find the total number of moles
  - the **mole fraction** of each gas — divide the equilibrium number of moles by the total number of moles
  - the partial pressure of each gas — multiply the mole fraction of each substance by the total pressure
- Below the table, write the expression for the equilibrium constant.
- Substitute the equilibrium concentrations into the equilibrium constant expression and calculate its value. At the same time, work out the units of K_p and include them in your answer.

Worked example 1

Phosphorus pentachloride decomposes on heating:

$$PCl_5(g) \rightleftharpoons PCl_3(g) + Cl_2(g)$$

When some phosphorus pentachloride was heated to 250°C in a flask, 69% of it dissociated and the total pressure in the flask was 2.0 atm. Calculate the value of the equilibrium constant, K_p.

Answer

	PCl_5	PCl_3	Cl_2
Initial moles	1	0	0
Change	−0.69	+0.69	+0.69
Equilibrium moles	1 − 0.69 = 0.31	0 + 0.69 = 0.69	0 + 0.69 = 0.69

Total number of moles at equilibrium = 0.31 + 0.69 + 0.69 = 1.69

Mole fraction	0.31/1.69 = 0.183	0.69/1.69 = 0.408	0.69/1.69 = 0.408
Partial pressure/atm	0.18 × 2.0 = 0.366	0.408 × 2.0 = 0.816	0.408 × 2.0 = 0.816

$$K_p = \frac{p(PCl_3)p(Cl_2)}{p(PCl_5)} = \frac{0.816 \times 0.816}{0.366} = 1.8 \text{ atm at } 250°C$$

e When there is only one reactant, you might not be told its initial amount. You must assume that it is 1 mol. You will be told the percentage that reacts.

Worked example 2 is more complicated because the stoichiometry is not 1:1.

Worked example 2

One of the important reactions by which the gaseous fuel methane, CH_4, is produced from coal is:

$$3H_2(g) + CO(g) \rightleftharpoons CH_4(g) + H_2O(g)$$

Hydrogen and carbon monoxide were mixed in a 3:1 ratio and allowed to reach equilibrium at a temperature of 1000 K. 65% of the hydrogen reacted and the total pressure was 1.2 atm. Calculate the value of the equilibrium constant, K_p.

Answer

	$3H_2$	CO	CH_4	H_2O
Initial moles	3	1	0	0
Change	−0.65 × 3 = −1.95	−1.95/3 = −0.65	+0.65	+0.65
Equilibrium moles	3 − 1.95 = 1.05	1 − 0.65 = 0.35	0 + 0.65 = 0.65	0 + 0.65 = 0.65

Total number of moles at equilibrium = 1.05 + 0.35 + 0.65 + 0.65 = 2.7

Mole fraction	1.05/2.7 = 0.389	0.35/2.7 = 0.130	0.65/2.7 = 0.241	0.65/2.7 = 0.241
Partial pressure/atm	0.389 × 1.2 = 0.467	0.130 × 1.2 = 0.156	0.241 × 1.2 = 0.289	0.241 × 1.2 = 0.289

$$K_p = \frac{p(CH_4)p(H_2O)}{p(H_2)^3 p(CO)} = \frac{0.289 \times 0.289}{(0.467)^3 \times 0.156} = 5.3 \text{ atm}^{-2}$$

$$\text{dimensions} = \frac{(\text{pressure})^2}{(\text{pressure})^4} = (\text{pressure})^{-2}$$

units of $K_p = \text{atm}^{-2}$

Worked example 3

When lightning passes through the air, some of the nitrogen reacts with oxygen to form nitric oxide, NO.

$$N_2(g) + O_2(g) \rightleftharpoons 2NO(g) \qquad \Delta H = +180 \text{ kJ mol}^{-1}$$

1 mol of air at 1 atm pressure contains 0.7809 mol of nitrogen and 0.2095 mol of oxygen. (Most of the rest is argon.) After a bolt of lightning, the air was found to contain 0.0018 mol of nitric oxide. Assuming that equilibrium had been reached at the temperature of the lightning, calculate the value of K_p at this temperature.

NOAA

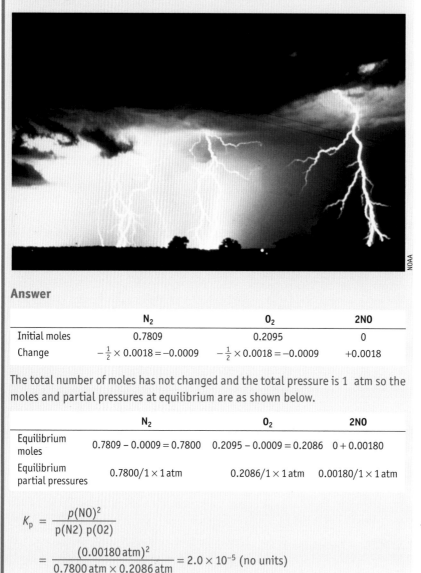

Answer

	N_2	O_2	2NO
Initial moles	0.7809	0.2095	0
Change	$-\frac{1}{2} \times 0.0018 = -0.0009$	$-\frac{1}{2} \times 0.0018 = -0.0009$	+0.0018

The total number of moles has not changed and the total pressure is 1 atm so the moles and partial pressures at equilibrium are as shown below.

	N_2	O_2	2NO
Equilibrium moles	$0.7809 - 0.0009 = 0.7800$	$0.2095 - 0.0009 = 0.2086$	$0 + 0.00180$
Equilibrium partial pressures	$0.7800/1 \times 1\,\text{atm}$	$0.2086/1 \times 1\,\text{atm}$	$0.00180/1 \times 1\,\text{atm}$

$$K_p = \frac{p(\text{NO})^2}{\text{p(N2) p(O2)}}$$

$$= \frac{(0.00180\,\text{atm})^2}{0.7800\,\text{atm} \times 0.2086\,\text{atm}} = 2.0 \times 10^{-5} \text{ (no units)}$$

Note that the units cancel, as there are two molecules of gas on each side of the equation.

Calculations using K_c and K_p

The first type of calculation is where you are given the value of K_p and the other partial pressures.

Worked example

The equilibrium constant, K_p, for the reaction

$N_2O_4(g) \rightleftharpoons 2NO_2(g$

is 3.52 atm. Calculate the total pressure required for an equilibrium mixture to have a mole fraction of N_2O_4 equal to 0.15.

The sum of the mole fractions of all the gases in a gaseous mixture is always 1.

Answer

mole fraction of N_2O_4 + mole fraction of $NO_2 = 1$

mole fraction of $N_2O_4 = 0.15$

so, the mole fraction of $NO_2 = 1 - 0.15 = 0.85$

Let total pressure = P_T

$$K_p = 3.52\,\text{atm} = \frac{P(NO_2)^2}{P(N_2O_4)} = \frac{(\text{mole fraction } NO_2 \times P_T)^2}{(\text{mole fraction } N_2O_4 \times P_T)}$$

$$= \frac{0.85^2 \times P_T{}^2}{0.15 \times P_T}$$

$$= \frac{0.7225 \times P_T}{0.15} = 4.817 \times P_T$$

$P_T = 3.52\,\text{atm}/4.817 = 0.73\,\text{atm}$

A second type of calculation is more difficult. This is when two reactant molecules react reversibly to form two product molecules. This calculation involves letting the fraction of one substance that reacts equal an unknown, z. A table is used as in the calculation of an equilibrium constant. The value of z is found by taking the square root of the equilibrium expression.

Worked example

The esterification of ethanol with ethanoic acid is represented by the equation:

$C_2H_5OH + CH_3COOH \rightleftharpoons CH_3COOC_2H_5 + H_2O$

The equilibrium constant K_c at 25°C is 4.0.

Calculate the percentage of ethanol that is converted to ester when 1.0 mol of ethanol is mixed with 1.0 mol of ethanoic acid in a propanone solvent of volume 1.0 dm^3 and allowed to reach equilibrium.

Answer

	C_2H_5OH	CH_3COOH	$CH_3COOC_2H_5$	H_2O
Initial moles	1.0	1.0	0	0
Change	$-z$	$-z$	$+z$	$+z$
Equilibrium moles	$(1-z)$	$(1-z)$	z	z
Equilibrium concentration/mol dm^{-3}	$(1-z)/1.0 = 1-z$	$(1-z)/1.0 = 1-z$	$z/1.0 = z$	$z/1.0 = z$

$$K_c = \frac{[CH_3COOC_2H_5][H_2O]}{[C_2H_5OH][CH_3COOH]}$$

$$= \frac{z \times z}{(1-z)(1-z)}$$

$$= \frac{z^2}{(1 - z)^2} = 4.0$$

Taking the square root of both sides:

$$= \frac{z}{(1 - z)} = \sqrt{4.0} = 2$$

$z = 2.0 - 2z$

$3z = 2.0$

$z = 0.67$, so 67% of the ethanol reacted

Heterogeneous equilibria

In all the examples so far, the reactants and products have been in the same phase. Some reversible reactions involve reactants and products in different phases.

There are three physical states — solid, liquid and gas.

- All gases mix completely. Therefore, a mixture of gases always forms a single phase in which any one part is identical with any other part.
- A solution of several solutes in water exists in a single phase also forming a homogeneous mixture.
- Two liquids mix to form either a single phase or, if they are immiscible, two layers — two liquid phases.
- Solids that dissolve in a solvent form a single liquid phase.
- A mixture of a solid and a gas forms two distinct phases.
- Mixtures of solids are usually in two different solid phases.

A heterogeneous mixture is one that exists in two or more different phases.

In a heterogeneous equilibrium reaction at least one substance is in a different phase from the others. An example of this is the reaction between carbon and steam to form carbon monoxide and hydrogen:

$$C(s) + H_2O(g) \rightleftharpoons CO(g) + H_2(g)$$

The three gases are in the same phase but carbon is in a different phase. The concentration of a solid, such as carbon, is constant and is determined only by its density. Its value is incorporated into the equilibrium constant and so is left out in the expression for K_c. For the reaction above, the equilibrium constant, K_c, is given by the expression:

$$K_c = \frac{[CO]_{eq}[H_2]_{eq}}{[H_2O]_{eq}}$$

Involatile solids have no vapour pressure and so they do not appear in the expression for the equilibrium constant, K_p. For the reaction:

$$3Fe(s) + 4H_2O(g) \rightleftharpoons Fe_3O_4(s) + 4H_2(g)$$

$$K_p = \frac{p(H_2)^4}{p(H_2O)^4}$$

where all partial pressures are equilibrium values.

Another heterogeneous reversible reaction is one that involves a solid and a solution – for example, the reaction between solutions of iron(II) ions and silver ions to form iron(III) ions and a precipitate of solid silver:.

$$Fe^{2+}(aq) + Ag^{+}(aq) \rightleftharpoons Fe^{3+}(aq) + Ag(s)$$

[Ag(s)] is left out of the expression for K_c, which becomes

$$K_c = \frac{[Fe^{3+}]_{eq}}{[Fe^{2+}]_{eq}\,[Ag^{+}]_{eq}}$$

Worked example 1

Ammonium hydrogensulfide, NH_4HS, decomposes when heated, according to the equation:

$NH_4HS(s) \rightleftharpoons NH_3(g) + H_2S(g)$ $K_p = 0.142\ atm^2$ at 50°C

a State the expression for K_p.

b Calculate the partial pressure of both gases at 50°C and hence the total pressure.

Answer

a $K_p = p(NH_3)p(H_2S) = 0.142$

b As the reaction produces NH_3 and H_2S in a 1:1 ratio, $p(H_2S) = p(NH_3)$

$K_p = p(NH_3)^2 = 0.142$

$p(NH_3) = \sqrt{0.142} = 0.377$ atm

$p(H_2S) = 0.377$ atm

total pressure = sum of partial pressures = 0.377 + 0.377 = 0.754 atm

Another heterogeneous equilibrium is that between solid calcium hydroxide and aqueous calcium and hydroxide ions:

$$Ca(OH)_2(s) + aq \rightleftharpoons Ca^{2+}(aq) + 2OH^{-}(aq)$$

The solid calcium hydroxide is in one phase and the dissolved ions and the solvent are in another. The expression for the equilibrium constant is:

$$K_c = [Ca^{2+}][OH^{-}]^2$$

Worked example 2

The value of K_c for dissolving calcium hydroxide is $5.5 \times 10^{-5}\ mol^3\,dm^{-9}$ at 25°C. Calculate the concentration of OH^- ions in a saturated solution.

Answer

$K_c = [Ca^{2+}][OH^{-}]^2 = 5.5 \times 10^{-5}$

let $[Ca^{2+}] = z$

ratio of OH^- to Ca^{2+} = 2:1, so $[OH^-] = 2z$

$K_c = z \times (2z)^2 = 5.5 \times 10^{-5}$

$4z^3 = 5.5 \times 10^{-5}$

$z = 0.024$

$[OH^-] = 2z = 0.048\ mol\,dm^{-3}$

Relationship between the equilibrium constant and the total entropy change

Calculation of equilibrium constant

As stated on p. 52, the relationship between the equilibrium constant and the total entropy change is:

$$\Delta S^{\ominus}_{total} = R \ln K$$

Thus:

$$\ln K = \Delta S^{\ominus}_{total}/R \qquad \text{or} \qquad K = e^{\Delta S/R}$$

where R is the gas constant, which has the value 8.31 J K^{-1} mol^{-1}

and

$$\Delta S^{\ominus}_{total} = \Delta S^{\ominus}_{system} + \Delta S^{\ominus}_{surr} = \Delta S^{\ominus}_{system} - \Delta H^{\ominus}/T$$

Worked example

Use the data below to calculate the value of the equilibrium constant at a temperature of 25°C (298 K) for the reaction:

$$AgCl(s) \rightleftharpoons Ag^+(aq) + Cl^-(aq)$$

	$S^{\ominus}$/J K^{-1} mol^{-1}	$\Delta H^{\ominus}_{soln}$/kJ mol^{-1}
AgCl(s)	+96	+66
Ag^+(aq)	+73	
Cl^-(aq)	+57	

Answer

$$\Delta S^{\ominus}_{system} = \Delta S^{\ominus}_{Ag^+(aq)} + \Delta S^{\ominus}_{Cl^-(aq)} - \Delta S^{\ominus}_{AgCl(s)} = +73 + 57 - 96 = +34\,\text{J K}^{-1}\,\text{mol}^{-1}$$

$$\Delta S^{\ominus}_{surr} = -(+66\,000/298) = -221\,\text{J K}^{-1}\,\text{mol}^{-1}$$

$$\Delta S^{\ominus}_{total} = +34 + (-221) = -187\,\text{J K}^{-1}\,\text{mol}^{-1}$$

$$\ln K = \Delta S^{\ominus}_{total}/R$$

$$K = e^{-187/R} = e^{-187/8.31} = 1.7 \times 10^{-10}\,\text{mol}^2\,\text{dm}^{-6}$$

The value from data books is 1.8×10^{-10}. This shows that the use of relative entropies of aqueous ions is fair.

Effect of a change in temperature on ΔS_{total} and hence on K

$$\Delta S^{\ominus}_{total} = \Delta S^{\ominus}_{system} + \Delta S^{\ominus}_{surr} = \Delta S^{\ominus}_{system} - \Delta H^{\ominus}/T$$

If $\Delta S^{\ominus}_{total}$ increases, the value of the equilibrium constant, K, also increases. Likewise, if the total entropy decreases, so does the value of K.

The magnitude of $\Delta H^{\ominus}/T$ always decreases as the temperature is increased.

Exothermic reactions: ΔH negative

$\Delta S^{\ominus}_{surr} = -\Delta H^{\ominus}/T$, so $\Delta S^{\ominus}_{surr}$ is positive. When the temperature is increased, $\Delta S^{\ominus}_{surr}$ becomes less positive. As $\Delta S^{\ominus}_{total} = \Delta S^{\ominus}_{system} + \Delta S^{\ominus}_{surr}$, the value of $\Delta S^{\ominus}_{total}$ decreases. This causes the value of the equilibrium constant, K, to decrease.

The value of K decreases with an increase in temperature.

Worked example

Nitrogen and hydrogen react together according to the equation:

$$N_2(g) + 3H_2(g) \rightleftharpoons 2NH_3(g)$$

Calculate the value of the equilibrium constant for the reaction at 527°C (800 K) and compare it with the value at 427°C (700 K).
Use the following data:

K_p at 427°C (700 K) = 7.76×10^{-5} atm^{-2}
$\Delta H = -92.4$ kJ mol^{-1}
$\Delta S_{system} = -214$ J K^{-1} mol^{-1}

Answer

$\Delta S^{\ominus}_{total}$ at 527°C (800 K) = $\Delta S^{\ominus}_{system} + \Delta S^{\ominus}_{surr} = -214 - (-92\,400/800)$
$= -98.5$ J K^{-1} mol^{-1}
$\ln K = \Delta S^{\ominus}_{total}/R = -98.5/8.31 = -11.9$
K_p at 527°C (800 K) = $e^{-11.9} = 6.79 \times 10^{-6}$ atm^{-2}, which is less than the value at 427°C (700 K)

This decrease is predicted by Le Chatelier's principle, as the reaction is exothermic.

Endothermic reactions: ΔH positive

$\Delta S^{\ominus}_{surr} = -\Delta H^{\ominus}/T$, so $\Delta S^{\ominus}_{surr}$ is negative. When the temperature is increased, $\Delta S^{\ominus}_{surr}$ becomes less negative. This causes $\Delta S^{\ominus}_{total}$ to increase and hence K to increase.

Worked example

Consider the reaction:

$AgCl(s) \rightleftharpoons Ag^+(aq) + Cl^-(aq)$

Use the data in the worked example on p. 68 to calculate the value of the equilibrium constant for this reaction at 100°C (373 K). Compare its value with that at 25°C (298 K).

Answer

$\Delta S^{\ominus}_{system}$ is unchanged at +34 J K^{-1} mol^{-1}
$\Delta H^{\ominus} = +66$ kJ mol^{-1}
$\Delta S^{\ominus}_{surr} = -(+66\,000/373) = -177$ J K^{-1} mol^{-1}
$\Delta S^{\ominus}_{total} = +34 + (-177) = -143$ J K^{-1} mol^{-1}
$\ln K = \Delta S^{\ominus}_{total}/R = -143/8.31 = -17.2$
K at 100°C = $e^{-17.2} = 3.4 \times 10^{-8}$ mol^2 dm^{-6}

This value is much larger than the value at 25°C.

This increase is predicted by Le Chatelier's principle, as the reaction is endothermic.

Summary

An increase in temperature has the following effects:

- For an exothermic reaction (ΔH negative) — $\Delta H/T$ becomes less positive which decreases the value of $\Delta S^{\ominus}_{surr}$. This causes both $\Delta S^{\ominus}_{total}$ and K to ***decrease***.
- For an endothermic reaction (ΔH positive) — $\Delta H/T$ becomes less negative. This causes both $\Delta S^{\ominus}_{total}$ and K to ***increase***.

A decrease in temperature has the opposite effects.

Estimating the extent of a reaction

The values of the entropy change and the equilibrium constant give an estimate of how complete a reaction is at a particular temperature.

ΔS_{total}/J K^{-1} mol^{-1}	K	Extent of reaction
Greater than +150	Greater than 10^8	Reaction almost complete
Between +60 and +150	Between 10^8 and 10^3	Reaction favours products
Between +60 and −60	Between 10^3 and 10^{-3}	Both reactants and products present in significant quantities
Between −60 and −150	Between 10^{-3} and 10^{-8}	Reaction favours reactants
More negative than −150	Less than 10^{-8}	Reaction not noticeable

***Table 3.2** Relationship between ΔS_{total}, K and extent of reaction*

When $\Delta S = 0$, $K_c = 1$

The values of ΔS_{total} for dissolving some ionic solids are listed in Table 2.10 (p. 47).

Group 2 hydroxides

$$Ba(OH)_2(s) + aq \rightleftharpoons Ba^{2+}(aq) + 2OH^-(aq) \qquad \Delta S_{total} = +62\,J\,K^{-1}\,mol^{-1}$$

ΔS_{total} is about +60, so barium hydroxide is fairly soluble.

$$Ca(OH)_2(s) + aq \rightleftharpoons Ca^{2+}(aq) + 2OH^-(aq) \qquad \Delta S_{total} = -106\,J\,K^{-1}\,mol^{-1}$$

ΔS_{total} is between +60 and +150, so calcium hydroxide is only very slightly soluble.

Group 2 sulfates

$$MgSO_4(s) + aq \rightleftharpoons Mg^{2+}(aq) + SO_4{}^{2-}(aq) \qquad \Delta S_{total} = +113\,J\,K^{-1}\,mol^{-1}$$

ΔS_{total} lies between +60 and +150, so magnesium sulfate is very soluble.

$$BaSO_4(s) + aq \rightleftharpoons Ba^{2+}(aq) + SO_4{}^{2-}(aq) \qquad \Delta S_{total} = -313\,J\,K^{-1}\,mol^{-1}$$

ΔS_{total} is more negative than –150, so barium sulfate is regarded as being totally insoluble. The digestive tract can be examined by giving the patient a 'barium meal' — a suspension of barium sulfate — and following this with a series of X-rays. Even though barium ions are extremely poisonous, [$Ba^{2+}(aq)$] is so very small that the meal is harmless.

Combustion of hydrogen

The equilibrium constant for the reaction:

$$2H_2(g) + O_2(g) \rightleftharpoons 2H_2O(g)$$

is 1.4×10^{53} at 200°C, so the reaction between hydrogen and oxygen goes virtually to completion.

The Haber process

$$N_2(g) + 3H_2(g) \rightleftharpoons 2NH_3(g)$$

The Haber process is carried out at a temperature of about 400°C and 200 atm pressure. The equilibrium constant, K_c, at this temperature is 0.26 $mol^{-2}\,dm^6$ and there is over 20% ammonia in the equilibrium mixture. As the industrial process does not take place in a closed system, equilibrium is not quite reached in the catalyst chamber and the actual percentage of ammonia leaving the catalyst chamber is only 15%.

Questions

1 Write the expressions for the equilibrium constant, K_c, for the following reactions:

a $CO(g) + 2H_2(g) \rightleftharpoons CH_3OH(g)$

b $NO(g) + \frac{1}{2}O_2(g) \rightleftharpoons NO_2(g)$

c $4NH_3(g) + 5O_2(g) \rightleftharpoons 4NO(g) + 6H_2O(g)$

d $C_2H_5Br + H_2O \rightleftharpoons C_2H_5OH + HBr$ (in ethanol solution)

e $[Fe(H_2O)_6]^{3+}(aq) + H_2O(l) \rightleftharpoons H_3O^+(aq) + [Fe(H_2O_5OH]^{2+}(aq)$

2 $K_c = 40\ mol^{-1}\ dm^3$ at 250°C for the reaction:

$PCl_3(g) + Cl_2(g) \rightleftharpoons PCl_5(g)$

Calculate the value of K_c, at 250°C, for the reaction:

$PCl_5(g) \rightleftharpoons PCl_3(g) + Cl_2(g)$

3 $K_c = 1.0$ at 1100 K for the reaction:

$CO(g) + H_2O(g) \rightleftharpoons CO_2(g) + H_2(g)$

a 1.5 mol of carbon monoxide and 1.5 mol of steam at 1100 K are mixed with 3.0 mol of carbon dioxide and 1.0 mol of hydrogen in a container of volume $100\ dm^3$. Calculate whether the system is in equilibrium. If it is not, explain in which direction the system will move in order to reach equilibrium.

b A mixture of the four gases at equilibrium at 500°C contained $0.010\ mol\ dm^{-3}$ carbon dioxide, $0.0040\ mol\ dm^{-3}$ steam and $0.035\ mol\ dm^{-3}$ carbon monoxide. Given that the value of K_c at 500°C = 0.20, calculate the concentration of hydrogen at equilibrium.

4 Sulfur trioxide was heated to 700°C in a vessel of volume $10\ dm^3$ and allowed to reach equilibrium:

$2SO_3(g) \rightleftharpoons 2SO_2(g) + O_2(g)$

The equilibrium mixture was found to contain 0.035 mol sulfur trioxide, 0.044 mol sulfur dioxide and 0.022 mol oxygen. Calculate the value of the equilibrium constant, K_c, at 700°C.

5 Partition experiments can be used to find the formula of the deep blue ammonia/copper(II) complex ion.

$100\ cm^3$ of a solution of copper sulfate containing 0.010 mol of $Cu^{2+}(aq)$ ions was added to $100\ cm^3$ of a solution of ammonia containing 0.100 mol of ammonia. The following reaction took place:

$Cu^{2+}(aq) + nNH_3(aq) \rightarrow [Cu(NH_3)_n]^{2+}(aq)$

This solution was shaken with $100\ cm^3$ of ethoxyethane and the dissolved ammonia became distributed between the aqueous and the organic layers. The partition constant of ammonia between water and ethoxyethane is 30.

$$\frac{[NH_3] \text{ in water}}{[NH_3] \text{ in ethoxyethane}} = 30$$

The organic layer was found to have an ammonia concentration of $0.00980\ mol\ dm^{-3}$.

Calculate:

a the concentration of ammonia in the aqueous layer

b the amount (in moles) of ammonia in each layer and hence the total amount (in moles) of free ammonia

c the amount (in moles) of ammonia that reacted with the copper ions

d the ratio of reacted ammonia to copper ions and hence the formula of the ammonia/copper(II) complex ion

6 1.00 mol of methane, CH_4, and 2.00 mol of steam were mixed in a vessel of volume $10\ dm^3$ and allowed to reach equilibrium at 1200 K:

$CH_4(g) + H_2O(g) \rightleftharpoons CO(g) + 3H_2(g)$

Analysis of the equilibrium mixture showed that 0.25 mol of methane was present.

a Write the expression for the equilibrium constant, K_c.

b Calculate the value of the equilibrium constant, K_c, at a temperature of 1200 K.

7 Methanol, CH_3OH, can be made by passing carbon monoxide and hydrogen over a heated zinc oxide and chromium(III) oxide catalyst:

$$CO(g) + 2H_2(g) \rightleftharpoons CH_3OH(g)$$

When 0.100 mol of carbon monoxide and 0.300 mol of hydrogen were heated to 400°C in a vessel of volume 10 dm^3, equilibrium was reached after 30% of the carbon monoxide had reacted. Calculate the value of the equilibrium constant, K_c.

8 The ester, dimethyl ethanedioate, is hydrolysed by water:

$$CH_3OOCCOOCH_3 + 2H_2O \rightleftharpoons HOOCCOOH + 2CH_3OH$$

11.8 g of dimethyl ethanedioate were mixed with 5.40 g of water and allowed to reach equilibrium. 75% of the ester reacted and the total volume was 15.0 cm^3. Calculate the value of the equilibrium constant, K_c, at the temperature of the experiment.

9 Write the expression for K_p for the following reactions:

a $SO_2Cl_2(g) \rightleftharpoons SO_2(g) + Cl_2(g)$

b $2NO(g) + O_2(g) \rightleftharpoons 2NO_2(g)$

c $2SO_2(g) + O_2(g) \rightleftharpoons 2SO_3(g)$

10 At 35°C, dinitrogen tetroxide, N_2O_4, is 15% dissociated in a flask at a pressure of 1.2 atm:

$$N_2O_4(g) \rightleftharpoons 2NO_2(g)$$

Calculate the value of the equilibrium constant, K_p.

11 Nitrogen and hydrogen were mixed in a 1:3 ratio and heated to 400°C over an iron catalyst at a pressure of 30 atm until equilibrium was reached. 15% of the nitrogen was converted to ammonia:

$$N_2(g) + 3H_2(g) \rightleftharpoons 2NH_3(g)$$

a Write the expression for K_p.

b Calculate its value under these conditions.

12 Hydrogen and iodine react reversibly to form hydrogen iodide:

$$H_2(g) + I_2(g) \rightleftharpoons 2HI(g)$$

The value of the equilibrium constant, K_c, at 420°C = 49.

Calculate the percentage of hydrogen that reacts when 1.0 mol of hydrogen and 1.0 mol of iodine reach equilibrium at 420°C in a vessel of volume 50 dm^3.

13 Write the expression for K_p for the following reactions:

a $CaCO_3(s) \rightleftharpoons CaO(s) + CO_2(g)$

b $H_2O(g) + C(s) \rightleftharpoons CO(g) + H_2(g)$

14 Write the expression for K_c for the reactions:

a $Cr_2O_7^{2-}(aq) + H_2O(l) \rightleftharpoons 2CrO_4^{2-}(aq) + 2H^+(aq)$

b $2CH_3COOH + HOCH_2CH_2OH \rightleftharpoons CH_3COOCH_2CH_2OOCCH_3 + 2H_2O$

15 Silver carbonate decomposes according to the equation:

$$Ag_2CO_3(s) \rightleftharpoons Ag_2O(s) + CO_2(g) \qquad K_p = 1.48 \text{ atm at } 500°C$$

Calculate the partial pressure of carbon dioxide at 500°C.

Applications of rate and equilibrium

Chapter 4

Predicting the direction of change

When both reactants and products are present in a vessel at a stated temperature, the system is either at equilibrium or not. The situation that exists can be determined by calculating the value of the concentration term (**reaction quotient**) and comparing it with the value of the equilibrium constant at that temperature:

- If the two are equal, the system is in equilibrium.
- If the two are not equal, the system is not at equilibrium.

Worked example

Hydrogen and iodine, both at a concentration of 0.0020 mol dm^{-3}, were mixed at 480°C with hydrogen iodide of concentration 0.0040 mol dm^{-3}. $K_c = 49$ at 480°C. Is the system at equilibrium?

Answer

The equation for the equilibrium reaction is:

$$H_2(g) + I_2(g) \rightleftharpoons 2HI(g)$$

$$\frac{[HI]^2}{[H_2][I_2]} = \frac{0.0040^2}{0.0020 \times 0.0020} = 4$$

The value of the concentration term is 4, which is less than the value of K_c (49), and so the system is not at equilibrium.

For the system in the worked example to reach equilibrium, the value of the concentration term must increase. This will happen if hydrogen and iodine react to form more hydrogen iodide. This will increase the value of [HI] and decrease the values of $[H_2]$ and $[I_2]$. Thus the reaction moves to the right-hand side until the value of the concentration term equals 49. The system is then in equilibrium, and so:

$$\frac{[HI]^2_{eq}}{[H_2]_{eq}[I_2]_{eq}} = 49 = K_c$$

To summarise:

- If K_c equals the concentration term (the fraction or the quotient), the system is at equilibrium. There is no change to the relative amounts of the reactants and products.
- If the concentration term is smaller than K_c, the system is not at equilibrium. The system will react to form more of the products until the concentration term equals K_c. This means that the position of equilibrium moves to the right.

- If the concentration term is greater than K_c, the system is not at equilibrium. The system will react to form more of the reactants until the concentration term equals K_c. This means that the position of equilibrium moves to the left.

e You should use the subscript 'eq' to distinguish between equilibrium concentrations and the initial concentrations, before equilibrium has been reached.

Altering equilibrium conditions

Le Chatelier's principle can be used to predict the direction of change of equilibrium position when conditions such as temperature, pressure or concentration are altered. However, the principle does not *explain* the direction of change. It also gives no information about whether the value of the equilibrium constant changes.

A proper understanding is based on the fact that the equilibrium constant equals the concentration (or partial pressure) term only when the system is at equilibrium:

- If K_c is equal to the concentration term, then the system is at equilibrium — there will be no change in the equilibrium position.
- If K_c is *not* equal to the concentration term, then the system is *not* at equilibrium — the position of equilibrium will change until the concentration term equals K_c.

e The position of equilibrium is defined by the ratio of the amounts of products to the amounts of reactants. If the position moves to the right, this ratio increases.

Effect of change in temperature

Effect on the equilibrium constant

The only factor that alters the value of the equilibrium constant of a particular reaction is temperature.

The value of the equilibrium constant depends on the temperature because of the relationships:

$$\Delta S_{total} = R \ln K \text{ where } R \text{ is the gas constant}$$

and

$$\Delta S_{total} = \Delta S_{system} + \Delta S_{surr} = \Delta S_{system} + (-\Delta H/T)$$

All exothermic reactions have a negative ΔH, so $-\Delta H/T$ is positive. If the temperature is increased, the bottom line gets bigger, which makes $-\Delta H/T$ smaller. Thus ΔS_{total} becomes smaller. This makes $\ln K$ and hence K smaller.

All endothermic reactions have a positive ΔH, so $-\Delta H/T$ is negative. If the temperature is increased, the bottom line gets bigger, which makes $-\Delta H/T$ less negative. Thus ΔS_{total} becomes bigger. This makes $\ln K$ and hence K bigger.

For exothermic reactions, an increase in temperature results in lower K_c and K_p values.

For endothermic reactions, an increase in temperature results in higher K_c and K_p values.

This is illustrated in Table 4.1.

Reaction	ΔH/kJ mol^{-1}	T/°C	K_p
$N_2(g) + 3H_2(g) \rightleftharpoons 2NH_3(g)$ Exothermic	−92	25	6.8×10^5 atm^{-2}
		125	43 atm^{-2}
		225	3.7×10^{-2} atm^{-2}
		325	1.7×10^{-3} atm^{-2}
		425	7.8×10^{-5} atm^{-2}
$N_2O_4(g) \rightleftharpoons 2NO_2(g)$ Endothermic	+58	25	0.24 atm
		80	4.0 atm
		100	46 atm
		200	350 atm

***Table 4.1** Effect of temperature on K_p*

Effect on position of equilibrium

Exothermic reactions

- An increase in temperature causes the value of ΔS_{total} to become smaller and hence K_c to become *smaller*. Therefore, the concentration term is bigger than the new value of K_c.
- The system reacts to make the concentration term smaller — which it does by forming more of the reactants — until the value of the concentration term equals the new value of K_c.
- Thus, the position of equilibrium shifts to the *left*.

A decrease in temperature has the opposite effect.

Endothermic reactions

- An increase in temperature causes the value of ΔS_{total} to become bigger and hence K_c to become *larger*. Therefore, the concentration term is smaller than the new value of K_c.
- The system reacts to make the concentration term larger — which it does by forming more of the products — until the value of the concentration term equals the new value of K_c.
- Thus, the position of equilibrium shifts to the *right*.

A decrease in temperature has the opposite effect.

The logic of the effect of temperature on the equilibrium is:

change in temperature → change in the value of K → alteration in the value of the concentration term to regain equality → movement of position of equilibrium

Worked example

The chloride complex of cobalt(II) ions, $CoCl_4^{2-}$, is blue and is in equilibrium with pink hydrated cobalt(II) ions, $Co(H_2O)_6^{2+}$ according to the equation:

$$CoCl_4^{2-} + 6H_2O \rightleftharpoons Co(H_2O)_6^{2+} + 4Cl^-$$

When a blue solution of mainly $CoCl_4^{2-}$ ions is cooled, the colour changes to pink. Deduce whether the reaction as written is exothermic or endothermic.

Answer

On cooling, the equilibrium shifts to the right. This means that the value of K_c is higher at the lower temperature. Therefore, the reaction must be exothermic.

Effect on the rate of reaching equilibrium

An increase in temperature always results in an increase in the rate of reaction. This is true whether the reaction is exothermic or endothermic. It happens because the molecules possess greater average kinetic energy, so more collisions have energy that is greater than the activation energy. This means that a greater proportion of the collisions result in a reaction.

In a reversible reaction, an increase in temperature increases the rate of the forward and back reactions, but does not do so equally. The endothermic reaction has higher activation energy and so its rate is increased more than that of the exothermic reaction. This is one reason why the position of equilibrium shifts in the endothermic direction when the temperature of the system is raised.

Because the rates of both reactions are increased, equilibrium is reached more rapidly.

Effect of change of pressure or volume of the container on gaseous reactions

Effect on the equilibrium constant

Altering the pressure of a gaseous system or the volume of the container has *no* effect on the value of either the total entropy change or the equilibrium constant.

Effect on the position of equilibrium

This depends on the number of gas molecules on each side of the equation.

Increase in pressure by decreasing the volume (at constant temperature)

In all the examples below assume that the pressure has been doubled by halving the volume of the container. This will cause the concentrations of all species to double.

The equilibrium involving hydrogen, iodine and hydrogen iodide is an example of a reaction in which the number of gas moles on the left equals the number on the right:

$$H_2(g) + I_2(g) \rightleftharpoons 2HI(g)$$

$$K_c = \frac{[HI]^2_{eq}}{[H_2]_{eq}[I_2]_{eq}}$$

- An increase in pressure (caused by a decrease in the volume of the container) has *no* effect on the value of K_c.
- The concentration of HI rises by a factor of 2, and so $[HI]^2$ increases by a factor of 4.
- The concentrations of both H_2 and I_2 rise by a factor of 2. Therefore, $[H_2]$ multiplied by $[I_2]$ increases by a factor of 4.

- Both the top and the bottom lines of the concentration term rise by the same factor. Therefore, its value does not change.
- Neither K_c nor the concentration term has altered, thus K_c still equals the concentration term. This means that the system is still in equilibrium, so there is no change to the position of equilibrium.

The equilibrium between nitrogen, hydrogen and ammonia is an example of a reaction in which the number of gas moles on the left is more than the number on the right:

$$N_2(g) + 3H_2(g) \rightleftharpoons 2NH_3(g)$$

$$K_c = \frac{[NH_3]^2_{eq}}{[N_2]_{eq}[H_2]^3_{eq}}$$

- An increase in pressure (caused by a decrease in the volume of the container) has *no* effect on the value of K_c.
- The concentration of ammonia doubles, so $[NH_3]^2$ rises by a factor of 4.
- $[N_2]$ multiplied by $[H_2]^3$ rises by a factor of 2×2^3. This is a greater increase than that of the top line of the concentration term.
- The concentration term becomes smaller. Therefore, it no longer equals K_c, and the system is no longer in equilibrium.
- The system reacts to make the concentration term bigger until it once again equals the unaltered value of K_c. It does this by hydrogen reacting with nitrogen to make more ammonia, so the position of equilibrium shifts to the right.

An increase in pressure does not alter K_c, but the concentration term is lowered as there are fewer gas moles on the right. Therefore, the system reacts to make more ammonia, until the concentration term equals K_c once more.

The equilibrium between dinitrogen tetroxide and nitrogen dioxide is an example of a reaction in which the number of gas moles on the left is less than the number on the right:

$$N_2O_4(g) \rightleftharpoons 2NO_2(g)$$

$$K_c = \frac{[NO_2]^2_{eq}}{[N_2O_4]_{eq}}$$

- An increase in pressure (caused by a decrease in the volume of the container) has *no* effect on the value of K_c.
- The concentration of NO_2 doubles, so $[NO_2]^2$ increases by a factor of 4.
- The concentration of N_2O_4 only doubles, so the top line of the concentration term increases more than the bottom line.
- The concentration term becomes bigger and no longer equals K_c, so the system is no longer in equilibrium.
- The system reacts to make the concentration term smaller until it once again equals the unaltered value of K_c. It does this by converting NO_2 into N_2O_4, so the position of equilibrium shifts to the left.

e N_2O_4 is a colourless gas and NO_2 is brown. When the pressure is increased on this system at equilibrium, the colour fades as NO_2 is converted into N_2O_4. However, as it is an endothermic reaction from left to right, an increase in temperature causes a darkening of the colour, as the equilibrium position is pushed to the right.

An increase in pressure does not alter K_c, but the concentration term is increased, as there are fewer gas moles on the left. Therefore, the system reacts to make more N_2O_4, until the concentration term equals K_c once more.

Addition of an inert gas

When pressure is altered by the addition of an inert gas at constant volume, there is no effect on the concentrations of the reactants or products. This is because the number of moles of the reacting species has not been altered and neither has the volume. Concentration is the number of moles divided by the volume. As neither has altered, the concentration remains the same.

Since neither the value of K_c nor the value of the concentration term changes, the system is still in equilibrium. The position of equilibrium does not change, even though the pressure has been increased.

Effect on the rate of reaching equilibrium

If the reaction is homogeneous, the rate of collision increases when the pressure is increased. This causes an increase in the rates of the forward and back reactions and so equilibrium is reached sooner.

This is not the case for gaseous reactions catalysed by a solid catalyst. The rate is determined by the number of **active sites** on the catalyst surface, so an increase in pressure does not alter the rate. The active sites are always occupied unless the pressure falls to an extremely low value. A typical mechanism is:

reactant gases + free active sites $\rightleftharpoons$ adsorbed reactants — fast step

adsorbed reactants $\rightleftharpoons$ adsorbed products — slow step

adsorbed products $\rightleftharpoons$ gaseous products + free active sites — fast step

Biological reactions catalysed by enzymes have similar mechanisms — extra substrate does not increase the rate of the reaction. Eating a Mars bar will not make you run faster, but getting 'psyched up' will, as this causes you to produce adrenaline, which triggers the production of more enzyme.

Effect of change of concentration of one species

Effect on the equilibrium constant

Altering the concentration of a reactant or product has no effect on the value of the equilibrium constant.

Effect on the position of equilibrium

The equilibrium involving sulfur dioxide, oxygen and sulfur trioxide is represented by the equation:

$$2SO_2(g) + O_2(g) \rightleftharpoons 2SO_3(g)$$

$$K_c = \frac{[SO_3]^2_{eq}}{[SO_2]^2_{eq}[O_2]_{eq}}$$

The percentage conversion of sulfur dioxide to sulfur trioxide can be increased by adding more oxygen. Addition of extra oxygen to the system in equilibrium does not alter the value of K_c but causes the concentration term to become smaller. Therefore, it no longer equals K_c. The system reacts, making more SO_3, until equality is regained. This means that a greater proportion of sulfur dioxide is converted to sulfur trioxide.

> ***Worked example***
> State and explain the colour change that occurs when alkali is added to a solution containing dichromate(VI) ions.
>
> **Answer**
>
> $Cr_2O_7^{2-}(aq) + H_2O(l) \rightleftharpoons 2CrO_4^{2-}(aq) + 2H^+(aq)$
> orange yellow
>
> The addition of alkali does not alter the value of K_c but it removes H^+ ions, making $[H^+]$ smaller. This reduces the value of the concentration term, which no longer equals K_c. Therefore, dichromate(VI) ions react with water to form chromate ions until the concentration term once again equals K_c. The equilibrium position shifts to the right, so the solution turns yellow.

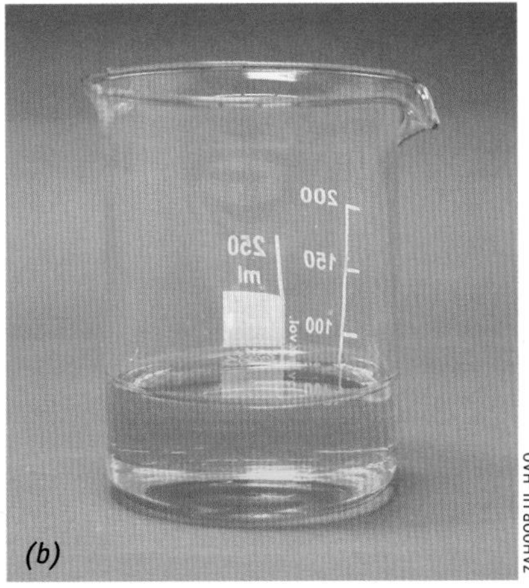

Acidified potassium dichromate (a) before and (b) after the addition of alkali

ZAHOOR UL-HAQ

Effect of adding a catalyst

- A catalyst has no effect on either the value of the equilibrium constant or the concentration term and so has no effect on the position of equilibrium.
- A catalyst speeds up both the forward and the back reactions equally. Therefore, equilibrium is reached more quickly.

Catalysts are used in exothermic industrial processes so that a faster rate can be obtained at a lower temperature. For an exothermic reaction this results in an increased yield, compared with the same reaction carried out at the same rate but at a higher temperature in the absence of a catalyst.

The cost of heating to a high temperature can always be partially recovered by the use of heat exchangers. High pressure is expensive, not only in the energy costs of compression, but also in the extra cost of a plant that will withstand the high pressure.

Industrial and pharmaceutical processes

The aim is to:

- maximise the percentage conversion of reactants to products (increase the equilibrium yield)
- make this amount of product as quickly as possible (fast rate of reaction)
- keep the costs as low as possible
- have as high an atom economy as possible

The chemical industry frequently uses a **continuous flow** method of production. The reactants are added continuously at one end of the plant and the products are removed continuously at the other. It is not truly an equilibrium system because it is not closed, but the reactants are in the presence of the catalyst for long enough for the principles of equilibrium to be useful when deciding on optimum conditions. This type of process is particularly effective when gases react in the presence of a solid catalyst.

The pharmaceutical industry uses a **batch process** — the reactants are added together in a reaction vessel. When the reaction is complete, the products are separated from any catalyst, the solvent and any unused reactants. The catalyst is often bonded to resin particles so that it can be removed easily from the reaction mixture and re-used.

The Haber process

The chemical reaction is:

$$N_2(g) + 3H_2(g) \rightleftharpoons 2NH_3(g) \quad \Delta H = -92\,\text{kJ mol}^{-1}$$

$$K_c = \frac{p(NH_3)^2_{eq}}{p(N_2)_{eq}\, p(H_2)^3_{eq}}$$

The hydrogen is obtained from methane by reacting it with steam at a high temperature. The overall reaction is:

$$CH_4(g) + 2H_2O(g) \rightleftharpoons CO_2(g) + 4H_2(g)$$
$$\Delta H = +165\text{ kJ mol}^{-1}$$

The nitrogen is obtained from the air.

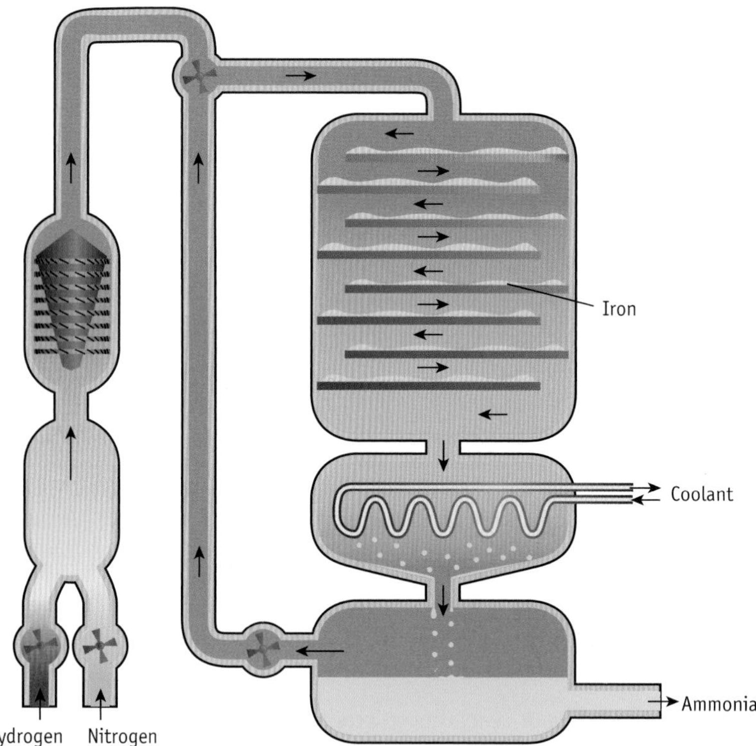

***Figure 4.1** Schematic diagram of the Haber process. Hydrogen and nitrogen are mixed and compressed. The mixture cycles through the reaction tower over trays of iron.*

The conditions are:

- temperature of 400°C to 450°C (673 K to 723 K)
- pressure of 200 atm
- catalyst — iron promoted by traces of aluminium and potassium oxides

Effect of temperature

$$\Delta S_{total} = \Delta S_{system} + \Delta S_{surr} = \Delta S_{system} - \Delta H/T$$

The reaction is exothermic, so the value of $-\Delta H/T$ and hence ΔS_{surr} is positive. Any increase in temperature will make the value of $-\Delta H/T$ less positive, which reduces the value of ΔS_{total}. As $\ln K = \Delta S_{total}$, any fall in the value of ΔS_{total} will cause a corresponding fall in the value of the equilibrium constant, K_p.

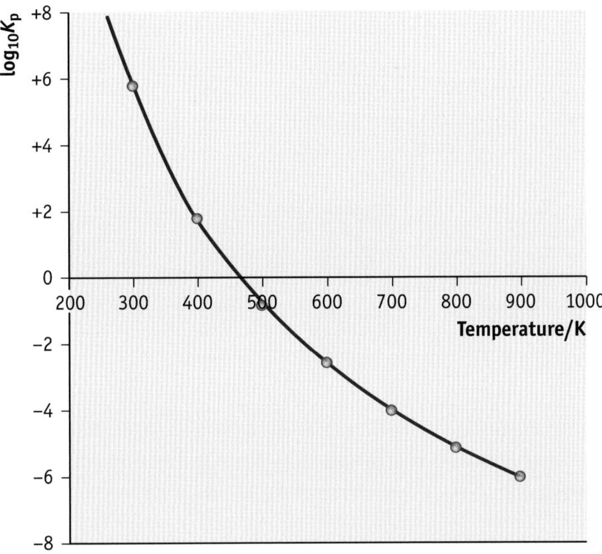

***Figure 4.2** Plot of $\log_{10} K_p$ against temperature for the Haber process equilibrium*

Figure 4.2 shows how the value of $\log K_p$ and hence K_p falls dramatically as the temperature is increased.

At first sight, it would appear that the Haber process should be carried out at room temperature, but then the rate would be so slow as to be effectively zero and almost no product would be formed. A high temperature would result in a low yield being achieved quickly; the rate at a lower temperature would be so slow that the reaction would not approach equilibrium. To overcome these problems, an iron catalyst is used. This allows the reaction to take place at a reasonable rate at a temperature of 700 K, which is a compromise temperature of a reasonable yield at an economically acceptable rate.

Effect of pressure

At 400°C, the equilibrium constant, K_p, is 3.9×10^{-4} atm^{-2}, so the yield at 1 atm pressure is low. Fritz Haber understood the principles of equilibrium and realised that high pressure would increase the yield. His co-worker, Carl Bosch, designed apparatus that could work at 200 atm pressure.

An increase in pressure does not alter the value of the equilibrium constant, but it causes the partial pressure expression to become smaller. This means that K_p no longer equals the partial pressure expression, so the system reacts making more ammonia until the partial pressure expression once again equals the value of the equilibrium constant. This means that the position of equilibrium shifts to the right.

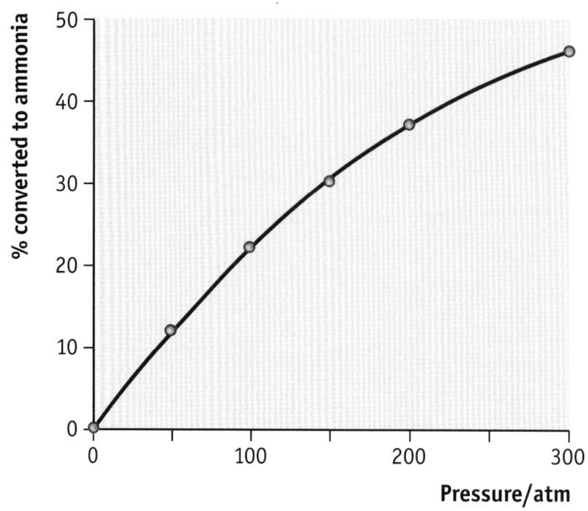

***Figure 4.3** Graph showing the effect of pressure on the percentage of reactants converted to ammonia at a temperature of 400°C*

As there are more molecules on the left-hand side of the equation, the bottom line of the partial pressure expression increases more than the top line. This makes K_p smaller.

The rate of reaction is not altered by increasing the pressure, as the rate is controlled by the availability of the active sites on the surface of the catalyst (see p. 78).

Atom economy

Even under these conditions only about 30% of the hydrogen is converted to ammonia. This is because the rate of reaction is still too slow for equilibrium to be reached in the catalyst chamber.

To obtain an economic overall yield with a high atom economy, the ammonia is removed by cooling the gases leaving the catalyst chamber. The ammonia liquefies and is separated from the unreacted nitrogen and hydrogen, which are then recycled through the catalyst chamber. In this way almost all the hydrogen is eventually converted to ammonia.

The choice of optimum conditions based on an understanding of the principles of kinetics and equilibrium is an example of 'How Science Works'.

Haber and Bosch were awarded the Nobel prize for their work, which enables nitrogenous fertilisers to be manufactured cheaply.

The Contact process

The crucial step in the manufacture of sulfuric acid is:

$$2SO_2(g) + O_2(g) \rightleftharpoons 2SO_3(g) \quad \Delta H = -196\,\text{kJ mol}^{-1}$$

$$K_c = \frac{p(SO_3)^2_{eq}}{p(SO_2)^2_{eq}\,p(O_2)_{eq}}$$

The conditions are:

- temperature of 425°C (698 K)
- pressure of 2 atm
- catalyst — vanadium(V) oxide, V_2O_5

Effect of temperature

The reaction is exothermic, so an increase in temperature decreases the value of ΔS_{surr}. This means that ΔS_{total} and K_p also decrease.

Figure 4.4 shows how $\log K_p$ varies with temperature.

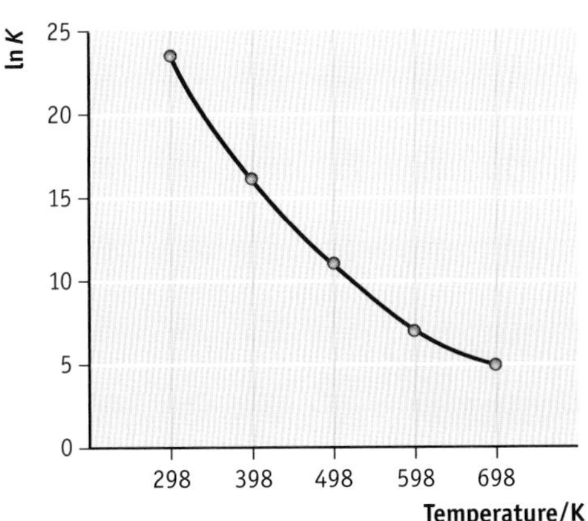

K_p at 298 K = 3×10^{24} atm^{-1}

K_p at 698 K = 6×10^{4} atm^{-1}

The value of K_p for this reaction, under the conditions of the Contact process, is about a billion times larger than that of the reaction in the Haber process.

To ensure that the reaction is fast enough to be economic at a temperature of 698 K, a catalyst of vanadium(V) oxide is used. A higher temperature would mean a lower yield and a lower temperature would mean an uneconomic rate. The temperature used is a compromise that produces a high yield quickly.

At a pressure of 2 atm and a temperature of 698 K, the equilibrium mixture contains 95% sulfur trioxide.

Effect of pressure

As there are more molecules on the left-hand side of the equation, an increase in pressure would drive the equilibrium to the right.

The reason for this is the same as for the Haber process.

However, the yield is high anyway and the use of high pressure is expensive in terms of both the energy required to compress the gases and the extra cost involved in making the plant able to withstand the high pressure. It does not make economic sense to increase the yield further by using high pressure.

However, a pressure greater than 1 atm must be used in order to drive the gases through the plant, so the air and sulfur dioxide are compressed to about 2 atm. Under these conditions, the equilibrium yield is over 95%.

Atom economy and pollution

It would be a waste of raw materials if unreacted sulfur dioxide were released into the atmosphere and it would also cause considerable atmospheric pollution. A higher atom economy is obtained by passing the gases from the catalyst chamber into concentrated sulfuric acid, which absorbs the sulfur trioxide from the equilibrium mixture. The unreacted sulfur dioxide and air are then passed back through another bed of catalyst where a further 95% conversion takes place. The result is that the gases released into the environment contain only slight traces of sulfur dioxide.

The sulfur trioxide reacts with the concentrated sulfuric acid to form oleum, but the sulfur dioxide does not react. Water is added carefully and sulfuric acid is formed.

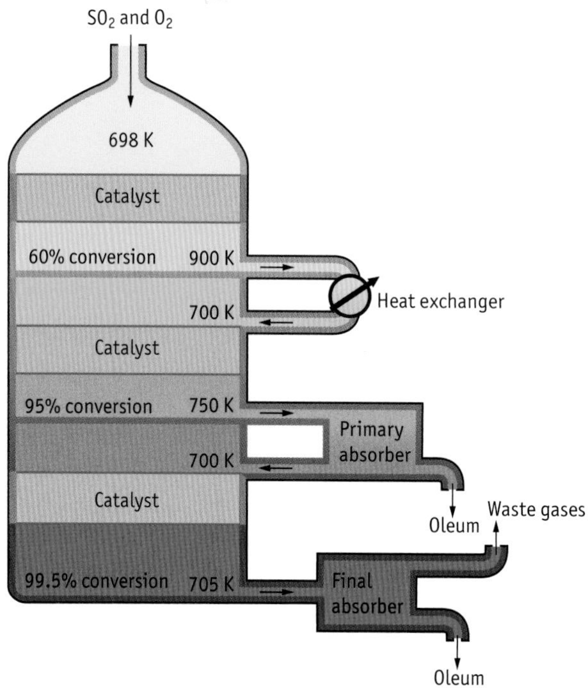

***Figure 4.5** A schematic representation of the Contact process*

The gases are passed through the first catalyst bed at a temperature of 698 K. As the reaction is exothermic, the gases heat up to about 900 K and the conversion is only 60%. The gases are then cooled to 700 K and passed back through another bed of catalyst. The conversion is now 95% and the mixture of sulfur trioxide, sulfur dioxide and oxygen is passed through a tower containing concentrated sulfuric acid. The sulfur trioxide forms oleum, $H_2S_2O_7$, and the remaining gases are passed through another catalyst bed. The overall conversion of sulfur dioxide is 99.5%.

The manufacture of hydrogen

Hydrogen is made by reacting methane with steam over a nickel catalyst at a temperature of 1000 K.

$$CH_4(g) + H_2O(g) \rightleftharpoons CO(g) + 3H_2(g) \quad \Delta H = +206 \text{ kJ mol}^{-1}$$

$$K_p = \frac{p(CO)_{eq}p(H_2)^3_{eq}}{p(CH_4)_{eq}p(H_2O)_{eq}}$$

This is a highly endothermic reaction and so ΔS_{surr} is negative.

At 298 K, $\Delta S_{surr} = -(+206\,000/298) = -691\,\text{J K}^{-1}\,\text{mol}^{-1}$ and $K_p \approx 10^{-20}\,\text{atm}^2$

At 700 K, $\Delta S_{surr} = -(+206\,000/700) = -294\,\text{J K}^{-1}\,\text{mol}^{-1}$ and $K_p \approx 2 \times 10^{-12}\,\text{atm}^2$

At 1000 K, $\Delta S_{surr} = -(+206\,000/1000) = -206\,\text{J K}^{-1}\,\text{mol}^{-1}$ and $K_p \approx 10\,\text{atm}^2$

As this is an endothermic reaction, the value of the equilibrium constant rises with increasing temperature. Thus the reaction is very much more favourable at the operating temperature of 1000 K than at lower temperatures.

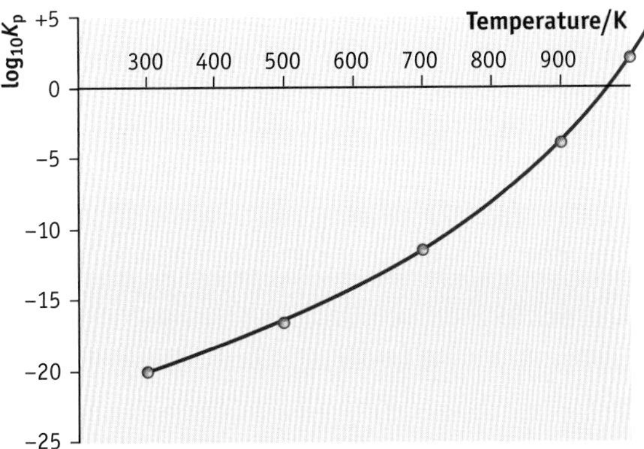

Figure 4.6 *Variation in $\log_{10} K_p$ with temperature for the reaction of methane with steam*

The manufacture of hydrogen requires a great deal of energy. The reaction is highly endothermic and consumes 34 000 kJ per tonne of hydrogen manufactured. In addition, there are considerable heat losses from the catalyst chamber. Almost a quarter of the methane used is required to produce the heat for the reaction.

The extraction of methane from methane hydrate

A possible future source of methane is the methane hydrate deposits (see page 253–254 of *Edexcel AS Chemistry*). These are found either in the oceans on some continental shelves or in the permafrost in the polar regions. Methane can be extracted by bringing the solid methane hydrate to room temperature and pressure. It has been estimated that the total oceanic methane hydrate reservoirs contain about eight times that in the known deposits of natural gas.

$$CH_4.6H_2O(s) \rightleftharpoons CH_4(g) + 6H_2O(l) \qquad \Delta H \text{ positive}$$

It can be seen from the equation that either a reduction in pressure or an increase in temperature will drive this equilibrium to the right, releasing methane gas from the solid methane hydrate.

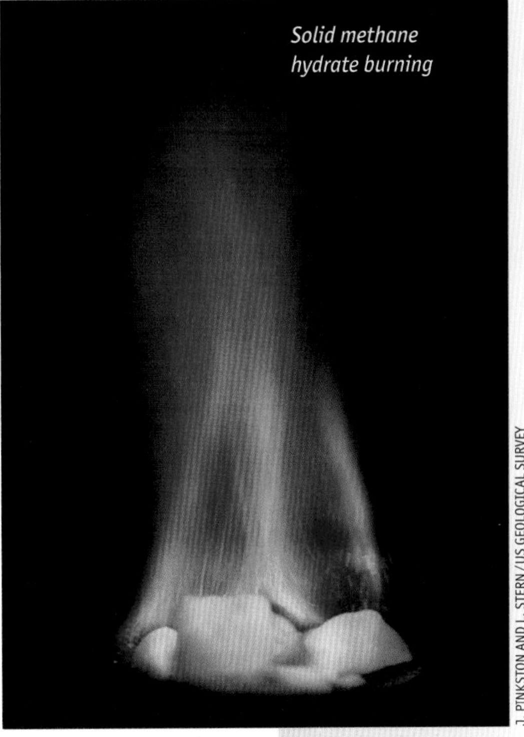

Solid methane hydrate burning

J. PINKSTON AND L. STERN/US GEOLOGICAL SURVEY

Pharmaceutical processes

As most pharmaceutical substances decompose when strongly heated, their manufacture is normally carried out between room temperature and 100°C. If energy is needed, it is usually provided by absorbing microwaves — in a sort of microwave oven.

Summary

- Enthalpy and entropy changes enable chemists to work out ΔS_{total} for a reaction and hence to decide whether the reaction is thermodynamically feasible — whether it will 'go'.
- An increase in temperature of an exothermic reaction will result in a decrease in equilibrium yield.
- An increase in temperature of an endothermic reaction will result in an increase in equilibrium yield.
- A decrease in temperature will result in a lower rate of reaction, so a catalyst and a compromise temperature may be used to optimise rate and yield.

- An increase in pressure will drive the equilibrium to the side of the equation with fewer gas molecules. However, the use of high pressure is expensive.
- Unreacted reactants are recycled to improve the atom economy, to conserve raw materials and to reduce environmental pollution.
- Many industrial processes are not true equilibrium systems because the products are removed.

Questions

1 0.20 mol of sulfur dioxide and 0.10 mol of oxygen were mixed with 2.0 mol of sulfur trioxide in a vessel of volume 20 dm^3 and heated to 425 °C.

$$2SO_2(g) + O_2(g) \rightleftharpoons 2SO_3(g)$$

The equilibrium constant $K_c = 1.7 \times 10^6\ mol^{-1}\,dm^3$ at 425 °C.

Is the system in equilibrium? If not, in which direction will the system move when a catalyst of vanadium(v) oxide is added?

2 When ammonium chloride is heated it decomposes into ammonia and hydrogen chloride:

$$NH_4Cl(s) \rightleftharpoons NH_3(g) + HCl(g) \quad \Delta H^\circ = +176\ kJ\,mol^{-1}$$

a Suggest the sign of ΔS°_{system}.

b Write the expression for K_p.

c At a temperature of 700 K, the value of the equilibrium constant is 50 atm^2. In terms of entropy and enthalpy, explain whether at 298 K the value will be more or less than the value at 700 K.

d Calculate the value of the total entropy change at 700 K and hence the value of ΔS°_{system}.
(The gas constant, R, = 8.31 $J\,K^{-1}\,mol^{-1}$)

3 Methanol is manufactured by passing carbon monoxide and hydrogen over a zinc oxide catalyst at a temperature of 300°C and a pressure of 300 atm:

$$CO(g) + 2H_2(g) \rightleftharpoons CH_3OH(g) \quad \Delta H^\circ = -90\ kJ\,mol^{-1}$$

K_p at 573 K = 5 x $10^{-4}\ atm^{-2}$

Explain, in terms of thermodynamics and kinetics, why these conditions are chosen.

4 Quicklime, CaO, is made by heating limestone, $CaCO_3$.

$$CaCO_3(s) \rightleftharpoons CaO(s) + CO_2(g) \quad \Delta H^\circ = +177\ kJ\,mol^{-1}$$
$$\Delta S^\circ_{system} = +158\ J\,K^{-1}\,mol^{-1}$$

a Write the expression for the equilibrium constant, K_p.

b State the units of K_p.

c Calculate the value of ΔS_{total} and hence K_p at:
(i) 700 K
(ii) 1200 K
(The gas constant, R, = 8.31 $J\,K^{-1}\,mol^{-1}$)

d At what temperature will the reaction change from being unfavourable to being favourable?

5 Explain how the atom economy of the Haber process is high, despite the equilibrium constant at the operating temperature having a low value.

6 Explain why, in the Contact process:

a the gases are cooled after a first pass through the catalyst bed before being passed through a second bed of vanadium(v) oxide

b why the gases are passed into concentrated sulfuric acid before being passed through the catalyst bed again

7 It has been suggested that a component of petrol, C_8H_{18}, could be made by reacting carbon with hydrogen. Comment on the feasibility of this reaction.

$$8C(s) + 9H_2(g) \rightleftharpoons C_8H_{18}(l) \quad \Delta H^\circ = -250\ kJ\,mol^{-1}$$
$$\Delta S^\circ_{system} = -860\ J\,K^{-1}\,mol^{-1}$$

8 When copper(II) chloride is dissolved in the minimum of water the following equilibrium is established:

$$Cu(H_2O)_6^{2+}(aq) + 4Cl^-(aq) \rightleftharpoons CuCl_4^{2-}(aq) + 6H_2O(l)$$

The reaction is exothermic and at 298 K the value of the equilibrium constant K_c is 5.3 $mol^{-4}\,dm^{12}$. The solution is green due to the mixture of blue hydrated copper(II) ions and yellow $CuCl_4^{2-}$ ions.

Explain, in terms of any changes to the equilibrium constant and the concentration term, what happens when:

a (i) concentrated hydrochloric acid is added
(ii) the solution is heated
(iii) a solution containing aqueous silver ions is added

b What colour change would you see when the concentrated hydrochloric acid is added?

9 Consider the equilibrium:

$$N_2(g) + 3H_2(g) \rightleftharpoons 2NH_3(g)$$

Explain why doubling the partial pressure of hydrogen has more effect on the position of the equilibrium than doubling the partial pressure of nitrogen.

10 Methane is trapped deep under the arctic sea in the form of solid methane hydrate $[CH_4(H_2O)_6]$. The equilibrium reaction for release of methane is:

$$[CH_4(H_2O)_6](s) \rightleftharpoons CH_4(g) + 6H_2O(s)$$

ΔH is endothermic.

a State and explain the effect of a decrease in pressure on this equilibrium at −40 °C.

b State and explain the effect of increasing the temperature of methane hydrate from −40 °C to −5 °C.

c If the partial pressure of methane at −40 °C is 0.86 atm, calculate the value of the equilibrium constant, K_p.

11 Consider the reaction in which ammonia is oxidised by air at 900 °C over a platinum catalyst at a pressure of 2 atm:

$$4NH_3(g) + 5O_2(g) \rightleftharpoons 4NO(g) + 6H_2O(g)$$

$\Delta H = -906\ kJ\ mol^{-1}$

Explain the effect on the equilibrium constant and on the position of equilibrium of:

a increasing the temperature

b increasing the pressure

c adding excess air

d removing the catalyst from the equilibrium mixture

12 Nitrogen and hydrogen react to form ammonia:

$$N_2(g) + 3H_2(g) \rightleftharpoons 2NH_3(g)$$

State and explain the effect, if any, on the equilibrium yield of ammonia of adding an inert gas, such as argon, to the equilibrium mixture:

a at constant volume and constant temperature

b at constant pressure and constant temperature

Acid–base equilibria

Chapter 5

Introduction

Acids were first defined in terms of their sour taste. When indicators such as litmus were discovered, an acid was thought of as any substance that turned litmus red. Later, the understanding of acidity led Arrhenius to define an acid as a substance that produces an excess of H^+ ions in aqueous solution. This definition is limited to aqueous solutions and was extended by Lowry and Brønsted to include non-aqueous solvents. Their definitions of acids and bases are given below.

An acid is a substance that gives a proton (H^+ ion) to a base.
A base is a substance that accepts a proton (H^+ ion) from an acid.

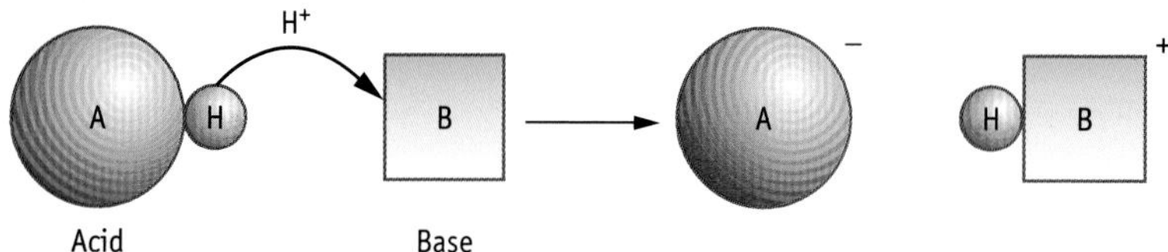

For an acid–base equilibrium, the changes are:

$$\text{acid} \underset{\text{gain of proton}}{\overset{\text{loss of proton}}{\rightleftharpoons}} \text{base}$$

The American chemist Lewis extended this idea further — for a base to be able to accept a proton, it must have a lone pair of electrons.

A base is a species that has a lone pair of electrons, which it uses to form a covalent bond. An acid is a substance that can accept the pair of electrons and form a bond.

The development of a theory from the very simple (based on observation) to the more complete (based on theory) is an example of 'How Science Works'

For example, an aluminium ion, Al^{3+}, is a Lewis acid as it accepts lone pairs of electrons from water molecules when the hydrated ion is formed.

Acid–base conjugate pairs

The reaction of an acid with a base can be written as the chemical equation:

$$HA + B \rightarrow BH^+ + A^-$$

For many acids this is a reversible reaction:

$$HA + B \rightleftharpoons BH^+ + A^-$$

- For the left-to-right reaction, HA is the acid and B is the base.
- For the right-to-left reaction, the acid is BH^+ and A^- is the base.

The acid, HA, and the base, A^-, derived from it by loss of a proton, are called an acid–conjugate base pair. The base, B, and the acid, BH^+, derived from it by acceptance of a proton, are also a base–conjugate acid pair.

The acid HCl reacts with the base H_2O:

$HCl(aq)$	+	$H_2O(l)$	$\longrightarrow$	$H_3O^+(aq)$	+	$Cl^-(aq)$
acid		base		conjugate acid		conjugate base

Cl^- is the conjugate base of the acid HCl; the conjugate acid of the base H_2O is H_3O^+.

Ammonia is a base and reacts with water, which acts as an acid giving a proton to the ammonia molecule:

$NH_3(aq)$	+	$H_2O(l)$	$\rightleftharpoons$	$OH^-(aq)$	+	$NH_4^+(aq)$
base		acid		conjugate base		conjugate acid

The relationship between conjugate pairs is:

- acid – $H^+ \rightarrow$ conjugate base
- base + $H^+ \rightarrow$ conjugate acid

Conjugate means joined together, here by the loss or gain of a proton.

Worked example

Concentrated sulfuric acid reacts with concentrated nitric acid:

$H_2SO_4 + HNO_3 \rightarrow H_2NO_3^+ + HSO_4^-$

Mark in the acid–base conjugate pairs.

Answer

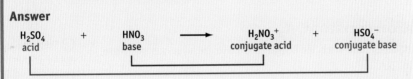

H_2SO_4	+	HNO_3	$\longrightarrow$	$H_2NO_3^+$	+	HSO_4^-
acid		base		conjugate acid		conjugate base

In this reaction, nitric acid is acting as a base, as it is protonated by the sulfuric acid, which is a stronger acid.

Some acid–base conjugate pairs are listed in Table 5.1 in order of decreasing acid strength and increasing base strength.

Table 5.1 *Acid–conjugate base pairs*

	Acid		Base	
	Name	Formula	Formula	Name
Strong acids	Sulfuric acid	H_2SO_4	HSO_4^-	Hydrogensulfate ion
Strong acids	Hydroiodic acid	HI	I^-	Iodide ion
Strong acids	Hydrobromic acid	HBr	Br^-	Bromide ion
Strong acids	Hydrochloric acid	HCl	Cl^-	Chloride ion
Strong acids	Nitric acid	HNO_3	NO_3^-	Nitrate ion
	Hydronium ion	H_3O^+	H_2O	Water
Weak acids	Hydrogensulfate ion	HSO_4^-	SO_4^{2-}	Sulfate ion
Weak acids	Hydrofluoric acid	HF	F^-	Fluoride ion
Weak acids	Ethanoic acid	CH_3COOH	CH_3COO^-	Ethanoate ion
Weak acids	Carbonic acid	H_2CO_3	HCO_3^-	Hydrogencarbonate ion
Weak acids	Ammonium ion	NH_4^+	NH_3	Ammonia
Weak acids	Hydrogencarbonate ion	HCO_3^-	CO_3^{2-}	Carbonate ion
Weak acids	Water	H_2O	OH^-	Hydroxide ion

Strong and weak acids and bases

A strong acid is an acid that is totally ionised in aqueous solution forming hydrated hydrogen ions, H_3O^+.

For example, hydrochloric acid, HCl(aq), is a strong acid:

$$HCl(aq) + H_2O(l) \rightarrow H_3O^+(aq) + Cl^-(aq)$$

A weak acid is an acid that is only very slightly ionised in aqueous solution.

For example, ethanoic acid is a weak acid:

$$CH_3COOH(aq) + H_2O(l) \rightleftharpoons H_3O^+(aq) + CH_3COO^-(aq)$$

A 0.1 mol dm^{-3} solution of ethanoic acid is only 1.3% ionised.

A strong base is totally ionised in aqueous solution forming hydroxide ions, OH^-.

For example, sodium hydroxide is a strong base:

$$NaOH(aq) \rightarrow Na^+(aq) + OH^-(aq)$$

A weak base is protonated to only a small degree in solution and so only forms a small proportion of hydroxide ions.

For example, ammonia is a weak base:

$$NH_3(aq) + H_2O(l) \rightleftharpoons NH_4^+(aq) + OH^-(aq)$$

Do not say that a weak acid is only partially ionised because this could mean that the ionisation is considerable but less than complete. Even HF, which is one of the strongest 'weak' acids, is only 5.7% ionised in a 0.1 mol dm^{-3} solution.

Acid and base equilibrium constants

Acid dissociation constant, K_a

A weak acid is in equilibrium with its conjugate base in aqueous solution. Consider a weak acid, HA:

$$HA + H_2O \rightleftharpoons H_3O^+ + A^-$$

The expression for the equilibrium constant (known in this context as the **acid dissociation constant, K_a**) is:

$$K_a = \frac{[H_3O^+][A^-]}{[HA]}$$

$[H_2O]$ is not included in this expression as the concentration of water, in aqueous solutions, is constant (p. 54).

The true expression of the equilibrium constant includes $[H_2O]$.

$$K_{true} = \frac{[H_3O^+][A^-]}{[HA][H_2O]}$$

1 dm^3 of water has a mass of 1000 g.

It contains $\frac{1000\,g}{18.0\,g\,mol^{-1}} = 55.6\,mol$.

Thus, the concentration of water in an aqueous solution is 55.6 mol dm^{-3}.

State symbols need not be written in equations in this topic, as all the substances are in solution.

Never include $[H_2O]$ in the expression for K_a.

This is a constant for all aqueous equilibria involving weak acids. Therefore, its value can be incorporated into the equilibrium expression:

$$K_{true} \times 55.6 = \frac{[H_3O^+][A^-]}{[HA]}$$

$$K_a = K_{true} \times 55.6$$

Base dissociation constant, K_b

An aqueous solution of a weak base is in equilibrium with its conjugate acid, for example:

$$NH_3 + H_2O \rightleftharpoons NH_4^+ + OH^-$$

The expression for the **base dissociation constant**, K_b, is:

$$K_b = \frac{[NH_4^+][OH^-]}{[NH_3]}$$

As with weak acids, $[H_2O]$ is omitted from the expression because its value is constant in aqueous equilibria involving weak bases.

Auto-ionisation of water and the pH scale

Water is amphoteric. It can act as both a base, as in its reaction with hydrogen chloride, or as an acid, as in its reaction with ammonia.

The amphoteric nature of water is even evident in the absence of another acid or base. One molecule of water can protonate another molecule of water:

H_2O	+	H_2O	$\rightleftharpoons$	H_3O^+	+	OH^-
acid		base		conjugate acid		conjugate base

The equilibrium constant for this reaction is given the symbol K_w. The equilibrium expression does not include the term $[H_2O]$ because its value is constant.

$$K_w = [H_3O^+][OH^-]$$

This is often written as $K_w = [H^+][OH^-]$.

K_w is also called the **ionic product** of water. Its value, at 25°C, is $1.0 \times 10^{-14}\,mol^2\,dm^{-6}$.

In any aqueous solution, the value of $[H^+] \times [OH^-]$ always equals K_w, the value of which is $1.0 \times 10^{-14}\,mol^2\,dm^{-6}$ at 25°C.

Thus $[OH^-] = \frac{1.0 \times 10^{-14}}{[H^+]}$

For an acidic solution containing $1.0 \times 10^{-3}\,mol\,dm^{-3}$ of $H^+(aq)$ ions,

$$[OH^-] = \frac{1.0 \times 10^{-14}}{1.0 \times 10^{-3}} = 1.0 \times 10^{-11}\,mol\,dm^{-3}$$

e The equation for a weak acid is sometimes written as $HA \rightleftharpoons H^+ + A^-$. So the expression for K_a is:

$$K_a = \frac{[H^+][A^-]}{[HA]}$$

$[H^+]$ can be regarded as being shorthand for $[H_3O^+]$.

pH scale

Hydrogen ion concentration varies over a huge range of values (by a factor of about a trillion), so a logarithmic scale of measurement was devised. To avoid negative numbers in most cases, the scale was defined as:

$$pH = -\log_{10}[H^+] \quad \text{(this is often written as } -\log[H^+]\text{)}$$

$[H^+] = 10^{-pH}$

pH equals the negative logarithm to the base 10 of the hydrogen ion (hydronium ion) concentration.

In practice, the pH scale runs from about –1 to just over 14.

Worked example

Calculate the pH of a solution in which $[H^+]$ is equal to:

a $10\,mol\,dm^{-3}$
b $0.10\,mol\,dm^{-3}$
c $1.23 \times 10^{-4}\,mol\,dm^{-3}$
d $4.56 \times 10^{-9}\,mol\,dm^{-3}$
e $7.89 \times 10^{-15}\,mol\,dm^{-3}$

Answer

a $pH = -\log 10 = -1.00$
b $pH = -\log 0.10 = 1.00$
c $pH = -\log 1.23 \times 10^{-4} = 3.91$
d $pH = -\log 4.56 \times 10^{-9} = 8.34$
e $pH = -\log 7.89 \times 10^{-15} = 14.10$

e If an answer to a question is a pH value, you should report it to two decimal places.

The hydrogen ion concentration can be calculated from the pH, using the expression:

$$[H^+] = 10^{-pH}$$

Worked example

Calculate the hydrogen ion concentration in solutions of pH:

a 3.50
b 7.00
c 12.85

Answer

a $[H^+] = 10^{-3.50} = 3.16 \times 10^{-4}\,mol\,dm^{-3}$
b $[H^+] = 10^{-7.00} = 1.0 \times 10^{-7}\,mol\,dm^{-3}$
c $[H^+] = 10^{-12.85} = 1.41 \times 10^{-13}\,mol\,dm^{-3}$

Neutrality

A neutral solution is one in which the concentrations of H^+ and OH^- ions are the same:

$$K_w = [H^+][OH^-] = 1.0 \times 10^{-14}\,mol^2\,dm^{-6} \text{ at } 25°C$$
$$[H^+] = [OH^-]$$
$$[H^+]^2 = 1.0 \times 10^{-14} \text{ or } [H^+] = \sqrt{(1.0 \times 10^{-14})} = 1.0 \times 10^{-7}\,mol\,dm^{-3}$$
$$pH = -\log 1.0 \times 10^{-7} = 7.00$$

In all aqueous solutions at 25°C, $[H^+] \times [OH^-] = 1.0 \times 10^{-14}$.

An acidic solution has $[H^+] > [OH^-]$. Therefore $[H^+] > 1.0 \times 10^{-7}$ and the solution has pH < 7.

An alkaline solution has $[H^+] < [OH^-]$. Therefore $[H^+] < 1.0 \times 10^{-7}$ and the solution has pH >7.

At 25°C, neutral pH = 7; acidic pH < 7; alkaline pH > 7.

pK_w, pOH and pK_a

The prefix 'p', in this context, means 'the negative log of'.

$pK_w = -\log K_w$ $\quad$ $pK_w = -\log(1.0 \times 10^{-14})$ $\quad$ $pK_w = 14.0$ at 25°C

$pOH = -\log[OH^-]$

The neutral value of 7 is only true when the solution is at 25°C.

As $K_w = [H^+][OH^-]$ and since $\log(a \times b) = \log a + \log b$:

$\log K_w = \log[H^+] + \log[OH^-]$

$pK_w = pH + pOH$ *or* $pH + pOH = 14$

Worked example

Calculate the concentration of hydroxide ions in a solution of pH 1.23.

Answer using method 1, which uses pH + pOH = 14

$pOH = 14 - pH = 14 - 1.23 = 12.77$

$[OH^-] = 10^{-12.77} = 1.70 \times 10^{-13}\,mol\,dm^{-3}$

Answer using method 2, which uses $[OH^-] = \dfrac{K_w}{[H^+]}$

$[H^+] = 10^{-1.23} = 0.0589\,mol\,dm^{-3}$

$$[OH^-] = \frac{1.00 \times 10^{-14}}{0.0589} = 1.70 \times 10^{-13}\,mol\,dm^{-3}$$

Note how even in very acidic solutions there are some OH^- ions present.

The acid dissociation constant, K_a, and pK_a are related by the expression:

$pK_a = -\log K_a$ $\quad$ or $\quad$ $K_a = 10^{-K_a}$

Worked example

a Calculate the value of K_a of a weak acid with $pK_a = 5.45$.

b Calculate the value of pK_a of a weak acid with $K_a = 2.17 \times 10^{-6}\,mol\,dm^{-3}$.

Answer

a $K_a = 10^{-5.45} = 3.55 \times 10^{-6}\,mol\,dm^{-3}$

b $pK_a = -\log K_a = -\log(2.17 \times 10^{-6})$

$= 5.66$

e The ionisation of water is the reverse of neutralisation of a strong acid with a strong base and so is endothermic. This means that at a higher temperature, the value of K_w is greater. At 37°C (normal blood temperature), the value of K_w is $2.4 \times 10^{-14}\,mol^2\,dm^{-6}$ and at 100°C it is $5.13 \times 10^{-13}\,mol^2\,dm^{-6}$. Therefore, at 37°C neutral pH is 6.8 and at 100°C it is 6.1.

The pH of acids, bases and salts

Introduction

Solutions of equal concentration

The pH of 0.10 mol dm^{-3} solutions of a number of acids, bases and salts are shown in Table 5.2.

Acid/base	pH	Salt	pH
Strong acid (e.g. HCl)	1.0	Salt of strong acid and strong base (e.g. NaCl)	7.0
Weak acid (e.g. CH_3COOH)	2.9	Salt of a strong acid and weak base (e.g. NH_4Cl)	5.1
Strong base (e.g. NaOH)	13.0	Salt of a weak acid and a strong base (e.g. CH_3COONa)	8.9
Weak base (e.g. NH_3)	11.1	Salt of a weak acid and weak base (e.g. CH_3COONH_4)	7.0

***Table 5.2** The pH of 0.1 mol dm^{-3} solutions of some acids, bases and salts*

Note how the pH of a solution of a weak acid or a weak base differs by approximately two from that of a strong acid or base (The rule of two — see p. 104.)

Dilution of solutions

The pH values resulting from diluting solutions of strong and weak acids and bases by factors of ten are shown in Table 5.3.

Concentration/ mol dm^{-3}	pH of a solution of HCl	pH of a solution of NaOH	pH of a solution of CH_3COOH	pH of a solution of NH_3
0.1	1	13	2.9	11.1
0.01	2	12	3.4	10.6
0.001	3	11	3.9	10.1
0.0001	4	10	4.4	9.6

***Table 5.3** Effect on pH of dilution*

Note: the pH of an acid increases as it is diluted; the pH of a base decreases as it is diluted.

When strong acids or strong bases are diluted, the pH changes by 1 unit for each ten-fold dilution.

When weak acids or weak bases are diluted, the pH changes by half a unit for each ten-fold dilution.

The pH of strong acids

A **strong acid**, such as nitric acid, HNO_3, is totally ionised in aqueous solution. Thus, for example, a nitric acid solution of concentration 0.123 mol dm^{-3} has a hydrogen ion concentration of 0.123 mol dm^{-3}.

$pH = -\log 0.123 = 0.91$

Worked example 1

Calculate the pH of a solution of HCl made by dissolving 4.56 g of hydrogen chloride, HCl (or 3.00 dm^3 of HCl gas) in water and making the solution up to a volume of 250 cm^3.

Answer

$$\text{amount of HCl} = \frac{4.56\,g}{36.5\,g\,mol^{-1}} = 0.125\,mol$$

(or $\frac{3.00\,dm^3}{24.0\,dm^3\,mol^{-1}} = 0.125\,mol$)

$[HCl] = \frac{mol}{volume} = \frac{0.125\,mol}{0.250\,dm^3} = 0.500\,mol\,dm^{-3}$

pH = $-\log 0.500 = 0.30$

Worked example 2

Calculate the pH of a $2.00\,mol\,dm^{-3}$ solution of hydrochloric acid.

Answer

$[H^+] = 2.00\,mol\,dm^{-3}$

pH = $-\log 2.00 = -0.30$

e Note that if a strong acid has a concentration greater than $1.00\,mol\,dm^{-3}$, it will have a negative pH.

Sulfuric acid, H_2SO_4, is only a strong acid in its first ionisation:

$$H_2SO_4 \rightarrow H^+ + HSO_4^-$$

The second ionisation is weak:

$$HSO_4^- \rightleftharpoons H^+ + SO_4^{2-}$$

and is suppressed by the H^+ ions from the first ionisation. So, a solution of sulfuric acid of concentration $0.10\,mol\,dm^{-3}$ has $[H^+]$ of just above 0.10 $mol\,dm^{-3}$, *not* $0.20\,mol\,dm^{-3}$ and hence its pH is very slightly less than 1.00.

The pH of strong bases

A **strong base** is totally ionised in aqueous solution. For example, a solution of a soluble base MOH of concentration $0.123\,mol\,dm^{-3}$ has a hydroxide ion concentration of $0.123\,mol\,dm^{-3}$.

A solution of a strong base $M(OH)_2$ of concentration $0.123\,mol\,dm^{-3}$ has a hydroxide ion concentration of $0.246\,mol\,dm^{-3}$, as there are two moles of OH^- ions per mole of base.

The pH can be worked out in one of two ways.

Method 1

- Using the expression pH + pOH = 14, calculate pOH and hence pH.

  If $[OH^-] = 0.123\,mol\,dm^{-3}$

  pOH = $-\log[OH^-] = -\log 0.123 = 0.91$

  pH = 14 − pOH = 14 − 0.91 = 13.09

Method 2

- Using the expression $[H^+] \times [OH^-] = 1.0 \times 10^{-14}\,mol^2\,dm^{-6}$, calculate $[H^+]$ and hence the pH.

  In the example above

  $[OH^-] = 0.123\,mol\,dm^{-3}$

$$[H^+] = \frac{1.0 \times 10^{-14}}{[OH^-]} = \frac{1.0 \times 10^{-14}}{0.123} = 8.13 \times 10^{-14}\,mol\,dm^{-3}$$

$$pH = -\log [H^+] = -\log 8.13 \times 10^{-14} = 13.09$$

Worked example

Calculate the pH of a $0.0444\,mol\,dm^{-3}$ solution of barium hydroxide, $Ba(OH)_2$.

Answer using method 1

1 mol of $Ba(OH)_2$ produces 2 mol of OH^- ions.

$[OH^-] = 2 \times 0.0444 = 0.0888\,mol\,dm^{-3}$

$pOH = -\log 0.0888 = 1.05$

$pH = 14 - pOH = 14 - 1.05 = 12.95$

Answer using method 2

1 mol of $Ba(OH)_2$ produces 2 mol of OH^- ions.

$[OH^-] = 2 \times 0.0444 = 0.0888\,mol\,dm^{-3}$

$$[H^+] = \frac{1.0 \times 10^{-14}}{0.0888} = 1.13 \times 10^{-13}$$

$pH = -\log [H^+] = -\log 1.13 \times 10^{-13} = 12.95$

e You should always check your calculation of pH to ensure it makes sense. An acid solution at 298 K cannot have a pH > 7; an alkaline solution cannot have a pH < 7. If you have obtained an impossible answer as a result of a calculation, then you should do the calculation again.

Titration of a strong acid with a strong base

In an experiment, $20.0\,cm^3$ of hydrochloric acid (a strong acid) of concentration $0.100\,mol\,dm^{-3}$ was titrated with a solution of sodium hydroxide (a strong base) of concentration $0.100\,mol\,dm^{-3}$.

The variation of pH with the volume of sodium hydroxide added can be estimated by calculating the pH at certain points.

pH at the start

The acid concentration is $0.100\,mol\,dm^{-3}$, so $pH = -\log 0.100 = 1.00$.

pH after the addition of $10.0\,cm^3$ sodium hydroxide

$0.0200 \times 0.100 = 0.00200$ mol of acid were present originally. Half the acid has reacted, so $\frac{1}{2} \times 0.00200 = 0.00100$ mol are present in $30.0\,cm^3$ of solution. Therefore,

$$[H^+] = \frac{0.00100}{0.0300} = 0.0333\,mol\,dm^{-3}$$

$$pH = -\log 0.0333 = 1.48$$

e You must convert the volume in cm^3 to dm^3 by dividing by 1000.

pH at equivalence point (after $20.0\,cm^3$ added)

All the acid has reacted and the solution contains sodium chloride which is neutral at pH = 7.

ANDREW LAMBERT PHOTOGRAPHY/SPL

Performing a titration

pH after the addition of 30.0 cm^3 of sodium hydroxide

Two-thirds of the sodium hydroxide has reacted, so 10.0 cm^3 did not react. Therefore, $0.100 \times 0.0100 = 0.00100$ mol of NaOH are present in 50.0 cm^3 of solution.

$$pOH = \frac{-\log 0.00100}{0.0500} = 1.70$$

$$pH = 14 - 1.70 = 12.30$$

The pH values at different points during this titration are given in Table 5.4.

Table 5.4 *pH during the titration of 20 cm^3 0.100 mol dm^{-3} HCl with 0.100 mol dm^{-3} NaOH*

Volume of NaOH added/cm^3	pH	Volume of NaOH added/cm^3	pH
0	1.00	20.0	7.00
10	1.48	20.1	10.40
15	1.85	21.0	11.39
19	2.59	30.0	12.30
19.9	3.60	40.0	12.52

These data can be presented as a graph (Figure 5.1), which is usually called a titration curve.

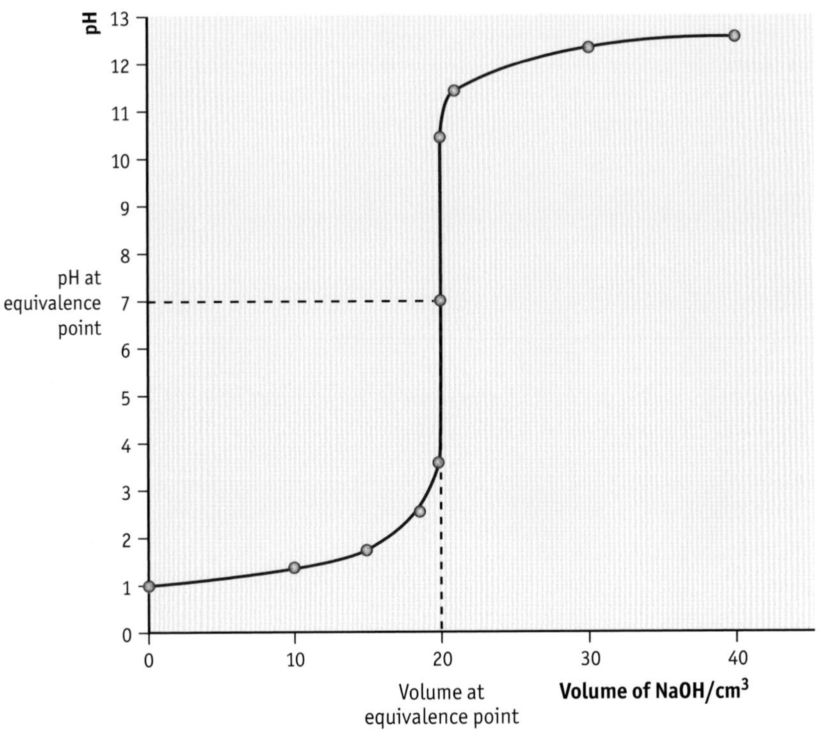

Figure 5.1 *Titration of a strong acid with a strong base*

Notice that the graph rises slowly to start with, then vertically at the equivalence point, before flattening off just below pH 13.

If the acid were added to 20 cm^3 of the alkali, the graph would have a similar shape but would be a mirror image. It would start at pH 13, fall slowly, and then just before the equivalence volume of 20.0 cm^3 of acid, it would plunge vertically from about pH 10 to pH 3. Finally, it would flatten off just above pH 1.

The pH of weak acids

A weak acid is in equilibrium with its conjugate base:

$$HA + H_2O \rightleftharpoons H_3O^+ + A^-$$

The rates of the forward and back reactions are so rapid, that the system is always in equilibrium. The equilibrium constant (the acid dissociation constant, K_a) is given by:

$$K_a = \frac{[H_3O^+][A^-]}{[HA]}$$

The reaction is sometimes simplified to:

$$HA \rightleftharpoons H^+ + A^-$$

$$K_a = \frac{[H^+][A^-]}{[HA]}$$

$[H_2O]$ is omitted from these expressions, as its concentration is effectively constant.

e In A-level answers either expression for K_a is acceptable, unless the equation in the question has H_3O^+ on the right-hand side, in which case $[H_3O^+]$ must be used.

The pH of a solution of a weak acid can be calculated using the same method as for K_c calculations.

Consider a $0.10\,mol\,dm^{-3}$ solution of a weak acid with $K_a = 1.0 \times 10^{-5}\,mol\,dm^{-3}$. Let x mol of HA ionise per dm^3:

	HA	H_3O^+	A^-
Initial concentration	0.10	0	0
Equilibrium concentration	$(0.10 - x)$	x	x

$$K_a = \frac{[H_3O^+][A^-]}{[HA]} = \frac{x^2}{0.10 - x} = 1.0 \times 10^{-5}\,mol\,dm^{-3}$$

This can be solved only by using the formula for a quadratic equation, which is outside the A-level Chemistry specification. However, if the value of x is much less than the initial concentration of the weak acid, the term $(0.10 - x)$ can be approximated to 0.10. The value of x can now be solved easily:

$$\frac{x^2}{0.10} = 1.0 \times 10^{-5}$$

$$x = \sqrt{(1.0 \times 10^{-5} \times 0.10} = 1.0 \times 10^{-3}\,mol\,dm^{-3}$$

e You should check that the approximation is fair. Look at the value of x that you have calculated and check that it is small compared with the initial value of [HA]. In this example, the value of x is 0.0010. This is small compared with 0.10, as x is only 1% of [HA].

Thus,

$$[H_3O^+] = 1.0 \times 10^{-3}\,mol\,dm^{-3}$$

$$pH = -\log 1.0 \times 10^{-3} = 3.00$$

In calculations involving weak acids, the following assumptions are made:

- The tiny amount of H_3O^+ ions from the auto-ionisation of water is ignored. (This is because the degree of ionisation of water is very small and is suppressed by the H_3O^+ ions formed from the ionisation of the weak acid, HA.)

- $[H_3O^+] = [A^-]$. (There is no other source of A^- ions and the tiny amount of H_3O^+ ions from the auto-ionisation of water has been ignored.)
- [HA] at equilibrium is equal to the initial concentration of the acid HA. (Only about 1% of the weak acid molecules are ionised, so this is a fair assumption.)

These assumptions allow the expression for K_a to be simplified:

$$K_a = \frac{[H_3O^+][A^-]}{[HA]} = \frac{[H_3O^+]^2}{[HA]_{initial}}$$

This simplified expression can be used to calculate the pH of a solution of a weak acid, given its concentration and the value of K_a.

e Always use the equilibrium constant expression in these and buffer solution calculations. It is safer than trying to remember a formula such as $pH = \frac{1}{2}pK_a - \frac{1}{2}\log c$ (where c = the concentration of the acid).

Worked example

Calculate the pH of a 0.135 mol dm^{-3} solution of ethanoic acid. (K_a for ethanoic acid = 1.8×10^{-5} mol dm^{-3} at 25°C)

Answer

Ethanoic acid ionises in water:

$$CH_3COOH + H_2O \rightleftharpoons H_3O^+ + CH_3COO^-$$

$$K_a = \frac{[H_3O^+][CH_3COO^-]}{[CH_3COOH]}$$

$$= \frac{[H_3O^+]^2}{0.135}$$

$$= 1.8 \times 10^{-5}\ \text{mol dm}^{-3}$$

$$[H_3O^+]^2 = 0.135 \times 1.8 \times 10^{-5}$$

$$= 2.43 \times 10^{-6}\ \text{mol}^2\ \text{dm}^{-6}$$

$$[H_3O^+] = \sqrt{(2.43 \times 10^{-6}}$$

$$= 0.00156\ \text{mol dm}^{-3}$$

$$pH = -\log[H_3O^+] = -\log 0.00156 = 2.81$$

Some questions give the pK_a value of the weak acid. In this case, pK_a must be converted to K_a.

$$K_a = 10^{-pK_a}$$

For example, pK_a for methanoic acid = 3.75

$$K_a = 10^{-3.75} = 1.78 \times 10^{-4}\ \text{mol dm}^{-3}$$

The strengths of some weak acids are shown in Table 5.5.

Weak acids have $K_a < 1$ and $pK_a > 0$; strong acids have $K_a > 1$, so $pK_a < 0$ (negative).

The pH of a solution can be measured using a pH meter (p. 185). If the pH and the concentration of a weak acid are known, the value of its acid dissociation constant, K_a, can be calculated.

Inorganic acid	pK_a	K_a/mol dm^{-3}
Hydrogensulfate ion, HSO_4^-	2.0	0.010
Hydrogen fluoride, HF	3.25	5.65×10^{-4}
Carbonic acid, H_2CO_3	6.38	4.17×10^{-7}
Ammonium ion, NH_4^+	9.25	5.62×10^{-10}
Hydrogen cyanide, HCN	9.40	3.98×10^{-10}
Hydrogencarbonate ion, HCO_3^-	10.32	4.79×10^{-11}
Organic acid		
Methanoic, HCOOH	3.75	1.78×10^{-4}
Ethanoic, CH_3COOH	4.76	1.75×10^{-5}
Chloroethanoic, $ClCH_2COOH$	2.86	1.32×10^{-3}
Propanoic, C_2H_5COOH	4.86	1.39×10^{-5}
Benzoic, C_6H_5COOH	4.20	6.31×10^{-5}
Phenol, C_6H_5OH	10.0	1.0×10^{-10}

Table 5.5 *Acid dissociation constants, K_a, and the pK_a values of some weak acids*

The smaller the value of K_a, the weaker is the acid. Thus, the larger the value of pK_a, the weaker is the acid.

Worked example

The pH of a 0.200 mol dm^{-3} solution of propanoic acid was found to be 2.78. Calculate the value of its acid dissociation constant, K_a.

Answer

Propanoic acid ionises according to the equilibrium:

$$C_2H_5COOH \rightleftharpoons H^+ + C_2H_5COO^-$$

pH = 2.78, so $[H^+] = 10^{-2.78} = 0.00166\ \text{mol dm}^{-3}$

$$K_a = \frac{[H^+][C_2H_5COO^-]}{[C_2H_5COOH]_{initial}}$$

$$= \frac{[H^+]^2}{[C_2H_5COOH]_{initial}}$$

$$= \frac{0.00166^2}{0.200} = 1.38 \times 10^{-5}\ \text{mol dm}^{-3}$$

$[C_2H_5COOH]_{initial}$ = the original concentration of the acid

e Note that in this worked example a simplified equilibrium expression was written, with H^+ being used instead of H_3O^+. Both are acceptable.

In the worked example above, the approximation:

$$[C_2H_5COOH]_{eq} \rightleftharpoons [C_2H_5COOH]_{initial}$$

need not have been made. The question gives the solution pH of 2.78, which equals a hydrogen ion concentration of 0.00166 mol dm^{-3}. Therefore, 0.00166 mol dm^{-3} of acid must have dissociated, leaving $[C_2H_5COOH] = 0.200 - 0.00166 = 0.19834$ mol dm^{-3}. This difference is so small that the true value of K_a obtained without using the approximation is 1.39×10^{-5} mol dm^{-3}, which is hardly any different from that calculated using the approximation.

However, with weak acids such as HF and HNO_2, which are stronger than propanoic acid, the approximation should not be made.

Worked example

The pH of a 0.100 mol dm^{-3} solution of hydrofluoric acid, HF, was found to be 2.14. Calculate the value of the acid dissociation constant, K_a, of hydrofluoric acid.

Answer

$[H^+] = 10^{-pH} = 10^{-2.14} = 0.00724\,mol\,dm^{-3}$

$[H^+] = [F^-]$

0.00724 mol dm^{-3} of H^+ ions were produced, so 0.00724 mol of HF must have dissociated.

$[HF] = [original] - [amount\ dissociated] = (0.100 - 0.00724) = 0.0928\,mol\,dm^{-3}$

$$K_a = \frac{[H^+][F^-]}{[HF]} = \frac{0.00724^2}{0.0928} = 5.65 \times 10^{-4}\,mol\,dm^{-3}$$

In the worked example above, if the approximation:

$[HF] = [original\ acid] = 0.100\,mol\,dm^{-3}$

had been made, the value of K_a would have been $5.24 \times 10^{-4}\,mol\,dm^{-3}$.

Diprotic acids

A **diprotic acid** can produce two H^+ ions per molecule. Sulfuric acid, H_2SO_4, is an example. It is a strong acid from its first ionisation:

$H_2SO_4 \rightarrow H^+ + HSO_4^-$

A 0.10 mol dm^{-3} solution of sulfuric acid produces 0.10 mol dm^{-3} of H^+ ions in its *first* ionisation.

The HSO_4^- ion is only a weak acid and so the second H^+ is only partially ionised:

$HSO_4^- \rightleftharpoons H^+ + SO_4^{2-}$

$$K_2 = \frac{[H^+][SO_4^{2-}]}{[HSO_4^-]} = 1.0 \times 10^{-2}\,mol\,dm^{-3}$$

The H^+ ions produced by the first ionisation suppress the second ionisation. The amount of H^+ produced by the second ionisation is only 0.0080 mol dm^{-3}.

pH of 0.10 mol dm^{-3} sulfuric acid $= -\log(0.10 + 0.008) = 0.97$

It would be quite wrong to assume that in a 0.10 mol dm^{-3} solution of H_2SO_4, $[H^+] = 2 \times 0.10 = 0.20\,mol\,dm^{-3}$ and hence that the pH $= -\log 0.20 = 0.70$.

e Although this calculation, which involves solving a quadratic equation, is beyond the A-level specification, candidates are expected to be able to explain why the $[H^+]$ in a solution of sulfuric acid is only slightly more than that caused by its first ionisation.

The pH of weak bases

A **weak base** is partially protonated in water. Ammonia is an example of a weak base:

$NH_3 + H_2O \rightleftharpoons NH_4^+ + OH^-$

$$K_b = \frac{[NH_4^+][OH^-]}{[NH_3]}$$

As with weak acids, the term $[H_2O]$ is omitted, as its value is constant.

The same type of assumptions can be made as with weak acids:

$[NH_4^+] = [OH^-]$ and $[NH_3]_{eq} = [NH_3]_{initial}$

Thus:

$$K_b = \frac{[OH^-]^2}{[NH_3]}$$

$$[OH^-]^2 = K_b \times [NH_3]$$

The value of K_b for ammonia at 25°C = 1.78×10^{-5} mol dm^{-3}, so the pH of a 0.100 mol dm^{-3} ammonia solution can be calculated:

$$[OH^-] = \sqrt{K_b \times [NH_3]} = \sqrt{1.78 \times 10^{-5} \times 0.100} = 0.00133 \text{ mol dm}^{-3}$$

$$pOH = -\log [OH^-] = -\log 0.00133 = 2.88$$

$$pH = 14 - pOH = 14 - 2.88 = 11.12$$

The strengths of some weak bases are shown in Table 5.6.

Substance	K_b/mol dm^{-3}	pK_b
Ammonia, NH_3	1.78×10^{-5}	4.75
Cyanide ion, CN^-	2.51×10^{-5}	4.60
Ethanoate ion, CH_3COO^-	5.75×10^{-10}	9.24

Table 5.6 *Dissociation constants, K_b, and the pK_b values of some weak bases*

Salts of weak bases

The cations in the salts of weak bases are their conjugate acids. For example, the ammonium ion, NH_4^+, is the conjugate acid of the weak base ammonia, NH_3.

A solution of ammonium chloride is totally ionised:

$$NH_4Cl \rightarrow NH_4^+ + Cl^-$$

The ammonium ions act as an acid, reacting reversibly with water to produce H_3O^+ ions, which make the solution acidic.

$$NH_4^+ + H_2O \rightleftharpoons H_3O^+ + NH_3$$

$$K_a = \frac{[H_3O^+][NH_3]}{[NH_4^+]}$$

$$K_a \text{ for } NH_4^+ = 5.62 \times 10^{-10} \text{ mol dm}^{-3}$$

Ⓔ Remember that a solution is acidic if $[H^+]$ or $[H_3O^+] > [OH^-]$. A neutral solution becomes acidic if H^+ ions are produced. NH_4^+ ions produce H^+ ions and so the solution has a pH < 7.

Using the usual assumptions for a weak acid:

$$[H_3O^+]^2 = K_a \times [\text{weak acid}] = K_a \times [NH_4^+]$$

For a 0.10 mol dm^{-3} solution of ammonium chloride:

$$[H_3O^+] = \sqrt{5.62 \times 10^{-10} \times 0.10} = 7.50 \times 10^{-6} \text{ mol dm}^{-3}$$

$$pH = -\log [H_3O^+] = -\log 7.50 \times 10^{-6} = 5.13$$

The pH is less than 7, so the solution is acidic.

Salts of weak acids

The anions in the salts of weak acids are their conjugate bases. For example, the ethanoate ion in sodium ethanoate is the conjugate base of ethanoic acid, which is a weak acid.

Sodium ethanoate is totally ionised:

$$CH_3COONa \rightarrow CH_3COO^- + Na^+$$

The CH_3COO^- ions react reversibly with water:

$$CH_3COO^- + H_2O \rightleftharpoons CH_3COOH + OH^-$$

The formation of OH^- ions makes the solution alkaline.

$$K_b = \frac{[CH_3COOH][OH^-]}{[CH_3COO^-]}$$

The value of K_b of a conjugate base can be found by using the formula:

$$K_a \times K_b = K_w$$

K_a for ethanoic acid = 1.75×10^{-5} mol dm^{-3}

So

$$K_b \text{ for the ethanoate ion} = \frac{1.0 \times 10^{-14}}{1.75 \times 10^{-5}} = 5.71 \times 10^{-10}\,\text{mol dm}^{-3}$$

The pH of a 0.10 mol dm^{-3} solution of sodium ethanoate can be calculated:

$$[OH^-]^2 = K_b[CH_3COO^-]$$

$$[OH^-] = \sqrt{5.71 \times 10^{-10} \times 0.10} = 7.56 \times 10^{-6}\,\text{mol dm}^{-3}$$

$$pOH = -\log[OH^-] = -\log 7.56 \times 10^{-6} = 5.12$$

$$pH = 14 - pOH = 14 - 5.12 = 8.88$$

The pH is greater than 7, so the solution is alkaline.

The weaker the acid, the stronger is its anion as a conjugate base, and the more alkaline is the solution of its salt. For example, carbonic acid, H_2CO_3, is a weaker acid than ethanoic acid, so a solution of its conjugate base sodium carbonate has a higher pH than a solution of sodium ethanoate.

Salts of strong acids and strong bases

The conjugate base of a strong acid, such as HCl, is too weak to react with water. Similarly, the conjugate acid of a strong base does not react, so the salts of strong acids and strong bases dissolve in water without any reaction taking place. Their solutions are neutral, pH 7.

$$Cl^- + H_2O \rightarrow \text{no reaction} \qquad Na^+ + H_2O \rightarrow \text{no reaction}$$

The reaction $NaCl + H_2O \rightarrow HCl + NaOH$ does *not* occur. In fact, the reverse reaction goes to completion.

ZAHOOR UL-HAQ

Universal indicator in various salt solutions (left to right: ammonium chloride, sodium chloride, sodium ethanoate and sodium carbonate

The rule of two

- The pH of a 0.1 mol dm^{-3} solution of a strong acid is 1 and that of a 0.1 mol dm^{-3} solution of a strong base is 13.
- The pH of a solution of a salt of a strong acid or a strong base is 7.

The 'rule of two' gives an approximate pH of weak acids, weak bases and their salts:

- pH of weak acid = pH of strong acid + 2 = 1 + 2 = 3
- pH of weak base = pH of strong base – 2 = 13 – 2 = 11
- pH of salt of weak acid and strong base = 7 + 2 = 9
- pH of salt of weak base and strong acid = 7 – 2 = 5
- pH of salt of weak acid and weak base ≈ pH of salt of strong acid +2 and strong base –2 = 7 + 2 – 2 = 7

e The rule for going from strong to weak is to add 2 pH units for a weak acid; subtract 2 pH units for a weak base.

Buffer solutions

A buffer solution is one that resists a change in pH when a small amount of acid or base is added.

- An **acid buffer solution** consists of a mixture of a weak acid and its conjugate base of *similar* concentration — for example, the weak acid ethanoic acid, CH_3COOH, and its salt sodium ethanoate, CH_3COONa.
- An **alkaline buffer solution** consists of a weak base and its conjugate acid of *similar* concentration — for example, the weak base ammonia, NH_3, and its salt ammonium chloride, NH_4Cl.

e Do not state that a buffer has constant pH. A buffer solution resists the change in pH, but it does not completely remove all the added H^+ or OH^- ions. Therefore, the pH does change, but only very slightly.

The crucial points are that the members of the acid–base conjugate pair must be at a similar concentration, which should be not less than 0.05 mol dm^{-3}.

Blood plasma has a pH of 7.4 and acts as a buffer solution. The pH is maintained mainly by the mixture of carbonic acid and hydrogencarbonate ions. Inside the red blood cells, the pH is 7.25. Here, the buffer is also carbonic acid–hydrogen-carbonate ions, but haemoglobin molecules are acidic and lower the pH.

It is important that the pH of blood plasma and of the fluid inside the blood cells does not alter significantly. If the pH changes inside a haemoglobin cell, its ability to absorb oxygen is altered. If a person hyperventilates, carbon dioxide is removed from the blood plasma and the pH alters, which causes the person to become unconscious.

Some foods are buffered to prevent deterioration. Decay due to bacteria may release acids, which, in the absence of a buffer, would result in a pH change that would cause the food to deteriorate.

Calculation of the pH of a buffer solution

Consider a buffer solution made up of a weak acid, HA, and its sodium salt, NaA. The salt is totally ionised:

$$NaA \rightarrow Na^+ + A^-$$

The weak acid is only slightly ionised:

$$HA \rightleftharpoons H^+ + A^-$$

$$K_a = \frac{[H^+][A^-]}{[HA]}$$

Ionisation of the weak acid is *suppressed* by the A^- ions from the totally ionised salt. This means that both [HA] and $[A^-]$ are fairly large and both are much larger than $[H^+]$. The following assumptions can be made:

- The number of H^+ ions from the auto-ionisation of water is so small in comparison with the H^+ ions from the ionisation of the weak acid that it can be ignored.
- The number of A^- ions from the totally ionised salt is much greater than the few A^- ions from the weak acid, so it can be assumed that $[A^-]$ = [salt] (the concentration of salt originally present).
- The ionisation of the weak acid is so suppressed that [HA] = [weak acid] (the concentration of weak acid originally present).

The equilibrium expression can, therefore, be simplified:

$$K_a = \frac{[H^+][A^-]}{[HA]} = \frac{[H^+][\text{salt}]}{[\text{weak acid}]}$$

The pH of a buffer solution can be calculated given K_a for the weak acid and the concentrations or amounts of the weak acid and its salt.

e In the calculation of the pH of a weak acid, the assumption was made that $[H^+] = [A^-]$. This is true only when the sole source of A^- ions is the weak acid. In a buffer solution this is not true because A^- ions are formed from the ionisation of the salt.

Worked example 1

Calculate the pH of a buffer solution made by adding 50 cm^3 of 0.100 mol dm^{-3} ethanoic acid to 50 cm^3 of 0.200 mol dm^{-3} sodium ethanoate. (K_a for ethanoic acid = 1.80×10^{-5} mol dm^{-3})

Answer

$$CH_3COONa \rightarrow CH_3COO^- + Na^+$$

$$CH_3COOH \rightleftharpoons H^+ + CH_3COO^-$$

$$K_a = \frac{[H^+][CH_3COO^-]}{[CH_3COOH]} = \frac{[H^+][\text{salt}]}{[\text{weak acid}]}$$

$$= 1.80 \times 10^{-5} \text{mol dm}^{-3}$$

$$[H^+] = K_a \times \frac{[\text{weak acid}]}{[\text{salt}]} = 1.80 \times 10^{-5} \times \frac{0.0500}{0.100}$$

$$= 9.00 \times 10^{-6} \text{ mol dm}^{-3}$$

$$pH = -\log[H^+] = -\log(9.00 \times 10^{-6}) = 5.05$$

Note that mixing the two solutions doubles the total volume. Therefore, the concentration of the weak acid was halved from 0.100 to 0.0500 mol dm^{-3} and that of the salt was halved from 0.200 to 0.100 mol dm^{-3}.

Worked example 2

Calculate the pH of a buffer solution made by adding 1.42 g of potassium methanoate, HCOOK, to 50.0 cm^3 of a 0.111 mol dm^{-3} solution of methanoic acid, HCOOH. (K_a for methanoic acid = 1.78×10^{-4} mol dm^{-3})

Answer

molar mass of potassium methanoate = 1 + 12 + (2 × 16) + 39.1 = 84.1 g mol^{-1}

amount of potassium methanoate = $\frac{1.42\,g}{84.1\,g\,mol^{-1}}$ = 0.0169 mol

$[HCOOK] = \frac{0.0169\,mol}{0.0500\,dm^3} = 0.338\,mol\,dm^{-3}$

$HCOOK \rightarrow HCOO^- + K^+$

$HCOOH \rightleftharpoons H^+ + HCOO^-$

$K_a = \frac{[H^+][HCOO^-]}{[HCOOH]} = \frac{[H^+][salt]}{[weak\ acid]} = 1.78 \times 10^{-4}\,mol\,dm^{-3}$

$[H^+] = K_a \times \frac{[weak\ acid]}{[salt]} = 1.78 \times 10^{-4} \times \frac{0.111}{0.338}$

$= 5.85 \times 10^{-5}\,mol\,dm^{-3}$

$pH = -\log[H^+] = -\log(5.85 \times 10^{-5}) = 4.23$

Extra care must be taken when both the volume and the concentration of the acid and of the salt are different.

Worked example 3

Calculate the pH of a solution made by mixing 30 cm^3 of a 0.10 mol dm^{-3} solution of benzoic acid with 20 cm^3 of a 0.20 mol dm^{-3} solution of sodium benzoate. (K_a for benzoic acid = 6.31 × 10^{-5} mol dm^{-3}]

Answer

amount (moles) of acid = 0.10 mol dm^{-3} × 0.030 dm^3 = 0.0030 mol

$[acid] = \frac{0.0030\,mol}{0.050\,dm^3} = 0.060\,mol\,dm^{-3}$

amount (moles) of salt = 0.20 mol dm^{-3} × 0.020 dm^3 = 0.0040 mol

$[salt] = \frac{0.0040\,mol}{0.050\,dm^3} = 0.080\,mol\,dm^{-3}$

$K_a = \frac{[H^+][A^-]}{[HA]}$

$[H^+] = K_a \times \frac{[acid]}{[salt]} = 6.31 \times 10^{-5} \times \frac{0.060}{0.080} = 4.73 \times 10^{-5}\,mol\,dm^{-3}$

$pH = -\log[H^+] = -\log(4.73 \times 10^{-5}) = 4.32$

A more complicated calculation involving buffers arises when an *excess* of weak acid is mixed with a strong alkali. All the alkali reacts with some of the acid, forming a salt of the weak acid and, therefore, creating a buffer solution. The amount of salt formed is equal to the amount of alkali. The total volume of the solution is the sum of the volumes of the two solutions that were mixed.

Worked example 4

Calculate the pH of a buffer solution made by mixing 60 cm^3 of 0.20 mol dm^{-3} ethanoic acid solution with 40 cm^3 of sodium hydroxide solution of concentration 0.10 mol dm^{-3}. (K_a for ethanoic acid = 1.80 × 10^{-5} mol dm^{-3})

Answer

amount of alkali taken = 0.10 × 0.040 = 0.0040 mol = amount of salt formed

amount of acid taken = 0.20 × 0.060 = 0.012 mol

amount of acid left = 0.012 – 0.0040 = 0.0080 mol in a total volume of 100 cm^3

$$[\text{salt}] = \frac{0.0040}{0.10} = 0.040\,\text{mol dm}^{-3}$$

$$[\text{weak acid}] = \frac{0.0080}{0.10} = 0.080\,\text{mol dm}^{-3}$$

$$CH_3COONa \rightarrow CH_3COO^- + Na^+$$

$$CH_3COOH \rightleftharpoons H^+ + CH_3COO^-$$

$$K_a = \frac{[H^+][CH_3COO^-]}{[CH_3COOH]} = \frac{[H^+][\text{salt}]}{[\text{weak acid}]}$$

$$= 1.80 \times 10^{-5}\,\text{mol dm}^{-3}$$

$$[H^+] = K_a \times \frac{[\text{weak acid}]}{[\text{salt}]} = 1.8 \times 10^{-5} \times \frac{0.080}{0.040}$$

$$= 3.6 \times 10^{-5}\,\text{mol dm}^{-3}$$

$$pH = -\log[H^+] = -\log(3.6 \times 10^{-5}) = 4.44$$

It is dangerous to use the Henderson- Hasselbalch formula when calculating the pH of a buffer solution:

$$pH = pK_a + \log\left(\frac{[\text{salt}]}{[\text{weak acid}]}\right)$$

as it is often mis-remembered.

e Always start the calculation of the pH of a buffer solution and also of a weak acid with the expression for the equilibrium constant. Then make the assumptions:

- for a buffer solution: [HA] = [weak acid] and $[A^-]$ = [salt]
- for a solution of a weak acid: [HA] = [weak acid] and $[H^+] = [A^-]$

Calculation of the composition of a buffer solution

Calculation of the composition of a buffer solution requires the use of the expression for K_a. There are two types of calculation. In the first, the amount of solid salt has to be calculated.

Worked example

Calculate the mass of sodium ethanoate, CH_3COONa, that has to be added to 100 cm^3 of a 1.00 mol dm^{-3} solution of ethanoic acid to make a buffer solution of pH = 4.38. (K_a for ethanoic acid = 1.8×10^{-5} mol dm^{-3})

Answer

$$[H^+] = 10^{-pH} = 10^{-4.38} = 4.17 \times 10^{-5}\,\text{mol dm}^{-3}$$

$$K_a = \frac{[H^+][CH_3COO^-]}{[CH_3COOH]} = \frac{[H^+][\text{salt}]}{[\text{weak acid}]} = 1.8 \times 10^{-5}\,\text{mol dm}^{-3}$$

$$[\text{salt}] = K_a \times \frac{[\text{weak acid}]}{[H^+]} = \frac{1.8 \times 10^{-5} \times 1.00}{4.17 \times 10^{-5}} = 0.432\,\text{mol}$$

molar mass of CH_3COONa = 12 + 3 + 12 + (2 × 16) + 23 = 82 g mol^{-1}

mass of sodium ethanoate required for 1 dm^3 = 0.432 mol × 82 g mol^{-1} = 35 g

mass for 100 cm^3 of solution = 3.5 g

In the second type of calculation, the volumes of the solutions of the weak acid and the salt have to be calculated. In this type of calculation there is no unique answer. It is the ratio of the volumes of the two solutions that is found.

Worked example

Calculate the relative volumes of a $1.00\,mol\,dm^{-3}$ solution of ethanoic acid and a $1.00\,mol\,dm^{-3}$ solution of sodium ethanoate that have to be mixed to give a solution of pH = 4.00. (K_a for ethanoic acid = $1.8 \times 10^{-5}\,mol\,dm^{-3}$)

Answer

$[H^+] = 10^{-pH} = 1.00 \times 10^{-4}\,mol\,dm^{-3}$

$$K_a = \frac{[H^+][CH_3COO^-]}{[CH_3COOH]} = \frac{[H^+][salt]}{[weak\ acid]} = 1.8 \times 10^{-5}\,mol\,dm^{-3}$$

$$\frac{[weak\ acid]}{[salt]} = \frac{[H^+]}{K_a} = \frac{1.00 \times 10^{-4}}{1.8 \times 10^{-5}} = 5.6$$

The ratio of the volumes of solutions of ethanoic acid to sodium ethanoate = 5.6:1, so, mix $56\,cm^3$ of the ethanoic acid solution with $10\,cm^3$ of the sodium ethanoate solution.

Mode of action of a buffer solution

Consider a buffer solution made of a weak acid, HA, and its sodium salt, NaA. The salt is totally ionised:

$$NaA \rightarrow Na^+ + A^-$$

The weak acid is only slightly ionised and its ionisation is *suppressed* by the A^- ions from the totally ionised salt:

$$HA \rightleftharpoons H^+ + A^-$$

$$K_a = \frac{[H^+][A^-]}{[HA]} = \frac{[H^+][salt]}{[weak\ acid]}$$

The acid produces a reservoir of HA molecules and the salt produces a reservoir of A^- ions.

A buffer solution maintains a nearly constant pH because the reservoir of *both* the weak acid, HA, and its conjugate base, A^-, are large *relative* to the small amount of H^+ or OH^- added.

When a *small* amount of H^+ ions is added, the ions react with the relatively large reservoir of A^- ions from the salt:

$$H^+ + A^- \rightarrow HA$$

The value of $[A^-]$ decreases slightly and that of [HA] increases slightly, but these changes are insignificant in relation to the original values of $[A^-]$ (from the totally ionised salt) and [HA] (from the almost un-ionised weak acid), which remain virtually unchanged. As nothing has changed significantly in the expression for K_a, the hydrogen ion concentration and hence the pH will not change greatly.

A weak acid on its own is not a buffer as $[A^-]$ is very small. When H^+ or OH^- are added, the change in $[A^-]$ is significant.

When a *small* amount of OH^- ions is added, the ions react with the relatively large reservoir of HA molecules of the weak acid:

$$OH^- + HA \rightarrow H_2O + A^-$$

The value of $[A^-]$ increases slightly and that of [HA] decreases slightly, but these changes are insignificant in relation to the original values of $[A^-]$ and [HA], which remain virtually unchanged. As nothing has changed significantly in the expression for K_a, the hydrogen ion concentration and hence the pH will not change greatly.

e The addition of OH^- ions can also be explained by stating that they drive the equilibrium $HA \rightleftharpoons H^+ + A^-$ to the right, by removal of the H^+ ions. This causes an increase in $[A^-]$. However, the increase is not significant because of the relatively large value of $[A^-]$ from the salt. [HA] also decreases slightly, but also by an insignificant amount. Therefore, the pH hardly changes.

The mode of action of a buffer can be summarised as:

- The salt is fully ionised, so suppresses the ionisation of the weak acid.
- The amounts of the weak acid *and* its conjugate base are large relative to the small additions of H^+ or OH^-.
- H^+ ions are removed by reaction with the conjugate base: $H^+ + A^- \rightarrow HA$
- OH^- ions are removed by reaction with the weak acid: $OH^- + HA \rightarrow H_2O + A^-$

Efficiency of a buffer solution

This can be tested by adding $0.10\,mol\,dm^{-3}$ hydrochloric acid in $5\,cm^3$ portions to $100\,cm^3$ of a buffer solution made by dissolving 0.1 mol of sodium ethanoate in $100\,cm^3$ of $1.0\,mol\,dm^{-3}$ ethanoic acid solution and measuring the pH after each addition. The results are shown in Table 5.7.

Table 5.7

Volume of $0.10\,mol\,dm^{-3}$ HCl added	pH
0	4.76
5	4.76
10	4.77
15	4.77
20	4.78

Acid–base indicators

In acid–base titrations, the **equivalence point** is the point at which enough alkali has been added from the burette to react with all the acid in the conical flask or when enough acid has been added from the burette to react with all the alkali in the conical flask. For a reaction with a 1:1 stoichiometry, this means an equal number of moles of acid and alkali. If the stoichiometry is 2:1, then the ratio of moles required for the equivalence point is also 2:1.

The pH at the equivalence point is not necessarily 7. This is because at this point the solution consists of the salt of the acid and the alkali. If both the acid and alkali are strong, the solution will be pH 7, but if either is weak, then it will not (p. 94).

The purpose of an indicator is to show when the equivalence point has been reached. Indicators are weak acids with the colour of the conjugate base being different from that of the weak acid molecule. Shorthand for an indicator molecule is HInd. It dissociates in water according to the equation:

$$\underset{\text{colour 1}}{HInd} \rightleftharpoons H^+ + \underset{\text{colour 2}}{Ind^-}$$

$$K_{ind} = \frac{[H^+][Ind^-]}{[HInd]}$$

- When acid is added, the equilibrium is driven to the left and the indicator appears as colour 1.
- When alkali is added, the OH^- ions react with the H^+ ions from the indicator and the equilibrium is driven to the right. The indicator turns colour 2.
- The colour at the equivalence point appears when [HInd] = $[Ind^-]$. At this point:

  $K_{ind} = [H^+]$ or $pH = pK_{ind}$

With most indicators, the eye can see either colour 1 or colour 2 only if at least 10% of that species is present in the mixture. Thus, the range over which the colour is seen to change is from a ratio of [HInd]:$[Ind^-]$ of just less than 10 to just over 0.1. This is a range of pH of approximately ±1 from the pK_{ind} value.

Indicator	pK_{ind}	pH range	Acid colour	Alkaline colour	Neutral colour
Methyl orange	3.7	3.1–4.4	Red	Yellow	Orange
Bromophenol blue	4.0	3.0–4.6	Yellow	Blue	Green
Bromocresol green	4.7	3.8–5.4	Yellow	Blue	Green
Methyl red	5.1	4.2–6.3	Red	Yellow	Orange
Bromothymol blue	7.0	6.0–7.6	Yellow	Blue	Green
Thymol blue*	8.9	8.0–9.6	Yellow	Blue	Green
Phenolphthalein	9.3	8.3–10.0	Colourless	Red	Pale pink

* Thymol blue also changes from red to yellow around a pH of 2

Table 5.8
Some common acid–base indicators

The correct choice of indicator depends on the strengths of the acid and base in the titration.

ANDREW LAMBERT PHOTOGRAPHY/SPL

Methyl orange in acid (left)and alkaline (right) solutions

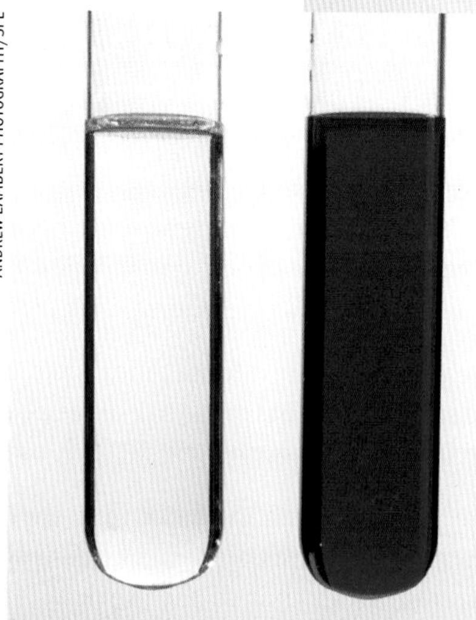

ANDREW LAMBERT PHOTOGRAPHY/SPL

Phenolphthalein in acid (left) and alkaline (right) solutions

Titration curves

A titration curve shows how the pH of a solution varies as the reagent in the burette is added. The shape of a titration curve depends upon the strength/weakness of the acid and base.

In order to sketch a titration curve, the following have to be estimated:

- the pH at the start
- the pH at the equivalence point
- the volume of liquid from the burette required to reach the equivalence point
- the pH range of the near vertical part of the graph
- the pH after excess reagent has been added from the burette (final pH)

The pH values at different points during a typical titration are shown in Table 5.9. The figures are only approximate, because they vary depending on the concentrations and on how weak the acids and bases are.

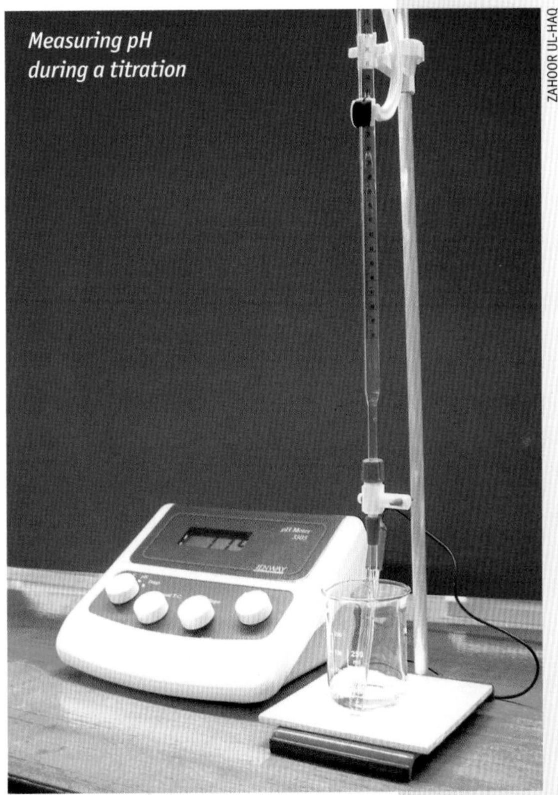

Measuring pH during a titration

ZAHOOR UL-HAQ

Table 5.9 *Variation of pH during different types of titration*

Reagent in conical flask	Reagent in burette	Initial pH	Equivalence point pH	Vertical range pH	Final pH
Strong acid	Strong base	1	7	3–11	Just < 13
Strong base	Strong acid	13	7	11–3	Just > 1
Weak acid	Strong base	3	9	7–11	Just < 13
Strong base	Weak acid	13	9	11–7	Just > 3
Strong acid	Weak base	1	5	3–7	Just < 11
Weak base	Strong acid	11	5	7–3	Just > 1

Red cabbage extract at (from left) pH 1, 3, 5, 7, 9, 11 and 13

ZAHOOR UL-HAQ

e Note that the rule of *two* (see p. 104) applies here:

- A weak acid starts or finishes at 2 pH units higher than that of a strong acid.
- A weak acid has an equivalence point 2 pH units higher than that of a strong acid.
- A weak base starts or finishes at 2 pH units lower than that of a strong base.
- A weak base has an equivalence point 2 pH units lower than that of a strong base.
- For both weak acids and bases, the vertical range is ± 2 pH units about the equivalence point pH.

The volume at the equivalence point has to be worked out by the usual titration method. In most questions the acids and bases react in a 1:1 ratio. For example, if a 0.10 mol dm^{-3} solution of a base is added to 20 cm^3 of 0.10 mol dm^{-3} acid, the equivalence point is at 20 cm^3 of added base. Note that if the acid had a concentration of 0.20 mol dm^{-3}, the equivalence point would be at 40 cm^3 of added base.

In the titration curves below:

- the starting volume in the flask is 20 cm^3
- the equivalence point is at 20 cm^3 of added reagent
- The left-hand graph of each pair shows the change in pH as base is added to acid; the right-hand graph shows the pH change as acid is added to base.

Strong acid–strong base titration

The variation of pH as sodium hydroxide solution is added to a strong acid, such as HCl, is given in Table 5.4 on p. 97.

The values are plotted in Figure 5.2. The graphs show the reaction between hydrochloric acid of concentration 0.10 mol dm^{-3} and sodium hydroxide of the same concentration.

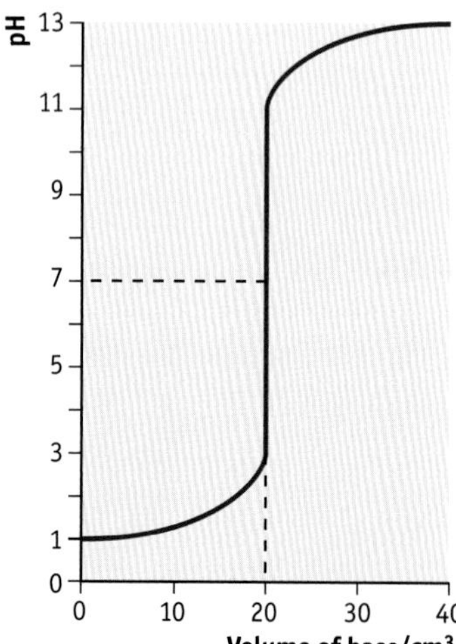

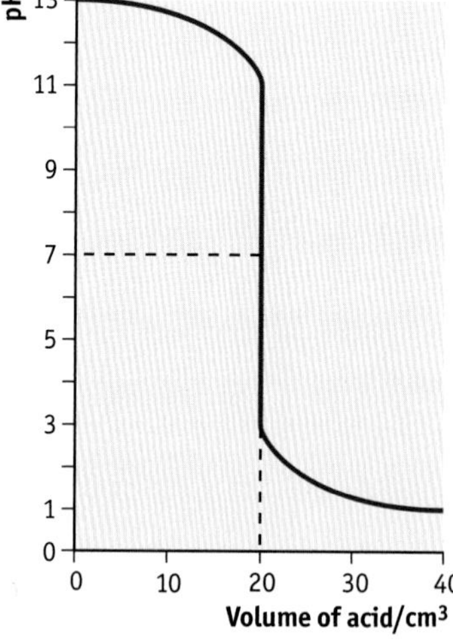

Figure 5.2 *Strong acid/strong base*

Weak acid–strong base titration

The graphs in Figure 5.3 show the variation of pH during the reaction between solutions of a weak acid, such as ethanoic acid, and a strong base, such as sodium hydroxide. The concentration of both reagents is 0.10 mol dm^{-3}.

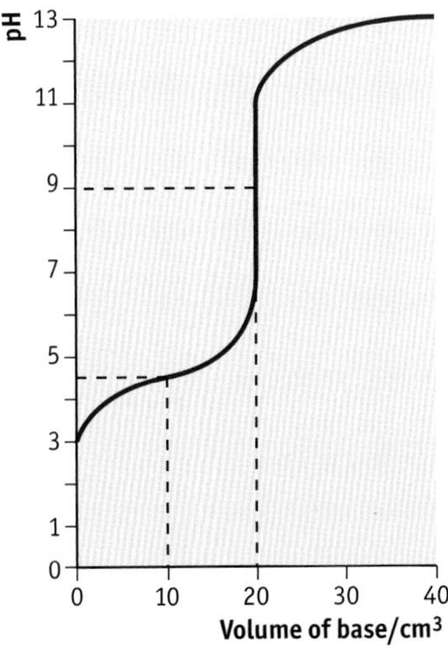

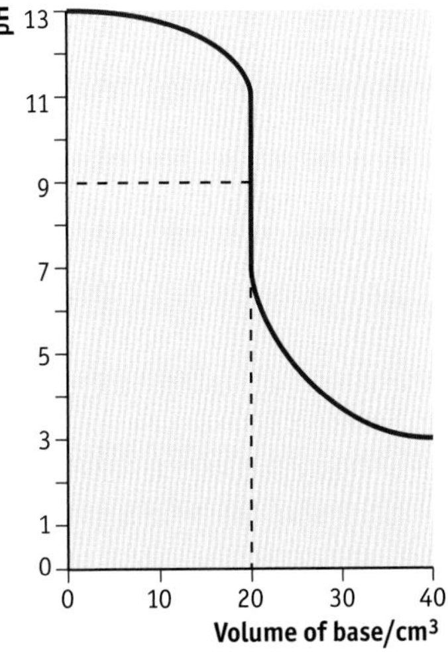

Figure 5.3 *Weak acid/ strong base*

Strong acid–weak base titrations

The graphs in Figure 5.4 show the variation of pH during the reaction between solutions of a strong acid, such as hydrochloric acid, and a weak base such as ammonia. The concentration of both reagents is $0.10\,mol\,dm^{-3}$.

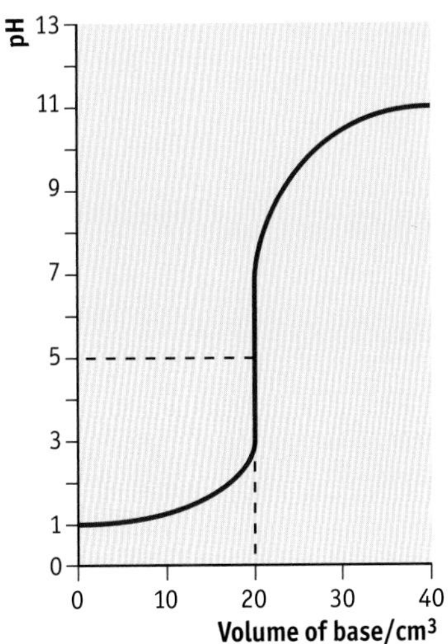

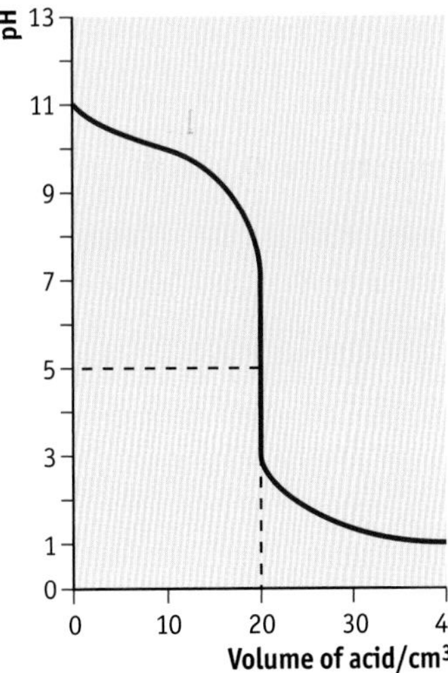

Figure 5.4 *Strong acid/ weak base*

Evaluation of K_a and K_b from titration curves

When a strong base is added to a solution of a weak acid, the point halfway to the equivalence point is when half the acid has been neutralised. For a weak acid, HA:

$$K_a = \frac{[H^+][A^-]}{[HA]}$$

At the half-neutralisation point, $[HA] = [A^-]$. Therefore, at this point, $K_a = [H^+]$ and so $pK_a = pH$. The pH at the half-neutralisation point can be read off the graph.

At the half-neutralisation point, the mixture is a buffer solution because both the weak acid and its conjugate base are present in equal and, therefore, significant, quantities.

The evaluation is similar for a weak base. If a strong acid is added to a weak base, the pH at the half-neutralisation point is equal to pK_b of the weak base, B.

$$B + H_2O \rightleftharpoons BH^+ + OH^-$$

$$K_b = \frac{[BH^+][OH^-]}{[B]}$$

At the half-neutralisation point, $[B] = [BH^+]$. Therefore, at this point $K_b = [OH^-]$, so $pK_b = pOH$. The pH at the half-neutralisation point can be read off the graph.

Choice of indicator

For an indicator to work, the entire range during which it changes colour must lie completely within the vertical section of the titration curve.

For a strong acid–strong base titration, the pH range of the indicator must lie completely within the range of pH 3–11. Therefore, all the indicators listed in Table 5.8 will give an accurate result. The indicators most usually chosen are methyl orange and phenolphthalein.

For a weak acid–strong base titration, the range must be completely within pH 7–11. Of the indicators in Table 5.8, only thymol blue and phenolphthalein will give the correct result.

For a weak base–strong acid titration, the colour change of the indicator must be completely within the range of pH 3–7. Methyl orange, bromophenol blue, bromocresol green and methyl red are all suitable indicators.

e People who are red–green colour blind have great difficulty seeing the end point of a titration using methyl orange. They should use screened methyl orange instead. This contains a blue dye, so the acid colour is purple (red plus blue), the alkaline colour is green (yellow plus blue) and the neutral colour is grey (red plus blue plus yellow).

Buffer region in titration curves

When a strong base is added to a weak acid, a buffer solution is produced when there are significant amounts of both the salt formed and unreacted weak acid present. This occurs in the fairly flat part of the curve between the addition of 5 cm^3 of strong base to within 5 cm^3 of the equivalence point.

Similarly, when a strong acid is added to a weak base, the fairly flat part of the curve occurs when there are significant amounts of both the salt formed and unreacted weak base present. This occurs in the region between the addition of 5 cm^3 of strong acid to within 5 cm^3 of the equivalence point.

Buffers are most resistant to pH changes when the amounts of the components of the acid–base conjugate pair are equal.

The buffered region would look much flatter if the scale of the x-axis were extended. This would then show clearly that the pH alters only slightly for small additions (less than 1 cm^3) of acid or alkali.

Enthalpy of neutralisation of acids

The standard enthalpy of neutralisation, ΔH_{neut}, is defined as the enthalpy change when an acid and a base react to form 1 mol of water under standard conditions of 1.0 mol dm^{-3} solutions, 1 atm pressure and a stated temperature, usually 298 K (25 °C).

For example, it is the molar enthalpy change for:

$$HCl(aq) + NaOH(aq) \rightarrow NaCl(aq) + H_2O(l)$$
$$\tfrac{1}{2}H_2SO_4(aq) + KOH(aq) \rightarrow \tfrac{1}{2}K_2SO_4(aq) + H_2O(l)$$
$$CH_3COOH(aq) + NaOH(aq) \rightarrow CH_3COONa(aq) + H_2O(l)$$

Strong acids neutralised by strong bases

Strong acids and strong bases are both totally ionised, so the equation for the reaction between them can be written ionically. For example, for the neutralisation of hydrochloric acid by sodium hydroxide the full ionic equation is:

$$H^+(aq) + Cl^-(aq) + Na^+(aq) + OH^-(aq) \rightarrow Na^+(aq) + Cl^-(aq) + H_2O(l)$$

The spectator ions can be crossed out, leaving the net ionic equation:

$$H^+(aq) + OH^-(aq) \rightarrow H_2O(l)$$

This equation represents the neutralisation of all strong acids by strong bases. Therefore, ΔH for all these reactions is approximately the same.

For a strong acid neutralised by a strong base, $\Delta H_{neut} = -57.2\,kJ\,mol^{-1}$

Weak acids neutralised by strong bases

The ionic equation given above does not apply to the neutralisation of a weak acid by a strong base because the acid is not totally ionised. The neutralisation can be regarded as the sum of two reactions:

$$HA(aq) \rightleftharpoons H^+(aq) + A^-(aq) \qquad \Delta H_1$$
$$H^+(aq) + OH^-(aq) \rightarrow H_2O(l) \qquad \Delta H_2$$

$$HA(aq) + OH^-(aq) \rightarrow A^-(aq) + H_2O(l) \qquad \Delta H_{neut}$$
$$\Delta H_{neut} = \Delta H_1 + \Delta H_2$$
where $\Delta H_2 = -57.2\,kJ\,mol^{-1}$

Even though the acid is only about 1% ionised, the removal of H^+ ions by the base drives the equilibrium to the right until all the acid has been neutralised.

- If ΔH_1 is endothermic, the value of ΔH_{neut} is *less* exothermic than ΔH_{neut} of a strong acid.
- If ΔH_1 is exothermic, the value of ΔH_{neut} is *more* exothermic than ΔH_{neut} of a strong acid.

Hydrofluoric acid, HF, is a weak acid. It ionises exothermically because of the small size of the F^- ion (the smallest anion that can exist in solution), which forms strong hydrogen bonds with water molecules.

Other weak acids ionise endothermically. The weaker the acid, the more endothermic is its ionisation. Some examples are given in Table 5.10.

Acid	K_a/mol dm^{-3}	Base	ΔH_{neut}/kJ mol^{-1}
HF	5.6×10^{-4}	NaOH	−68.6
CH_3COOH	1.8×10^{-5}	NaOH	−55.2
H_2S	8.9×10^{-7}	NaOH	−32.2
HCN	4.9×10^{-10}	NaOH	−11.7

***Table 5.10** Enthalpy of neutralisation of some weak acids*

The enthalpy of ionisation of ethanoic acid is only +2 kJ mol^{-1}. This is why its enthalpy of neutralisation is so close to that of a strong acid.

The structure of acids

Acids can be classified according to three types. In the first type of acid, a hydrogen atom is joined to an oxygen atom in a molecule. The oxo-acids (e.g. sulfuric, nitric, carbonic) and all the organic acids fall into this category.

In addition to the oxygen attached to the hydrogen, an oxo-acid also contains one or more X=O groups. For example, the structure of sulfuric acid is:

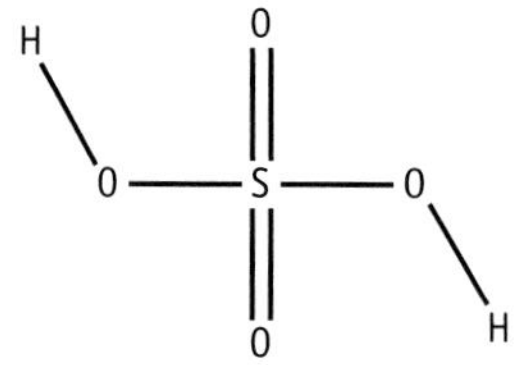

H
O — S — O
O O
H

The structure of nitric acid is:

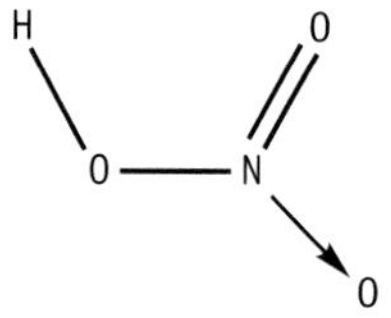

H O
O — N
O

The structure of carbonic acid is:

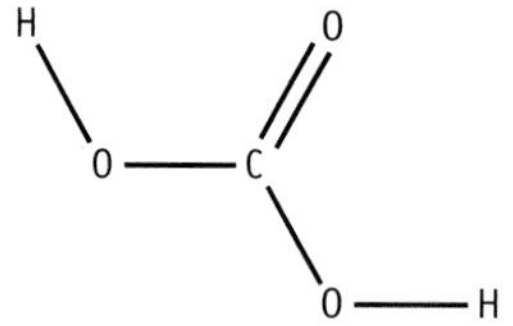

H O
O — C
O — H

In the second type of acid, a hydrogen atom is joined to an electronegative atom other than oxygen. Well-known examples of this type of acid are HF, HCl, HBr and HI. The strength of these acids depends mainly on the strength of the H–X bond: the stronger the bond, the weaker the acid. This is why HF is the weakest acid of the hydrogen halide acids and HI is the strongest. The bond enthalpies are:

H–F: +562 kJ mol^{-1}

H–I: +299 kJ mol^{-1}

H_2S and HCN are other examples of this type of acid.

e In oxo-acids containing an element that can have more than one oxidation state, the lower the oxidation state the weaker the acid is. For example, sulfuric acid (oxidation state of sulfur = +6) is a stronger acid than sulfurous acid (oxidation state of sulfur = +4)

The third type of acid is a hydrated metal ion. The polarising power of the cation, especially small 3+ ions such as Al^{3+} or Fe^{3+}, draws electrons away from the H–O bond in the surrounding water molecules. This means that a water molecule can be deprotonated easily:

$$[Al(H_2O)_6]^{3+} + H_2O \rightleftharpoons [Al(H_2O)_5OH]^{2+} + H_3O^+$$

H_3O^+ ions are produced and so the solution becomes acidic.

Questions

1 Hydrogen chloride gas reacts with water. Write the equation for the reaction and use it to explain why HCl can be classified as a Brønsted–Lowry acid and why the solution is acidic.

2 When ethanoic acid reacts with ethanol in the presence of concentrated sulfuric acid, the first step is the reaction between sulfuric acid and ethanoic acid:

$$CH_3COOH + H_2SO_4 \rightarrow CH_3COOH_2^+ + HSO_4^-$$

Identify the acid–base conjugate pairs.

3 State the conjugate bases of the following acids:

a HCN
b NH_3
c $HClO_3$
d OH^-
e $[Fe(H_2O)_6]^{3+}$

4 Write the formulae for the conjugate acids of the following bases:

a NH_3
b CH_3NH_2
c OH^-
d HNO_3

5 Explain why the pH of pure water is not always 7.

6 At 25 °C, $pK_w = 14$. Calculate the $[OH^-]$ of the following solutions and state whether the solutions are acidic or alkaline:

a $[H^+] = 1.0 \times 10^{-2}\,mol\,dm^{-3}$
b $[H^+] = 2.2 \times 10^{-7}\,mol\,dm^{-3}$
c $[H^+] = 3.3 \times 10^{-10}\,mol\,dm^{-3}$

7 At 25 °C, $K_w = 1.0 \times 10^{-14}\,mol^2\,dm^{-6}$. Calculate the pH of the following solutions in which:

a $[H^+] = 4.4 \times 10^{-5}\,mol\,dm^{-3}$
b $[H^+] = 5.5 \times 10^{-9}\,mol\,dm^{-3}$
c $[OH^-] = 6.6 \times 10^{-2}\,mol\,dm^{-3}$
d $[OH^-] = 7.7 \times 10^{-11}\,mol\,dm^{-3}$

8 Calculate the concentration of hydrogen ions in solutions with the following pH values:

a pH = 1.33 **b** pH = 7.00 **c** pH = 13.67

9 Calculate the ratio of $[H^+]$ to $[OH^-]$ ions in solutions of pH:

a 7 **b** 10 **c** 3

10 Calculate the pH of the following solutions:

a $0.200\,mol\,dm^{-3}$ hydrobromic acid, HBr, which is a strong acid
b $0.200\,mol\,dm^{-3}$ lithium hydroxide, LiOH, which is a strong base
c $0.0500\,mol\,dm^{-3}$ strontium hydroxide, $Sr(OH)_2$, which is a strong base

11 Calculate the pH of a $2.0\,mol\,dm^{-3}$ solution of a strong acid such as nitric acid, HNO_3. Then calculate its pH when it is diluted 10 times, 100 times and 1 000 000 times.

12 Chloric(I) acid, HOCl, is a weak acid with $K_a = 3.02 \times 10^{-11}\,mol\,dm^{-3}$.

a Write the equation for its ionisation in water and hence the expression for the acid dissociation constant, K_a.
b Calculate the pH of a $0.213\,mol\,dm^{-3}$ solution of chloric(I) acid.

13 Nitrous acid, HNO_2, is also called nitric(III) acid. It is a weak acid. A 0.200 mol dm^{-3} solution of HNO_2 has a pH of 2.02. Calculate the value of its acid dissociation constant, K_a.

14 Propanoic acid is a weak acid with K_a = 1.35×10^{-5} mol dm^{-3}. A solution of the acid has a pH = 3.09. Calculate the concentration of the solution.

15 Hydroxyethanoic acid, $CH_2(OH)COOH$, has a pK_a = 3.83. Calculate the pH of a 1.05 mol dm^{-3} solution of this weak acid.

16 Methylamine is a weak base, with $K_b = 4.36 \times 10^{-4}$ mol dm^{-3}:

$$CH_3NH_2 + H_2O \rightleftharpoons CH_3NH_3^+ + OH^-$$

a Give the expression for K_b.

b Calculate the pH of 0.200 mol dm^{-3} methylamine solution.

17 a Define 'buffer solution'.

b Calculate the pH of a buffer solution made by mixing 100 cm^3 of 1.00 mol dm^{-3} ethanoic acid solution with 5.65 g of sodium ethanoate, CH_3COONa. (K_a for ethanoic acid = 1.8×10^{-5} mol dm^{-3})

18 Calculate the pH of a buffer solution made by adding 50 cm^3 of 2.00 mol dm^{-3} sodium hydroxide solution to 150 cm^3 of 1.00 mol dm^{-3} ethanoic acid solution. (K_a for ethanoic acid = 1.8×10^{-5} mol dm^{-3})

19 What volume of 1.00 mol dm^{-3} propanoic acid, pK_a = 4.87, is needed to make a buffer solution of pH 4.50 with 50 cm^3 of 1.00 mol dm^{-3} sodium propanoate solution?

20 What mass of calcium ethanoate, $Ca(CH_3COO)_2$, must be added to 100 cm^3 of 1.25 mol dm^{-3} ethanoic acid solution to make a buffer solution of pH = 5.00? (K_a for ethanoic acid = 1.80×10^{-5} mol dm^{-3})

21 What volume of 1.00 mol dm^{-3} sodium hydroxide must be added to 100 cm^3 of 1.00 mol dm^{-3} ethanoic acid solution to make a buffer solution of pH = 4.44? (K_a for ethanoic acid = 1.80×10^{-5} mol dm^{-3})

22 Write the equation for the reaction, if any, of the following ions with water and state whether their solutions would be neutral, acidic or alkaline:

a $CH_3NH_3^+$

b CN^-

c CO_3^{2-}

d I^-

23 a Sketch the titration curve obtained when 40 cm^3 of 0.20 mol dm^{-3} hydrochloric acid is added to 40 cm^3 of 0.10 mol dm^{-3} ammonia solution. Use Table 5.8 (p. 110) to select a suitable indicator. Justify your choice.

b Mark on your curve a place where the solution is acting as a buffer.

24 20 cm^3 of a 0.100 mol dm^{-3} solution of a weak acid, HA, was placed in a conical flask. Sodium hydroxide solution of concentration 0.100 mol dm^{-3} was added in portions and the pH of the stirred solution read after each addition. The readings obtained are given in the table below.

Volume of NaOH/cm^3	pH	Volume of NaOH/cm^3	pH
0	2.9	19.9	7.0
2.5	3.9	20.1	10.4
5.0	4.3	20.5	11.1
15.0	5.2	21.0	11.4
17.5	5.6	25.0	12.0
19.0	6.0	30.0	12.3
19.5	6.3	35.0	12.4

a Plot a graph of pH (*y*-axis) against the volume of sodium hydroxide (*x*-axis).

b Use the graph to find a value for the acid dissociation constant, K_a, of the weak acid, HA.

c Estimate the pH of a 0.050 mol dm^{-3} solution of the salt, NaA.

d Use the data in Table 5.8 (p. 110) to select a suitable indicator for this titration. Justify your choice.

25 Thymol blue can be regarded as a weak acid of formula HThy. Use the data in Table 5.8 (p. 110) to explain the colour changes that take place when dilute hydrochloric acid, followed by an excess of sodium hydroxide solution, is added to a solution of thymol blue.

26 **a** Explain why the standard enthalpies of neutralisation of hydrobromic acid and hydrochloric acid by aqueous sodium hydroxide are both $-57\ kJ\ mol^{-1}$.

b Ethanoic acid is only about 1% ionised in solution, yet its enthalpy of neutralisation is $-55\ kJ\ mol^{-1}$. Explain why its value is so similar to that of hydrobromic and hydrochloric acids.

c Explain why the very weak acid, hydrocyanic acid, HCN, has a much lower exothermic enthalpy of neutralisation than that for a strong acid.

27 Draw the structural formula, showing all the bonds, of each of the following:

a chloric(V) acid, $HClO_3$

b hydrocyanic acid (hydrogen cyanide), HCN

c sulfurous acid (sulfuric(IV) acid), H_2SO_3

28 Explain why, in the absence of its salt (e.g. NaOCl), a solution of the weak acid HOCl does not act as a buffer solution when small amounts of either H^+ ions or OH^- ions are added

29 Use the internet as a resource and write a short article on either blood as a buffer solution during exercise or the role of buffers in preventing the deterioration of food. Identify the sources that you use.

Isomerism

Chapter 6

Isomers are different compounds that have the same molecular formula.

e Structural and geometric isomerism are covered in the AS course.

Structural isomerism

Structural isomers are compounds with the same molecular formula but different structural formulae.

Structural isomers can be divided into three categories: carbon-chain, positional and functional group.

Carbon-chain isomerism

In **carbon-chain isomerism**, the difference between the isomers is the length of the carbon chain. For example, if the compound contains four carbon atoms, they can be arranged with two different chain lengths:

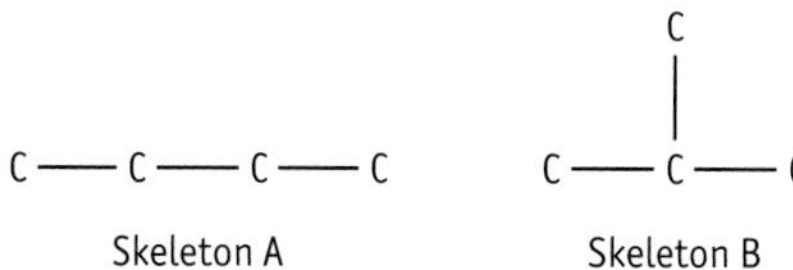

Note that the carbon skeleton below is the same as that of A above, because they both contain a chain of four carbon atoms:

Positional isomerism

Positional isomers have the same functional group in different locations on the carbon skeleton. For example, there are two isomers of molecular formula C_3H_8O. In one isomer (propan-1-ol) the –OH group is bonded to an end carbon atom; in the other (propan-2-ol) it is bonded to the middle carbon atom:

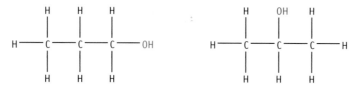

Functional-group isomerism

In **functional-group isomerism**, the isomers are members of different homologous series and, therefore, have different functional groups. For example, there are two isomers of molecular formula C_2H_6O. One isomer is the alcohol, ethanol, CH_3CH_2OH; the other is the ether, methoxymethane, CH_3OCH_3.

Worked example

Draw and name the structural isomers of C_4H_9Cl.

Answer

There are two ways of arranging the four carbon atoms:

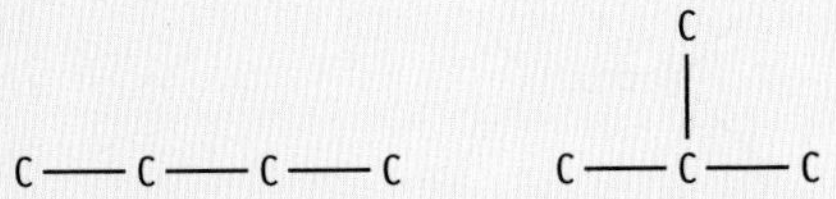

C—C—C—C

Skeleton A

C—C(—C)—C

Skeleton B

There are two positions on the four-carbon skeleton for the chlorine atom:

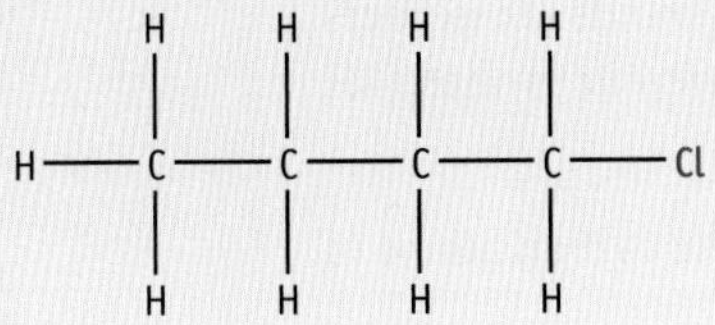

1-chlorobutane

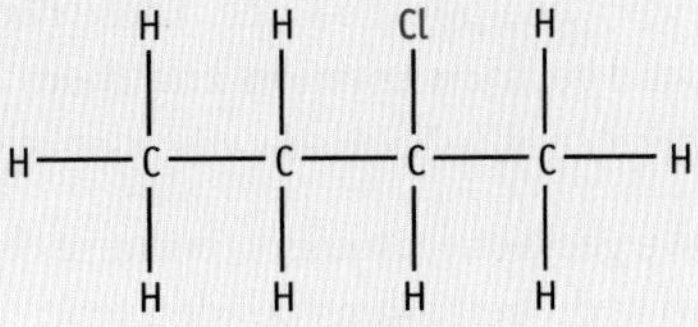

2-chlorobutane

There are also two positions on the three-carbon skeleton for the chlorine atom:

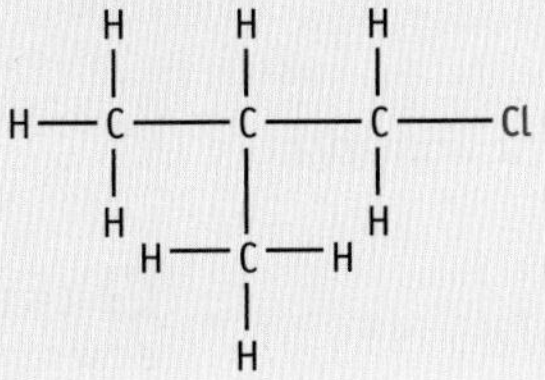

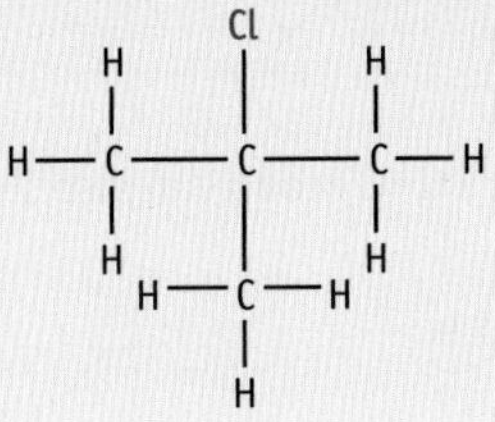

Stereoisomerism

Stereoisomers are compounds with the same structural formula but which have the atoms arranged differently in space.

There are two types of stereoisomerism — geometric and optical.

Geometric isomerism

Geometric isomerism is also called *cis–trans* isomerism. In organic chemistry, it is caused by the presence of a functional group that restricts rotation. For example, a C=C group consists of a σ-bond, which lies along the axis between the two carbon atoms and a π-bond which is above and below that axis:

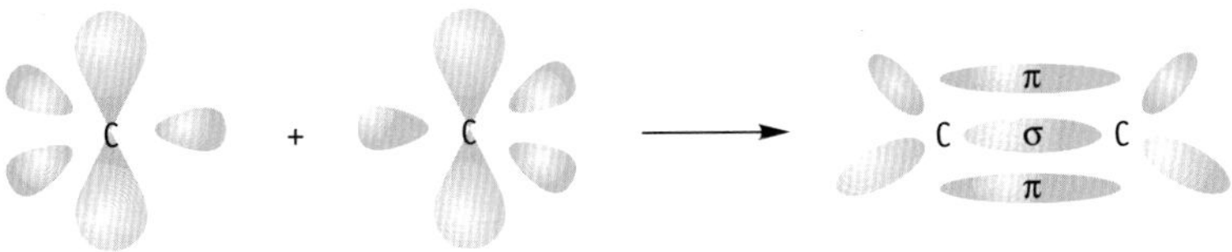

For rotation round the σ-bond to occur, the π-bond would have to break and then reform. The energy required to do this is far too great for this to occur at room temperature.

Alkenes exhibit geometric isomerism if there are different groups on each carbon atom of the C=C bond. The simplest example is but-2-ene, which has two geometric isomers:

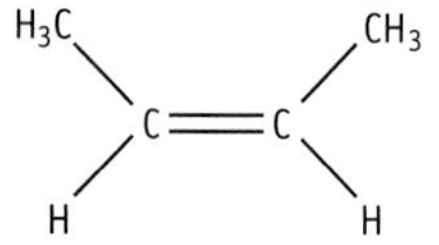

cis-but-2-ene

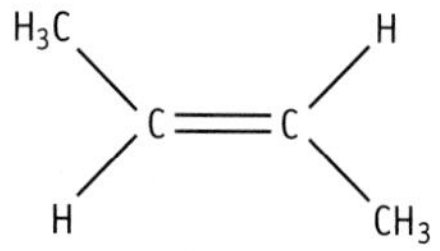

trans-but-2-ene

In *cis*-but-2-ene, the two $-CH_3$ groups are on the same side of the double bond; in *trans*-but-2-ene they are on opposite sides. These two compounds are isomers because the double bond restricts rotation and there are different groups (–H and $-CH_3$) on each of the double-bonded carbon atoms.

But-1-ene does not have geometric isomers, because one of the carbon atoms in the C=C group has two hydrogen atoms bonded to it:

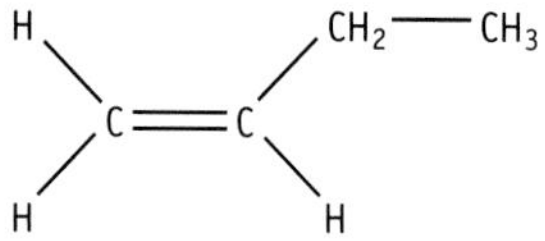

More complex compounds with a C=C group can also show geometric isomerism, for example, but-2-enoic acid:

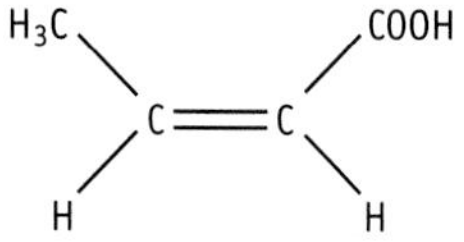

cis-but-2-enoic acid

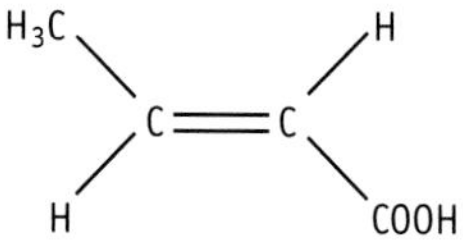

trans-but-2-enoic acid

e It is advisable to draw correct bond angles around the C=C bond.

Geometric isomers have the same chemical properties, but their biological properties may differ. *Cis*-retinal occurs in receptor cells in the retina of the human eye. When it absorbs a photon of visible light, the π-bond breaks. The molecule reforms as the *trans*-isomer. This change in shape causes a nerve impulse to be sent to the brain.

Geometric isomerism can occur in cyclic compounds in which rotation is not possible. When chlorine adds to cyclohexene, one of two possible geometric isomers is formed:

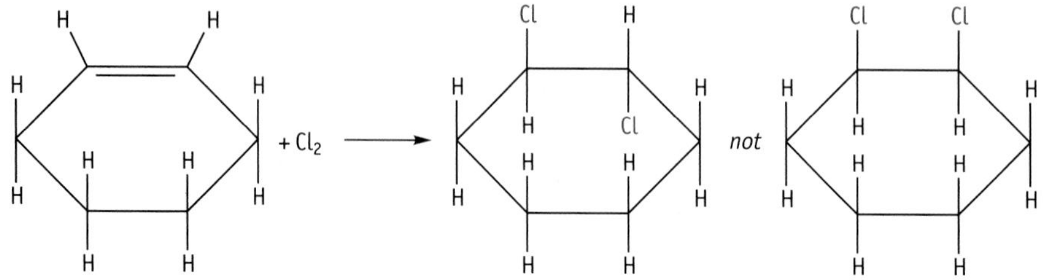

$+ Cl_2 \longrightarrow$ ... *not* ...

Geometric isomerism also occurs in some transition metal complexes. For example, platinum(II) forms planar complexes with four ligands. The complex $[PtCl_2(NH_3)_2]$ exists as two geometric isomers:

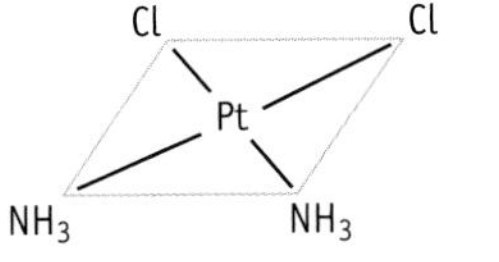

cis-diamminedichloroplatinum(II)

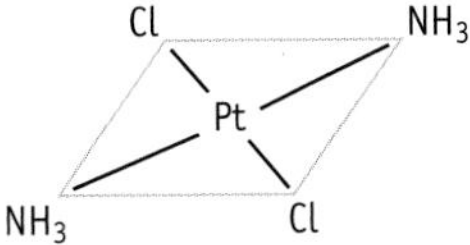

trans-diamminedichloroplatinum(II)

The *cis*-isomer (known as cisplatin) and similar *cis*-platinum complexes are used in the treatment of cancer. The complexes inhibit cell division and cancer cells are particularly susceptible. The treatment, known as chemotherapy, also causes hair loss because cisplatin stops the regrowth of hair. The *trans*-isomer has no biological activity.

Cisplatin blocks DNA replication because it is the correct shape to bond to the base guanine in DNA. The *trans*-form cannot do this.

The *cis–trans* method of naming geometric isomers breaks down in more complex compounds. For example, there are two geometric isomers of the unsaturated acid, 2-methylbut-2-enoic acid, $CH_3CH{=}C(CH_3)COOH$. One has the common name tiglic acid and can be obtained from the seeds of the croton tree; the other is called angelic acid and can be obtained from the root of the plant *Angelica archangelica*:

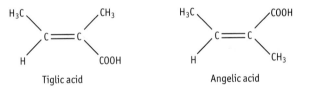

Tiglic acid

Angelic acid

The *cis–trans* naming system cannot be used in this case. This is because in tiglic acid, the left-hand $-CH_3$ group is *cis* to the right-hand $-CH_3$ but *trans* to the $-COOH$ group. The **E / Z system** is used instead. This system depends on the priorities of the groups attached to the carbon atoms of the double bond. The priority is equal to the sum of the atomic numbers of the group attached. The priority numbers for the some common groups are:

- $-H = 1$
- $-CH_3 = 9$
- $-C_2H_5 = 17$
- $-COOH = 23$

The prefix *Z* is given to the isomer with the two higher priority groups on the same side of the double bond; *E* is given to the isomer with the two higher priority groups on opposite sides. Thus, tiglic acid is *E*-2-methylbut-2-enoic acid and angelic acid is *Z*-2-methylbut-2-enoic acid.

e One way of remembering what *E and* Z stand for is 'Zee meanz on zee zame zide'.

Angelic acid has a melting temperature of 46°C; tiglic acid has a melting temperature of 64°C. The *Z*-isomer, angelic acid, does not pack well because the $-CH_3$ and the $-COOH$ groups are on the same side of the double bond. This means that the intermolecular forces are weaker and the melting temperature lower.

Angelic acid was first isolated in 1862 by L. Buchner of Buchner flask and funnel fame.

ARCO IMAGES GMBH/ALAMY

Tiglic and angelic acids are produced in defensive secretions by many beetles, including those of the genus Pterostichus.

Optical isomerism

Compounds that show **optical isomerism** do not have a plane (or axis or centre) of symmetry. They are said to be **chiral.** Such compounds have two isomers, which are mirror images. The isomers are called **enantiomers.**

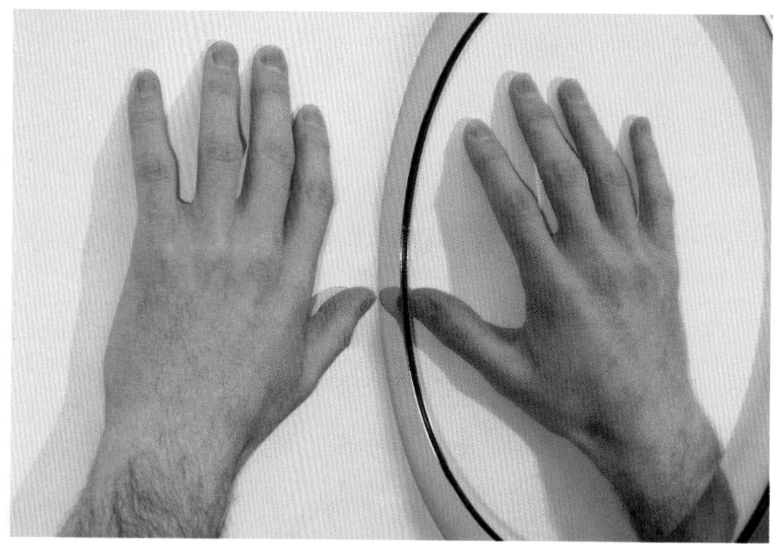

A left hand and its reflection

A left hand is different from a right hand, yet looks like a right hand when reflected in a mirror.

A chiral centre in a molecule or ion causes it to have two optical isomers, which are called enantiomers.

An enantiomer is an isomer that is non-superimposable on its mirror image.

Non-superimposable means that it is impossible to put one beside the other in such a way that their shapes are the same.

The most common cause of chirality in organic chemistry is when a carbon atom has four different groups or atoms attached to it. For example, the compound CHFClBr is chiral:

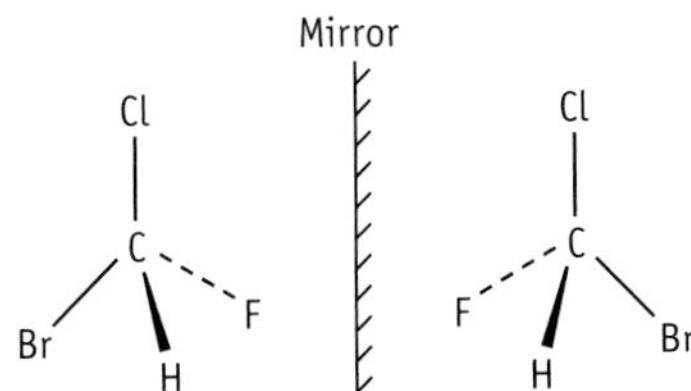

Note that the two isomers are drawn as mirror images. The wedges and dashes are meant to give an idea of the three-dimensional shapes of the molecules.

Enantiomers of CHFClBr

ZAHOOR UL-HAQ

Lactic acid is produced when milk goes sour and in muscles as a result of anaerobic respiration. Its formula is $CH_3CH(OH)COOH$ and its systematic name is 2-hydroxypropanoic acid. It contains a chiral carbon atom that has –H, –OH, $–CH_3$ and –COOH attached. The presence of these four different groups means that the substance exists as two optical isomers:

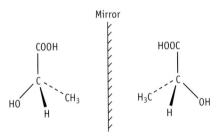

Mirror

COOH, C, HO, CH_3, H | HOOC, C, H_3C, OH, H

Always make sure that the bonds are drawn from the central carbon atom to the correct atom in the group, for example, to the oxygen atom of the –OH group and to the carbon atom of the $–CH_3$ group.

Enantiomers have identical chemical properties and the same boiling temperatures and solubilities. They differ in two ways:

- Uniquely among chemical compounds, they rotate the plane of polarisation of plane-polarised light.
- Optical isomers often have different biochemical reactions.

This property is optical and provides the origin of the name 'optical isomerism'.

Glucose, $CHO(CHOH)_4CH_2OH$ is one of 16 optical isomers. Glucose is the only one of the 16 that can be metabolised by humans.

Plane-polarised light

Light waves have peaks and troughs in all planes. When ordinary light is passed through a piece of Polaroid, the light that comes out only has peaks and troughs in a single plane. This light is said to be polarised.

A solution of one enantiomer rotates the plane of polarisation of **plane-polarised light** in a clockwise direction (+); the other enantiomer rotates it in an anticlockwise direction (–).

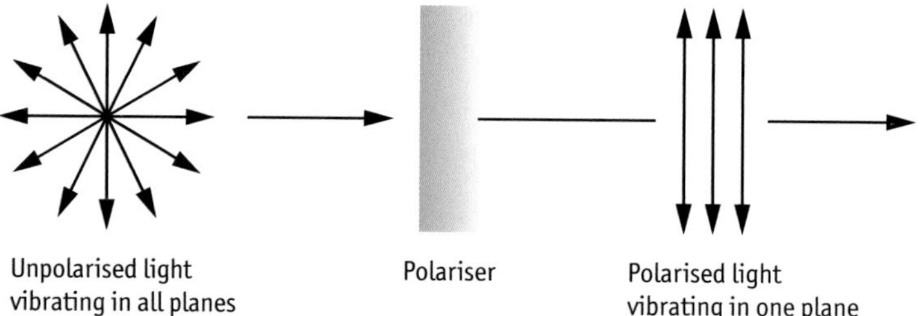

***Figure 6.1** Unpolarised and polarised light*

Lactic acid produced in muscles is a crystalline solid that melts at 26°C and rotates the plane of polarisation clockwise. Lactic acid obtained from the action of microorganisms on milk sugar (lactose) is a crystalline solid that also melts at 26°C, but rotates the plane of polarisation of plane-polarised light in an anticlockwise direction.

A solution containing equimolar amounts of the two enantiomers is called a racemic mixture. It does not rotate the plane of polarisation of plane-polarised light.

The extent by which an enantiomer rotates the plane of polarisation can be measured using a **polarimeter**.

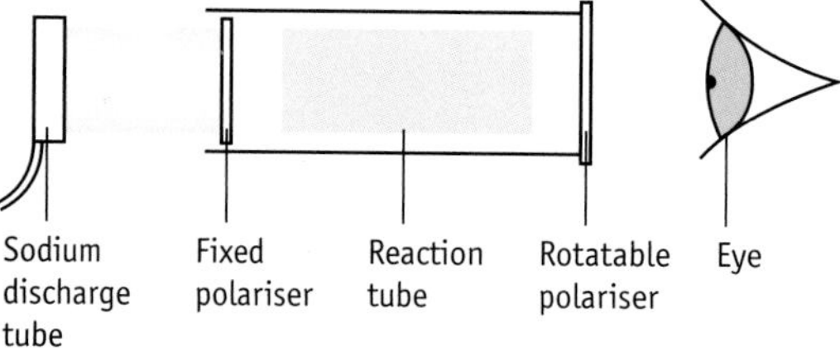

***Figure 6.2** A polarimeter*

The angle through which the plane of polarisation is rotated depends on:

- the nature of the enantiomer
- the concentration of the enantiomer in the solution

The second factor is useful in determining the change in concentration during a reaction. For example, the rate of the hydrolysis of sucrose can be followed using a polarimeter:

$$C_{12}H_{22}O_{11} + H_2O \rightarrow C_6H_{12}O_6 + C_6H_{12}O_6$$

sucrose — glucose — fructose

Sucrose rotates the plane of polarisation in one direction and the mixture of glucose and fructose rotates it in the other direction.

The rate of hydrolysis of one enantiomer of a halogenoalkane can also be followed using a polarimeter:

$$CH_3CHClC_2H_5 + OH^- \rightarrow CH_3CH(OH)C_2H_5 + Cl^-$$

one enantiomer — a racemic mixture

The rotation of the plane of polarisation of plane-polarised light drops from the original value to zero as the reaction progresses.

Biological differences

Chemical reactions are not stereospecific. Both enantiomers of lactic acid have identical chemical properties. However, most biochemical reactions take place with only one of the two enantiomers. For example, ibuprofen, $(CH_3)_2CHCH_2C_6H_4CH(CH_3)COOH$, is chiral and the tablets available commercially are a racemic mixture. Only the (–) form is active as a painkiller and anti-inflammatory drug. In the 1950s, the drug thalidomide was introduced as a cure for morning sickness in pregnant women. Sadly, it had not been tested properly on a range of pregnant mammals. About 10 000 children, whose mothers had been taking thalidomide, were born with severe abnormalities – for example, vestigial limbs. It was found subsequently that one enantiomer suppressed morning sickness and the other caused the birth defects.

Two chiral centres

If a molecule has two different chiral centres, there are four possible optical isomers. Let one chiral centre be called A. It has two mirror image structures, one of which will rotate the plane of polarisation of plane-polarised light clockwise, +A; the other will rotate it anticlockwise, –A. Let the second chiral centre be called B. The B centre also has two mirror image structures, +B and –B.

The four optical isomers are: +A with +B, –A with –B and +A with –B, –A with +B.

The first isomer rotates the plane of polarisation of plane-polarised light clockwise, the second rotates it anticlockwise by the same amount and the third and fourth each rotate it slightly, depending on the extent to which the A and B chiral centres each rotate the plane.

e If a molecule has two identical chiral centres, the isomer with a plane of symmetry will not be optically active because one chiral centre will rotate the plane of polarisation of plane-polarised light clockwise and the other chiral centre will rotate it equally anticlockwise.

To summarise:

- Optical activity occurs in organic compounds when four different groups are attached to the same carbon atom. This carbon atom is the chiral centre of the molecule.
- A chiral centre results in two optical isomers (enantiomers) that are non-superimposable mirror images of each other.
- Enantiomers rotate the plane of polarisation of plane-polarised light in opposite directions.
- An equimolar mixture of enantiomers is called a racemic mixture. It has no effect on plane-polarised light.

Chirality and mechanisms

Nucleophilic substitution of halogenoalkanes

Nucleophilic substitution of halogenoalkanes can proceed either through an S_N1 or an S_N2 mechanism (see p. 20). There are two pieces of evidence that enable the elucidation of the correct mechanism. One is from the kinetics of the reaction. An S_N1 mechanism is zero order with respect to the nucleophile and first order with respect to the halogenoalkane; an S_N2 mechanism is first order with respect to both the nucleophile and the halogenoalkane.

S_N1: rate = k[halogenoalkane]^1 × [nucleophile]^0

S_N2: rate = k[halogenoalkane]^1 × [nucleophile]^1

The other piece of evidence is the effect of the product of the reaction on plane-polarised light.

If a single optical isomer of a halogenoalkane is reacted with hydroxide ions and the reaction has an S_N2 mechanism, the result will be a single optical isomer of the product alcohol. The OH^- ion attacks from the side opposite to the halogen and thus a single optical isomer with an inverted stereostructure is obtained:

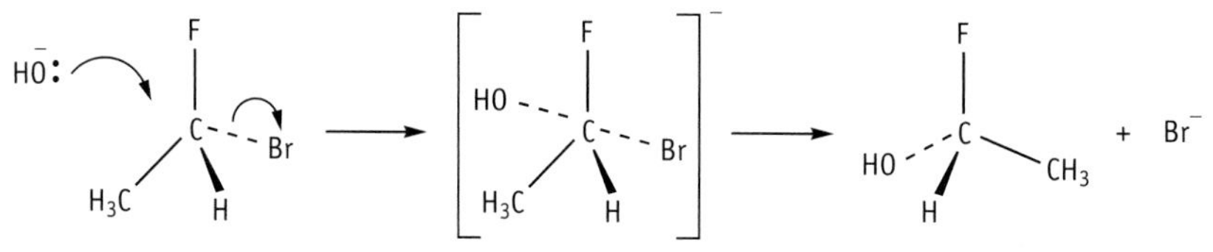

However, if the mechanism is S_N1, the result is optically inactive because the racemic mixture of the two enantiomers is produced:

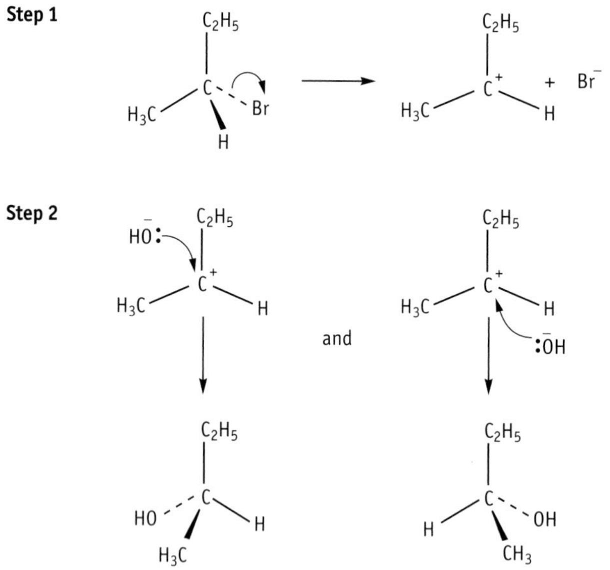

The first step is the loss of the halide ion from the halogenoalkane. The resulting carbocation has three pairs of bonding electrons and no lone pairs, so its shape is triangular planar around the positive carbon atom. The nucleophile can then attack from the top (left-hand side of the two step 2 mechanisms), producing one enantiomer, or from the bottom (right-hand side of step 2), producing the other. As the chance of attack from the top is identical to that from the bottom, the two optical isomers (enantiomers) are produced in equal amounts. This is the definition of a racemic mixture.

Nucleophilic addition to carbonyl compounds

The production of a racemic mixture in nucleophilic addition to aldehydes is evidence for the type of mechanism. This is explained in Chapter 7.

Questions

1 a Draw and name the four structural isomers of $C_3H_6Cl_2$.

b One of these exists as optical isomers. Identify which and draw the structural formulae of these two isomers, showing clearly how they differ.

2 Explain why 1-chloropropene, $CH_3CH{=}CHCl$, exists as two geometric isomers whereas 3-chloropropene, $CH_2{=}CHCH_2Cl$, does not.

3 The complex ion $[Cr(NH_3)_4(Cl)_2]^+$ has four ammonia molecules and two chloride ions attached to the central chromium ion by dative covalent bonds. Its shape is octahedral and similar to that of the hydrated magnesium ion $[Mg(H_2O)_6]^{2+}$. Draw the two geometric isomers of the chromium complex ion. Mark in the bond angles on your diagram.

4 Geraniol is a perfume found in rose petals. Its formula is:

$$(H_3C)_2C{=}CH(CH_2)_2C(CH_3){=}CHCH_2OH$$

Explain why geraniol exists as two geometric isomers.

5 Name the compounds:

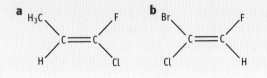

a $H_3C(H)C{=}C(F)Cl$ (H₃C and F on the same side; H and Cl on the same side)

b $Br(Cl)C{=}C(F)H$ (Br and F on the same side; Cl and H on the same side)

6 Draw the two optical isomers of $CH_3CH(OH)C_2H_5$ to show clearly the way in which they differ.

7 Limonene is a chiral molecule found in the oil of citrus fruits. Its structure is:

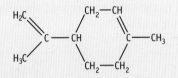

$$H_2C{=}C(CH_3){-}CH\langle CH_2{-}CH{=}C(CH_3){-}CH_2{-}CH_2\rangle$$

Redraw the structure, marking the chiral centre with an asterisk, *.

8 Alanine, 2-aminopropanoic acid, is an amino acid found in many proteins. It is chiral.

a Explain the meaning of the term 'chiral'.

b Draw the structures of the two optical isomers of alanine.

c Describe how optical activity is detected experimentally.

9 Tartaric acid, 2,3-dihydroxybutanedioic acid, has the formula HOOCCH(OH)CH(OH)COOH.

a Draw the full structural formula of this molecule, marking each chiral centre with an asterisk, *.

b Explain why there are only three optical isomers of this compound.

10 How would you demonstrate the difference in properties of two optical isomers?

11 Search the web for information on cisplatin and write approximately 100 words about its discovery, uses and problems.

Carbonyl compounds

Chapter 7

Introduction

Carbonyl compounds contain the >C=O group. The carbon atom is bonded to the oxygen by a σ-bond and a π-bond. Oxygen is more electronegative than carbon, so the bonding electrons are pulled towards the oxygen atom making it δ^- and the carbon δ^+:

$$>\overset{\delta^+}{C}=\overset{\delta^-}{O}$$

The polar nature of the bond makes the carbon atom susceptible to attack by nucleophiles.

A nucleophile is a species with a lone pair of electrons that is used to form a bond with a δ+ atom.

There are two types of carbonyl compounds: aldehydes and ketones.

Aldehydes

- Aldehydes have a hydrogen atom bonded to the carbon of the C=O group.
- The formulae of aldehydes can be represented by RCHO, where R is a hydrogen atom, an alkyl group or benzene ring, for example H, $-CH_3$, $-C_2H_5$ or $-C_6H_5$.
- Aliphatic aldehydes have the general formula $C_nH_{2n+1}CHO$, where n is 0, 1, 2, 3 etc.

n	Formula	Name	Boiling temperature/°C
0	HCHO	Methanal	−21
1	CH_3CHO	Ethanal	20
2	CH_3CH_2CHO	Propanal	49
3	$CH_3CH_2CH_2CHO$	Butanal	76
3	$(CH_3)_2CHCHO$	Methylpropanal	62

Never write the formula of an aldehyde as RCOH.

If you are asked for the structural formula in an exam, you must show the bonding in the –CHO group. For example, the structural formula of ethanal is:

H_3C—C(—H)=O

Ketones

- Ketones do *not* have a hydrogen atom bonded to the carbon of the group.
- The formulae of ketones can be represented as RCOR′, where R and R′ are alkyl or benzene ring groups such as $-CH_3$, $-C_2H_5$ or $-C_6H_5$. The simplest ketone is propanone, which contains three carbon atoms.

Formula	Name	Boiling temperature/°C
CH_3COCH_3	Propanone	56
$CH_3CH_2COCH_3$	Butanone	80
$CH_3CH_2COCH_2CH_3$	Pentan-3-one	102
$CH_3COCH_2CH_2CH_3$	Pentan-2-one	102
$CH_3COCH(CH_3)_2$	Methylbutanone	94

Propanone used to be called acetone.

If you are asked for the structural formula in an exam, you must show the bonding in the carbonyl group. For example, the structural formula of propanone is:

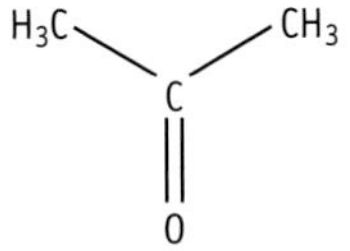

H_3C—C(=O)—CH_3

General structural formulae

The general structural formulae of aldehydes and ketones can be represented by:

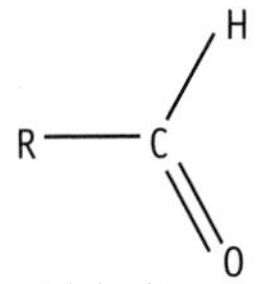

R—C(H)=O
Aldehyde

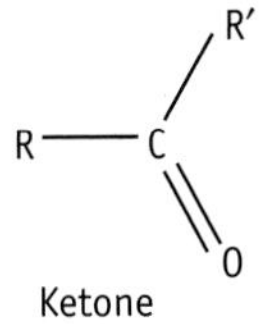

R—C(R′)=O
Ketone

Geometry around the C=O group

The carbon atom has two single bonds, one double bond and no lone pairs, so the electrons in the three bonds repel each other and take up the position of maximum separation. This is a planar triangular shape with bond angles of 120°:

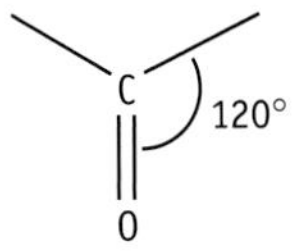

C=O, 120°

This planar shape makes it easy for nucleophiles to attack the carbon atom from either above or below, and so a single optical isomer is never obtained by addition to carbonyl compounds.

The polarity of the $>\overset{\delta+}{C}=\overset{\delta-}{O}$ group is not cancelled out by the other two groups attached to the carbon atom. Therefore aldehyde and ketone molecules are polar.

Physical properties

Boiling point

Methanal is a gas. The other carbonyl compounds are liquid at room temperature.

The molecules are polar, so dipole–dipole forces as well as instantaneous induced dipole–induced dipole (dispersion) forces exist between them. Alkanes and alkenes are not polar, so their intermolecular forces are weaker and their boiling points lower than those of aldehydes and ketones with the same number of electrons in the molecule. Intermolecular hydrogen bonding is not possible in carbonyl compounds because neither aldehydes nor ketones have a hydrogen atom that is sufficiently δ^+. Therefore, they have boiling points that are lower than those of alcohols, which do form intermolecular hydrogen bonds.

Solubility

The lower members of both series are soluble in water. This solubility is due to hydrogen bonding between the lone pair of electrons in the δ^- oxygen in the carbonyl compound and the δ^+ hydrogen in a water molecule.

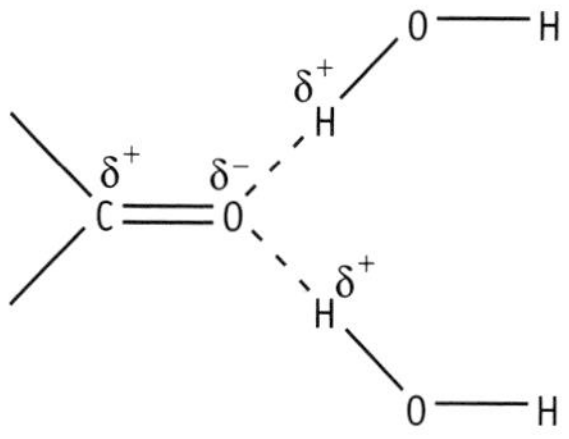

O—H
δ+
H
δ+ δ−
C=O
δ+
H
O—H

Propanone is an excellent solvent for organic substances.

Smell

- The lower members of the homologous series of aldehydes have pungent odours.
- Ketones have much sweeter smells than aldehydes.
- Complex aldehydes and ketones are often used as perfumes. Citral, an ingredient of lemon grass oil, is an aldehyde; β-ionone is a ketone that smells of violets and is used in perfume:

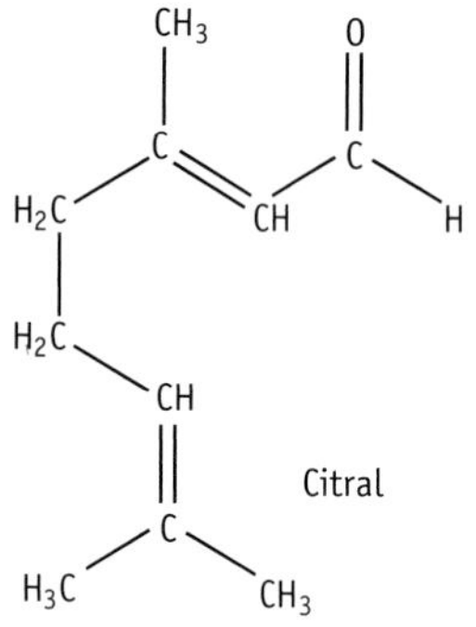

CH3 O C C H2C CH H H2C CH C H3C CH3

Citral

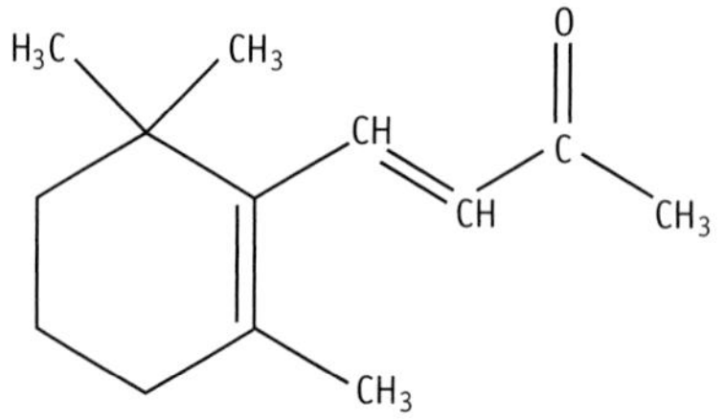

H3C CH3 O CH C CH CH3 CH3

β-ionone

Laboratory preparation

Aldehydes

Aldehydes are prepared by the partial oxidation of a *primary* alcohol. The usual oxidising agent is a solution of potassium (or sodium) dichromate(VI) in dilute sulfuric acid. The temperature must be below the boiling point of the alcohol and above that of the aldehyde. In this way the aldehyde is boiled off as it is formed and, therefore, cannot be oxidised further. If the mixture were to be heated under reflux, the aldehyde would be further oxidised to a carboxylic acid.

Ethanal can be prepared as follows:

- Heat ethanol in a flask to about 60°C, using an electric heater.
- Add a solution of potassium dichromate(VI) in dilute sulfuric acid slowly from a tap funnel.
- As it distils off, collect the ethanal in a flask surrounded by iced water.

e A Bunsen flame must *not* be used as both ethanol and ethanal are highly flammable.

The reaction is exothermic and so heat is generated. This maintains the temperature above the boiling point of ethanal, so it boils off before it can be oxidised further.

The preparation of ethanal is represented by the equation:

$$CH_3CH_2OH + [O] \rightarrow CH_3CHO + H_2O$$

e [O] can be used in equations for oxidation reactions in organic chemistry, apart from those involving oxygen gas. [H] can be used for all reduction reactions in organic chemistry, apart from when hydrogen gas is the reducing agent. Equations containing [O] or [H] must still balance.

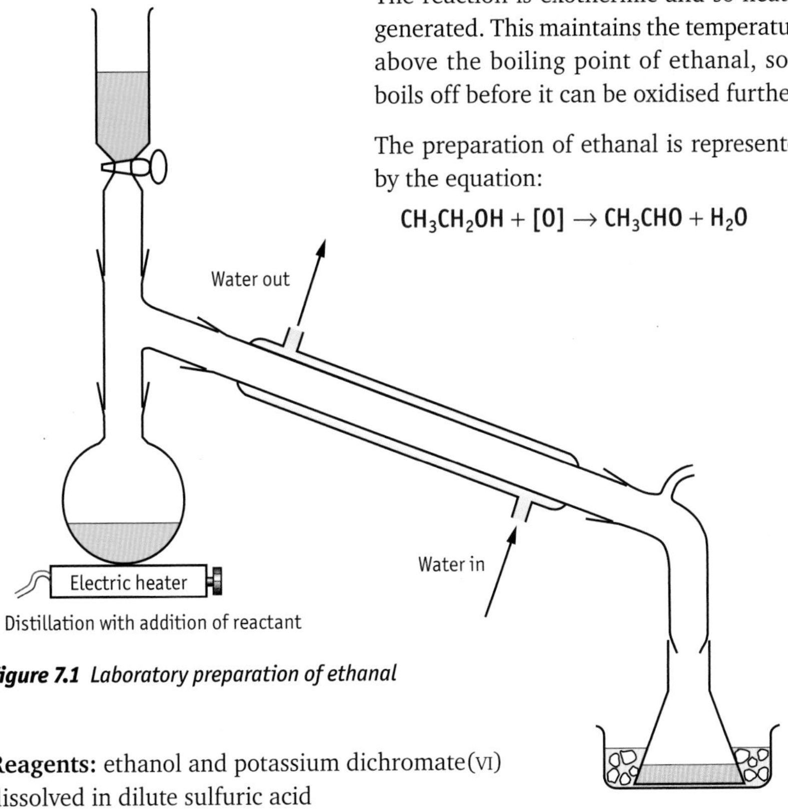

Water out

Water in

Electric heater

Distillation with addition of reactant

***Figure 7.1** Laboratory preparation of ethanal*

e The iced water keeps the collecting flask cool and so prevents evaporation of the ethanal.

Reagents: ethanol and potassium dichromate(VI) dissolved in dilute sulfuric acid

Conditions: a temperature of 60°C; collect the ethanal as it distils off

Observation: the orange colour of potassium dichromate(VI) changes to green because chromium(III) ions are formed.

Ethanal can be purified by re-distilling it, using a water bath as a source of heat and collecting the fraction that boils in the range 20–23°C.

Ketones

Ketones are prepared by the oxidation of *secondary* alcohols. If acidified potassium dichromate(VI) is used as the oxidising agent, the ketone is not further oxidised.

Propanone can be prepared as follows:

- Place propan-2-ol and a solution of potassium dichromate(VI) in dilute sulfuric acid in a round-bottomed flask.
- Fit a reflux condenser.
- Heat, using an electric heater, so that the mixture boils for about 15 minutes.
- Remove the reflux condenser and set up the apparatus for distillation. Distil off the propanone (boiling point 56°C) from any unreacted propan-2-ol (boiling point 82°C).

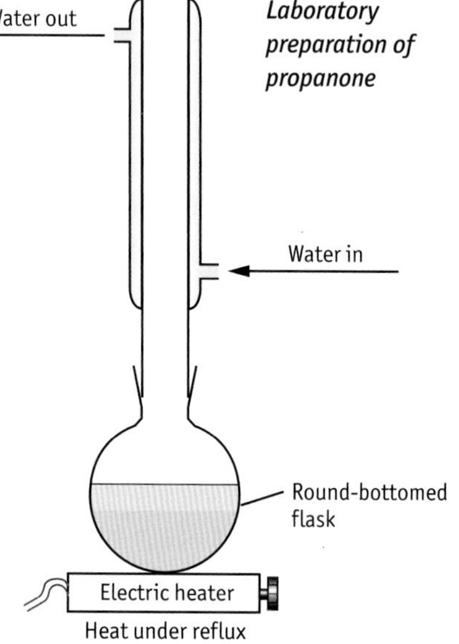

***Figure 7.2** Laboratory preparation of propanone*

The preparation of propanone is represented by the equation:

$$CH_3CH(OH)CH_3 + [O] \rightarrow CH_3COCH_3 + H_2O$$

Reagents: propan-2-ol and potassium dichromate(VI) dissolved in dilute sulfuric acid

Conditions: heat under reflux for about 15 minutes, and then distil off the propanone

Observation: the orange colour of potassium dichromate(VI) changes to green because chromium(III) ions are formed

Aromatic ketones

Ketones that have a benzene ring attached to the carbonyl group, such as phenylethanone, $C_6H_5COCH_3$, can be prepared using the **Friedel–Crafts reaction** (p. 246).

Reactions of aldehydes and ketones

- Both aldehydes and ketones undergo **nucleophilic addition** reactions because of the polar nature of the C=O bond.
- Aldehydes and ketones both undergo **addition/elimination** reactions.
- Aldehydes can be oxidised to carboxylic acids. Therefore, aldehydes are reducing agents.
- Ethanal and methylketones react with iodine in alkali to form a precipitate of iodoform, CHI_3. This is called the **iodoform reaction**.

Nucleophilic addition

Alkenes and carbonyl compounds contain π-bonds. However, C=C is non-polar whereas C=O is polar. The electron cloud above and below the σ-bond in the C=C group is an area of negative charge that is attacked by electrophiles.

The electron cloud in the C=O group is distorted, with the carbon atom being δ^+. This δ^+ carbon can be attacked by nucleophiles. The δ^- oxygen is not attacked by electrophiles, such as Br_2 or HBr, because the oxygen atom is itself strongly electronegative.

The $-CH_3$ group is electron releasing, so the carbon in methanal, HCHO (which has no $-CH_3$ group), is more δ^+ than the carbon of the C=O group in ethanal. The carbonyl carbon in propanone is even less δ^+ than that in ethanal. The reactivity in nucleophilic addition is: HCHO > CH_3CHO > CH_3COCH_3.

Reaction with hydrogen cyanide

The reaction between ethanal and hydrogen cyanide takes place only at about pH 8. This pH provides the catalyst, which is the cyanide ion, CN^-.

$$CH_3CHO + HCN \rightarrow CH_3CH(OH)CN$$

The product is 2-hydroxypropanenitrile.

Hydrogen cyanide, HCN, is a very weak acid and produces few CN^- ions in solution. Some base is added to produce the nucleophilic CN^- ions that attack the δ^+ carbon atom:

$$HCN + OH^- \rightleftharpoons :CN^- + H_2O$$

The mechanism of this reaction is:

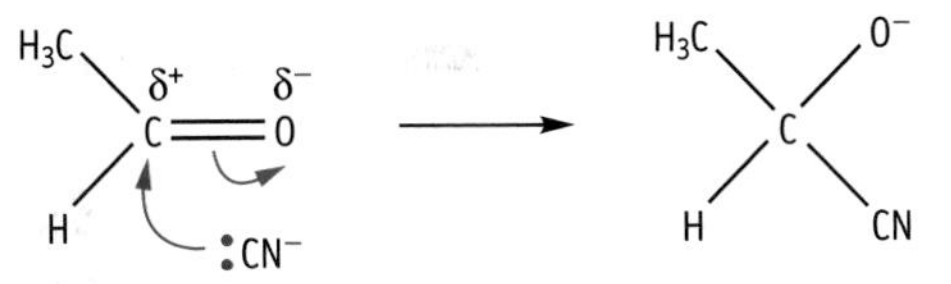

H₃C, δ+, δ−, C=O, H, :CN⁻ → H₃C, O⁻, C, H, CN

The lone pair of electrons on the carbon atom of the CN^- ion form a bond with the δ^+ carbon atom — as shown by the red curly arrow. At the same time, the π-electrons in the C=O group move to the oxygen atom — as shown by the blue curly arrow.

The anion formed in the first step removes a proton from a HCN molecule to form the organic product and another CN^- ion:

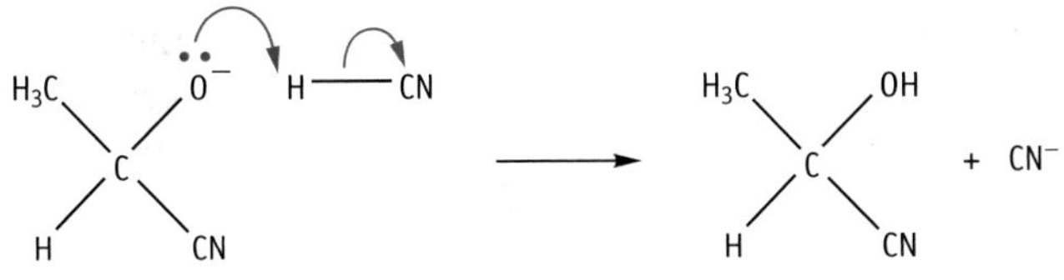

H₃C, O⁻, H—CN, C, H, CN → H₃C, OH, C, H, CN + CN⁻

The $-O^-$ group in the intermediate donates a lone pair of electrons to the hydrogen in a HCN molecule (red arrow) as the σ-bond between the H and the CN breaks (blue arrow). This regenerates CN^- ions, which catalyse the reaction.

The conditions for this reaction are very important. If the pH is too low, there are not enough CN^- ions for the first step to take place; if the pH is too high there are not enough HCN molecules for the second step.

The same type of reaction with the same mechanism occurs with ketones, but at a slower rate — for example, with propanone:

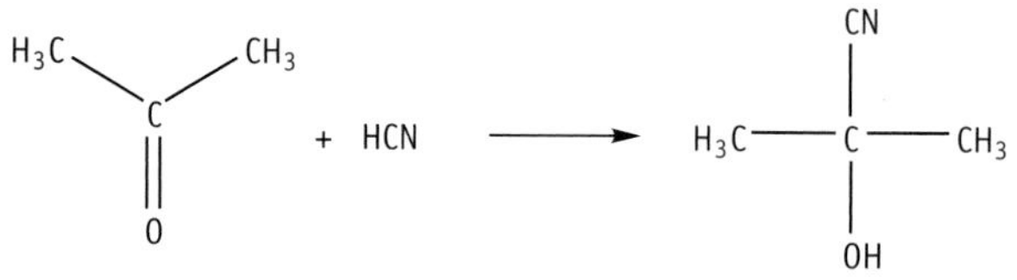

$$(CH_3)_2C{=}O + HCN \longrightarrow (CH_3)_2C(OH)CN$$

The product is 2-hydroxy-2-methylpropanenitrile.

Stereochemistry of addition

When hydrogen cyanide adds on to an aldehyde or to an asymmetric ketone, the product is a racemic mixture. This is because the carbonyl compound is planar around the >C=O group and, therefore, the cyanide ion can attack from above or below the plane.

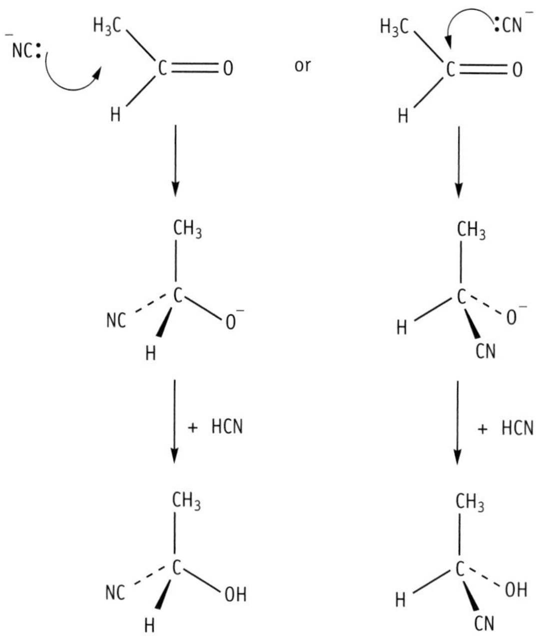

Figure 7.3 *Stereochemistry of addition to carbonyl compounds*

Thus, the addition of hydrogen cyanide to ethanal gives the racemic mixture of both enantiomers of 2-hydroxypropanenitrile.

Uses of nitriles in synthesis

Hydroxynitriles produced in this way can be hydrolysed to hydroxycarboxylic acids — for example:

$$CH_3CH(OH)CN + 2H_2O + H^+ \rightarrow CH_3CH(OH)COOH + NH_4^+$$

The reaction mixture is heated under reflux. The product is the racemic mixture of the two enantiomers of 2-hydroxypropanoic acid, which is also called lactic acid. Hydroxynitriles can also be reduced to hydroxyamines, for example:

$$CH_3CH(OH)CN + 4[H] \rightarrow CH_3CH(OH)CH_2NH_2$$

e A common error, when explaining the production of a racemic mixture, is to say that the *intermediate* is planar and can be attacked from either side. This is not true, as the intermediate is tetrahedral. It is the starting aldehyde or ketone that is planar at the reaction site and so can be attacked (by the cyanide ion) from above or below the plane.

Suitable reducing agents for this reaction are:

- lithium tetrahydridoaluminate(III) (lithium aluminium hydride), $LiAlH_4$, in dry ether solution
- hydrogen gas with a platinum catalyst
- sodium in ethanol

Reduction

Aldehydes and ketones can be reduced by lithium tetrahydridoaluminate(III), $LiAlH_4$.

This compound acts as a source of H^- ions. The reaction takes place in two distinct steps:

- The first step is the addition of H^- to the δ^+ carbon atom. In this step the reagents must be kept dry. It is carried out in ether solution.
- The second step is the addition of an aqueous solution of an acid, which protonates the O^- formed in the first step.

The result is that the carbonyl compound is *reduced* to an alcohol. For example, ethanal is reduced to ethanol:

$$\underset{\text{ethanal}}{CH_3CHO} + 2[H] \rightarrow \underset{\text{ethanol}}{CH_3CH_2OH}$$

The mechanism is almost the same as that for the addition of HCN. Step 1 requires curly arrows from the H^- ion to the carbon of the C=O group and from the π-bond to the oxygen atom. Step 2 requires a curly arrow from the $-O^-$ group to the H^+ of the acid.

When aldehydes are reduced, *primary* alcohols are formed.

Reduction of ketones takes place by a similar two-step reaction. The equation for the reduction of propanone is:

$$\underset{\text{propanone}}{CH_3COCH_3} + 2[H] \rightarrow \underset{\text{propan-2-ol}}{CH_3CH(OH)CH_3}$$

When ketones are reduced, *secondary* alcohols are formed.

$LiAlH_4$ is a reducing agent that reacts specifically with polar π-bonds. Thus, it reduces C=O in aldehydes, ketones, acids and acid derivatives, and C≡N in nitriles, but does *not* reduce the π-bond in C=C groups, which are non-polar.

Both C=C and C=O groups can be reduced by hydrogen and a platinum catalyst.

To summarise:

- Aldehydes and ketones are reduced to alcohols by:
  - $LiAlH_4$ in dry ether
  - H_2/Pt
- Alkenes are reduced by H_2/Pt only.

Addition–elimination reactions

Both aldehydes and ketones react with compounds containing an H_2N- group. The lone pair of electrons on the nitrogen atom acts as a nucleophile and forms a bond with the δ^+ carbon atom in the C=O group. However, instead of an H^+ ion adding on to the O^- formed, the substance loses a water molecule and a C=N bond is formed.

The general equation is:

$$>C{=}O + H_2N{-}X \rightarrow >C{=}N{-}X + H_2O$$

Since the nitrogen atom has a lone pair of electrons, the bond angles around the nitrogen in the product are 120°. This means that in this type of reaction aldehydes (other than methanal) and asymmetric ketones produce a mixture of two geometric isomers:

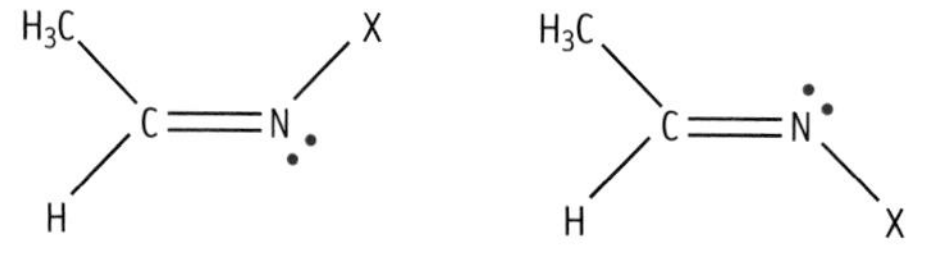

One compound that reacts in this way is 2,4-dinitrophenylhydrazine, in which the 'X' group is:

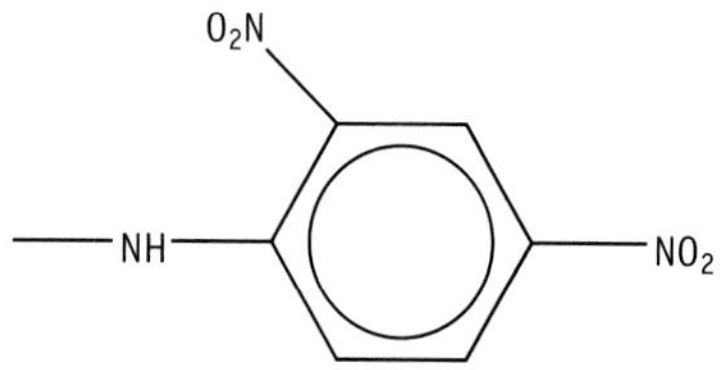

The full formula for 2,4-dinitrophenylhydrazine is:

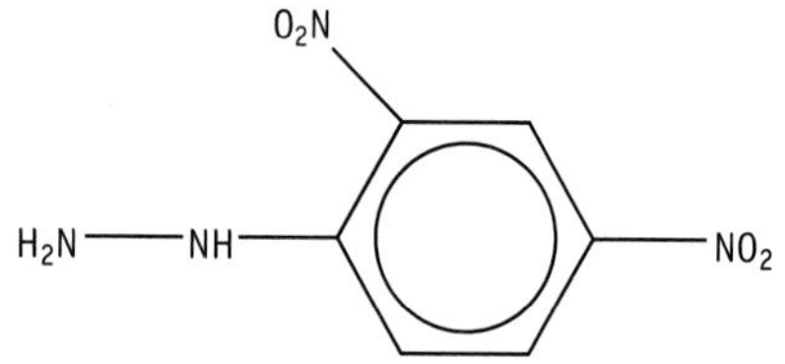

The equation for the reaction between ethanal and 2,4-dinitrophenylhydrazine is:

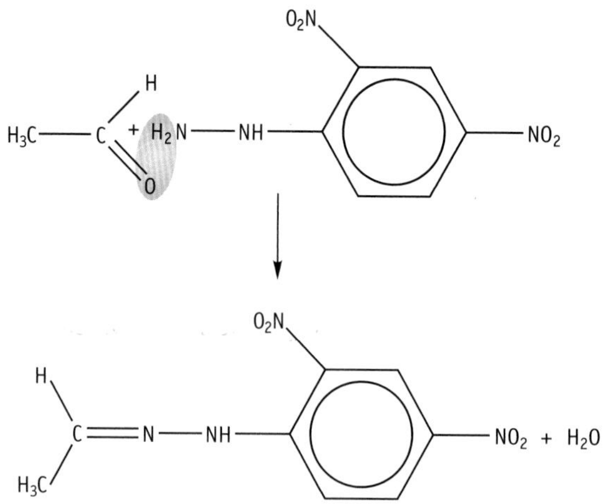

The equation for the reaction between propanone and 2,4-dinitrophenylhydrazine is:

$$(CH_3)_2C{=}O + H_2N{-}NH{-}C_6H_3(NO_2)_2 \rightarrow (CH_3)_2C{=}N{-}NH{-}C_6H_3(NO_2)_2 + H_2O$$

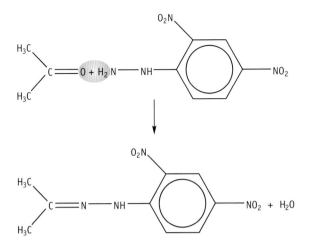

The importance of the reaction of carbonyl compounds with 2,4-dinitrophenylhydrazine is that the product is insoluble. Therefore, this reaction can be used as a test for the presence of a carbonyl group.

A solution of 2,4-dinitrophenylhydrazine is called Brady's reagent.

Test for a carbonyl group in a compound

Add a solution of 2,4-dinitrophenylhydrazine (**Brady's reagent**) to the suspected carbonyl compound:

- Simple aldehydes and ketones give yellow precipitates.
- Aromatic aldehydes (e.g. benzaldehyde, C_6H_5CHO) and ketones (e.g. phenylethanone, $C_6H_5COCH_3$) give orange precipitates.

The identity of the carbonyl compound can be found by the following method:

- React the carbonyl compound with a solution of 2,4-dinitrophenylhydrazine.
- Filter off the precipitate.
- Recrystallise the precipitate using the minimum amount of hot ethanol.
- Dry the purified product and measure its melting temperature.
- Refer to a data book and compare this melting temperature with those of 2,4-dinitrophenylhydrazine derivatives of aldehydes and ketones.

Oxidation

Aldehydes (but *not* ketones) are readily oxidised. If the reaction is carried out in acid or neutral solution the product is a carboxylic acid:

$$RCHO + [O] \rightarrow RCOOH$$

If the reaction is carried out in alkaline solution (Fehling's solution or Tollens' reagent), the product is the carboxylate anion:

$$RCHO + [O] + OH^- \rightarrow RCOO^- + H_2O$$

There are a number of suitable oxidising agents:

- Potassium manganate(VII) oxidises aldehydes in acidic, neutral or alkaline solution. In acidic solution, the purple manganate(VII) is reduced to give colourless Mn^{2+}. In neutral or alkaline solution, it is reduced to a brown precipitate of MnO_2.

- Orange potassium dichromate(VI) in acidic solution is reduced to green Cr(III) on heating with an aldehyde.
- Fehling's solution (a mixture of copper(II) sulfate and sodium potassium tartrate in alkali) is reduced from a deep-blue solution to a red precipitate of copper(I) oxide, Cu_2O, when warmed with an aldehyde. (Benedict's solution is a similar mixture and also produces a red precipitate of copper(I) oxide when warmed with an aldehyde.)
- Tollens' reagent (a solution made by adding a few drops of sodium hydroxide to silver nitrate solution and then dissolving the precipitate in dilute ammonia) is reduced to give a silver mirror on warming with an aldehyde.

e The reaction with Fehling's solution does not work with aldehydes of large molecular mass because they are too insoluble in water.

Test to distinguish an aldehyde from a ketone

An aldehyde can be distinguished from a ketone by the production of:

- a red precipitate on warming with Fehling's solution
- a silver mirror on warming with Tollens' reagent

In both reactions, the aldehyde is oxidised to the salt of a carboxylic acid — for example, with ethanal:

$$\underset{\text{ethanal}}{CH_3CHO} + [O] + OH^- \rightarrow \underset{\text{ethanoate ion}}{CH_3COO^-} + H_2O$$

Ketones do not undergo these reactions. Therefore, if these tests are carried out on a ketone, Fehling's solution remains blue and Tollens' reagent remains colourless.

Iodoform reaction

Ethanal and methyl ketones undergo the iodoform reaction with iodine in alkali, a complicated process in which the hydrogen atoms of the $CH_3C{=}O$ group are replaced by iodine atoms. The alkali present in the reaction mixture then causes the C–C bond to break and a pale yellow precipitate of **iodoform** (triiodomethane), CHI_3, is formed.

Sodium hydroxide solution is added to iodine solution to form iodate(I) ions (IO^-):

$$I_2 + OH^- \rightarrow IO^- + I^-$$

These substitute into the $-CH_3$ group next to the C=O group, forming a $CI_3C{=}O$ group. The electron-withdrawing effect of the three halogen atoms and the oxygen atom weaken the σ-bond between the two carbon atoms and this breaks, forming iodoform:

$$CH_3COR \xrightarrow{IO^-(aq)} CI_3COR \xrightarrow{OH^-(aq)} CHI_3 + RCOO^-$$

The overall equation is:

$$CH_3COR + 3I_2 + 4NaOH \rightarrow CHI_3 + RCOONa + 3NaI + 3H_2O$$

where R is a hydrogen atom (ethanal) or an organic group such as $-CH_3$, $-C_2H_5$ or $-C_6H_5$.

To carry out the iodoform reaction, the organic substance is warmed with either a mixture of iodine and sodium hydroxide solution or with a solution of potassium iodide in sodium chlorate(I).

The iodoform reaction also works with ethanol and methyl secondary alcohols. These compounds have a $CH_3CH(OH)$ group and are oxidised by the iodate(I) ions to the CH_3CO group, which then reacts as described above.

Summary

Test for a carbonyl group

Add a few drops of the organic compound to a solution of 2,4-dinitrophenylhydrazine. A yellow or orange precipitate indicates the presence of a carbonyl group.

Reactions of aldehydes

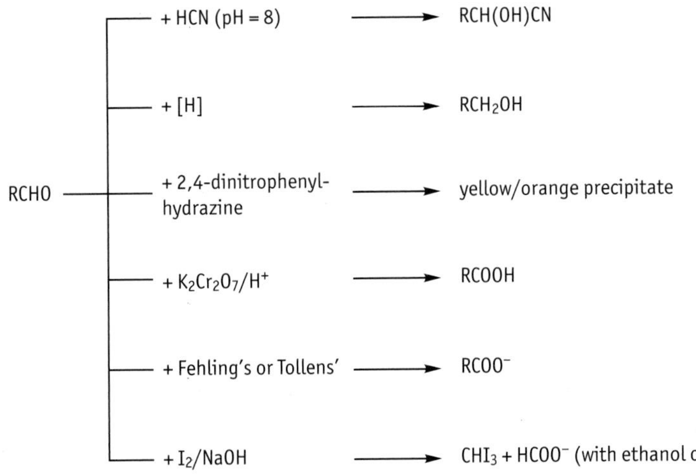

Reactions of ketones

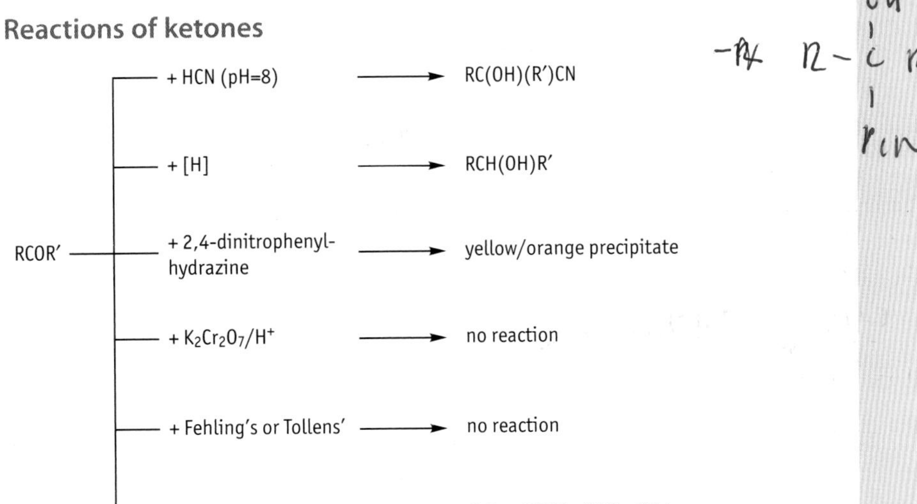

Tests to distinguish between an aldehyde and a ketone

Warm the substance with:

- Fehling's or Benedict's solution — aldehydes give a red precipitate; ketones do not alter the blue colour of the Fehling's solution
- Tollens' reagent — aldehydes give a silver mirror; ketones do not

Iodoform test

If a pale yellow precipitate is obtained when a substance is gently warmed with a solution of iodine and sodium hydroxide, the substance contains either a $CH_3CH(OH)$ or CH_3CO group.

- If the substance is an aldehyde, it can only be ethanal.
- If the substance is a primary alcohol it can only be ethanol.

Questions

1 What is the difference in structure between an aldehyde and a ketone?

2 Write the formulae of the ketone isomers of molecular formula, $C_5H_{10}O$.

3 When propanal is prepared from propan-1-ol, the mixture is kept at a temperature between the boiling temperatures of propanal and propan-1-ol. Explain why this is necessary.

4 Explain why ethanal is soluble in water but pentanal is almost insoluble.

5 Explain, in terms of the relative strengths of all the intermolecular forces present, why butanone has a boiling temperature between those of butan-1-ol and pentane, all three substances having a similar number of electrons.

6 **a** Write the equation for the reaction between hydrogen cyanide and propanone.

b Explain why this reaction does not take place unless a trace of alkali is added.

7 Write the mechanism for the reaction of propanone with lithium aluminium hydride in dry ether followed by addition of acid.

8 Explain why the product of the addition of hydrogen cyanide to propanone gives a product that has no effect on the plane of polarisation of plane-polarised monochromatic light.

9 Explain why the product of the addition of hydrogen cyanide to butanone gives a product that has no effect on the plane of polarisation of plane-polarised monochromatic light.

10 Give the formula of the organic products of the reaction, if any, between propanal and each of the following:

a 2,4-dinitrophenylhydrazine solution

b potassium dichromate(VI) in dilute sulfuric acid

c Fehling's solution

d Tollens' reagent

e a mixture of iodine and aqueous sodium hydroxide

In each case, state the observations that would be made.

11 A compound X of molecular formula $C_4H_{10}O$ was heated with a solution of potassium dichromate in sulfuric acid. The product, Y, was distilled off as it formed.

The compound Y gave a yellow precipitate with 2,4-dinitrophenylhydrazine. When Y was warmed with Fehling's solution, the Fehling's solution remained blue.

a Give the structural formulae of X and Y.

b What would you observe if X and Y were each warmed with a solution of iodine in aqueous sodium hydroxide?

c Write the equation for the reaction of Y with iodine in aqueous sodium hydroxide.

d A different compound, Z, also has the molecular formula $C_4H_{10}O$. When oxidised in the same way as X, it produced a product that also gave a yellow precipitate with 2,4-dinitrophenylhydrazine, but which reacted with warm Fehling's solution, producing a red precipitate.

Give two possible structural formulae for Z and name them.

12 Write the formula of the organic product of the reaction of $CH_2{=}CHCH_2CHO$ with each of the following:

a lithium tetrahydridoaluminate(III) in dry ether followed by the addition of dilute acid

b hydrogen gas with a platinum catalyst

13 Give details of how you would prepare a sample of butanone (boiling temperature 80°C) from butan-2-ol (boiling temperature 100°C). Include in your answer a diagram of the apparatus that you would use.

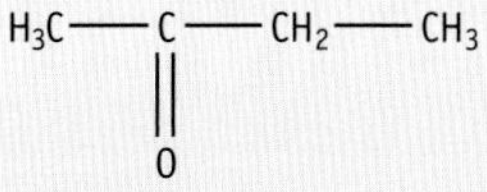

$H_3C—C(=O)—CH_2—CH_3$ $CH_3CH(OH)CH_2CH_3$

butanone butan-2-ol

14 a Define the terms 'nucleophile' and 'oxidation'.

b State the type of reaction between ethanal and each of the following:

(i) hydrogen cyanide

(ii) Fehling's solution

(iii) 2,4-dintrophenylhydrazine

15 Describe how you would identify the compound that gave an orange precipitate with 2,4-dintrophenylhydrazine and no silver mirror with Tollens' reagent.

Carboxylic acids and their derivatives

Chapter 8

Introduction

Carboxylic acids and their derivatives contain the group:

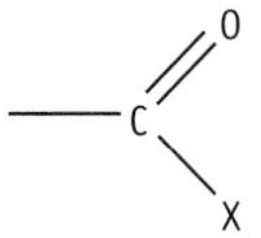

$$-C(=O)-X$$

The nature of X varies as shown in Table 8.1.

Table 8.1 *Carboxylic acids and their derivatives*

Type of compound	Structure of X	Example: Name	Example: Formula
Carboxylic acid	O–H	Ethanoic acid	CH_3COOH
Ester	O–R*	Ethyl ethanoate	$CH_3COOC_2H_5$
Acid chloride	Cl	Ethanoyl chloride	CH_3COCl
Amide	NH_2	Ethanamide	CH_3CONH_2

* R stands for an alkyl group (e.g. CH_3, C_2H_5) or a benzene ring

The reactivity of the C=O group is considerably modified by the presence of the X group, so much so that, unlike aldehydes and ketones, carboxylic acids and their derivatives do not react with 2,4-dinitrophenylhydrazine.

The C=O group is polarised with the less electronegative carbon atom δ^+ and the more electronegative oxygen atom δ^-.

The mechanism for many of the reactions is for a nucleophile to attack the δ^+ carbon atom. This is then followed by the loss of the X group as an X^- ion:

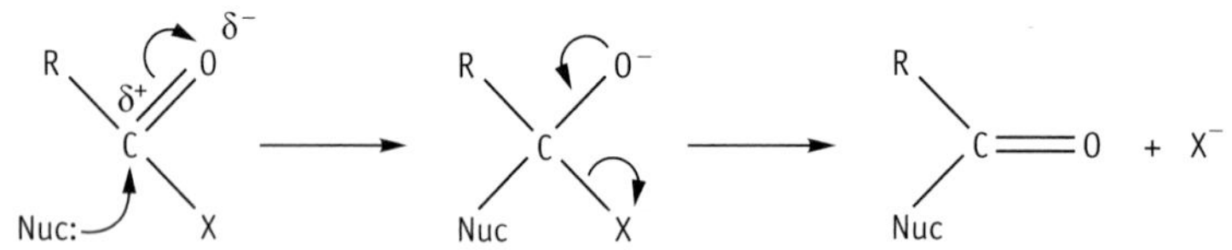

R, δ^+C=Oδ^-, X + Nuc: → R, Nuc, C–O^-, X → R(Nuc)C=O + X^-

This type of reaction is called an **addition–elimination** reaction. The ease of reaction depends on:

- the strength of the C–X bond. The weakest of these bonds is the C–Cl bond, so acid chlorides are the most susceptible to nucleophilic attack.
- the stability of the leaving group, X^-. The NH_2^- ion is too strong a base to be produced in aqueous solutions. This means that amides do not undergo addition–elimination reactions.

Carboxylic acids and their derivatives in nature

Carboxylic acids are found in a number of fruits and vegetables:

- Citric acid, $HOOCCH_2C(OH)(COOH)CH_2COOH$, is a tribasic acid and has the skeletal formula:

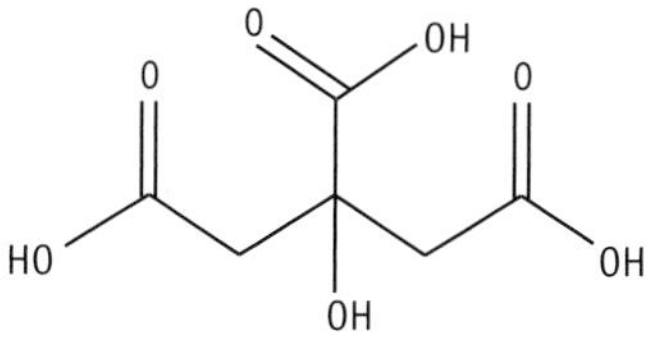

  It is present in citrus fruits, particularly lemons and limes. It is an antioxidant and is used in the food industry as an additive to give flavour and, mixed with citrate, as a buffer. Its E number is E330. In the form of citrate, it is an intermediate in the Krebs cycle in cells and is, therefore, involved in the metabolism of fats, proteins and carbohydrates.
- Malic acid, $HOOCCH_2CH(OH)COOH$, is present in apples (Latin *malum*) and grapes and gives them their sharp taste. Its systematic name is 2-hydroxybutanedioic acid.
- Ethanedioic acid (old name oxalic acid), HOOCCOOH, is found in rhubarb leaves.

Esters are also common in nature. Fats and vegetable oils are esters of propane-1,2,3-triol (glycerol), $CH_2OHCH(OH)CH_2OH$, and a variety of organic acids:

- Cows' milk contains an ester of the acids 2-hydroxypropanoic (lactic) acid, $CH_3CH(OH)COOH$ and butanoic acid, $CH_3CH_2CH_2COOH$; goats' milk contains an ester of octanoic (caproic) acid, $CH_3(CH_2)_5COOH$. The word 'caproic' comes from the Latin *caper* for goat, as does the Zodiac sign Capricorn. Solid animal fats are esters of acids, for example stearic acid, $C_{17}H_{35}COOH$, which is saturated.
- Vegetable and fish oils are esters containing unsaturated carboxylic acids such as $C_{17}H_{29}COOH$, $CH_3CH_2CH{=}CHCH_2CH{=}CHCH_2CH{=}CH(CH_2)_7COOH$, which is an omega-3 acid. It is called omega-3 because the first double bond starts at the third carbon from the end furthest from the –COOH group, omega being the last letter in the Greek alphabet.

Carboxylic acids

Carboxylic acids contain the group:

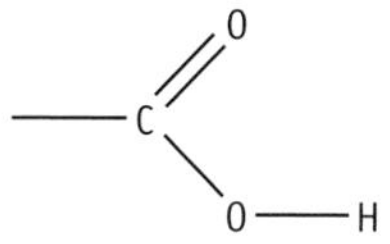

This is called the **carboxyl** group and consists of the *carb*onyl group, C=O, and the hydr*oxyl* group, OH.

The names of some of the members of this homologous series, together with their melting and boiling temperatures, are given in Table 8.2.

Table 8.2 *Carboxylic acids*

Formula	Name	Melting temperature/°C	Boiling temperature/°C	K_a/mol dm^{-3}
HCOOH	Methanoic acid	8	101	1.78×10^{-4}
CH_3COOH	Ethanoic acid	17	118	1.75×10^{-5}
CH_3CH_2COOH	Propanoic acid	−21	141	1.39×10^{-5}
$CH_3CH_2CH_2COOH$	Butanoic acid	−7	164	1.51×10^{-5}
$CH_3CH(CH_3)COOH$	Methylpropanoic acid	−47	154	1.41×10^{-5}

Never write the formula of butanoic acid as C_3H_7COOH because this could also represent its isomer, methyl-propanoic acid.

The melting temperatures of carboxylic acids do not fit the usual pattern of an increase with the number of electrons in the molecule. The reason for this is that with methanoic and ethanoic acids there is considerable hydrogen bonding. Pairs of acid molecules (dimers) are formed:

```
      O----H——O
     //         \
R——C             C——R
     \         //
      O——H----O
```

Methanoic acid is also called formic acid; ethanoic acid is also called acetic acid.

This occurs to a lesser extent with acids that have more carbon atoms, as the hydrogen bonding is inhibited by the zigzag chains of carbon atoms.

When the solid acid is melted, many hydrogen bonds are broken and so the boiling temperatures of the acids have the pattern that is expected from the increasing number of electrons: methanoic, 24 electrons; ethanoic, 32 electrons; propanoic, 40 electrons and so on. Methylpropanoic acid has fewer points of contact between adjacent molecules than does the straight chain butanoic acid, so the dispersion (temporary induced dipole–induced dipole) forces are weaker. This means that its melting and boiling temperatures are lower than those of butanoic acid, which is as expected with a branched-chain compound.

Names of carboxylic acids

The names of carboxylic acids are derived from the number of carbon atoms in the chain, *including* the carbon atom of the –COOH group. This carbon atom is regarded as the first carbon atom. Therefore, $ClCH_2CH_2COOH$ is called 3-chloro-propanoic acid and $CH_3CH(OH)COOH$ is called 2-hydroxypropanoic acid.

The old name for 2-hydroxypropanoic acid is lactic acid, which is still used in biochemistry.

Worked example

Name the compound that has the formula $CH_3CHClCH(CH_3)CH{=}CHCOOH$.

Answer

The carbon chain is six carbon atoms long (hex-). There is:

- a double bond on carbon number 2
- a $-CH_3$ group on carbon number 4
- a chlorine atom on carbon number 5

Its name is 5-chloro-4–methylhex-2-enoic acid.

Some organic acids have two –COOH groups. Examples are:

- HOOCCOOH (sometimes written $H_2C_2O_4$), which is ethanedioic acid
- HOOCCH=CHCOOH which exists as two geometric isomers, *cis*-butenedioic acid and *trans*-butenedioic acid

Aromatic acids contain a benzene ring.

The formula of benzenecarboxylic acid (benzoic acid) is:

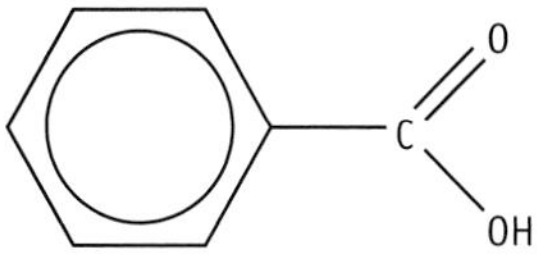

The formula of benzene-1,4-dicarboxylic acid (terephthalic acid) is:

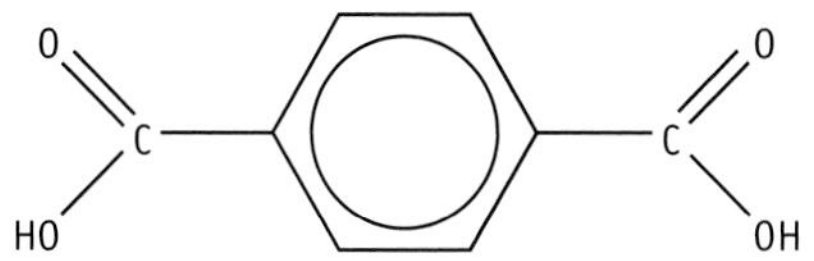

Physical properties

- The acids in the homologous series $C_nH_{2n+1}COOH$ are liquids up to $n = 9$.
- All carboxylic acids have strong smells. Vinegar is a dilute solution of ethanoic acid. The smell of rancid butter comes from butanoic acid formed by the action of bacteria on butter fat.
- Hydrogen bonding with water molecules enables those carboxylic acids with only a few carbon atoms to dissolve in water. Methanoic acid and ethanoic acid mix with water in all proportions. The solubility in water decreases as the hydrocarbon chain lengthens. This is because the chain is hydrophobic. Conversely, solubility in lipids increases as the chain lengthens.

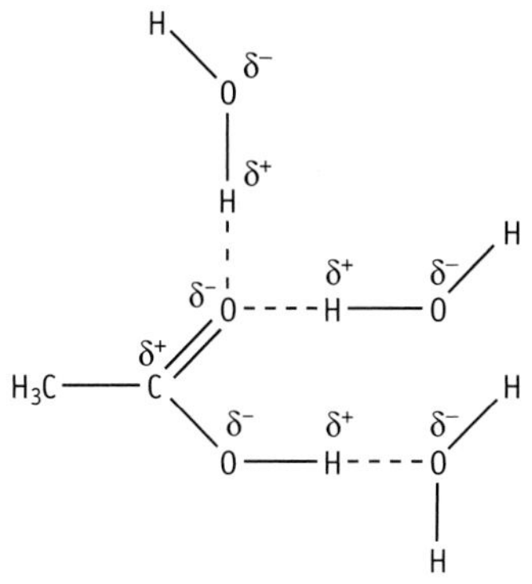

- Benzoic acid, C_6H_5COOH, is a solid that melts at 122°C. It is sparingly soluble in cold water and soluble in hot water. This makes water a suitable solvent for purifying benzoic acid by recrystallisation (p. 295).

Preparation

Carboxylic acids can be prepared by:

- the oxidation of a primary alcohol. When heated under reflux with acidified potassium dichromate(VI), the primary alcohol is oxidised to a carboxylic acid. For example, propan-1-ol is oxidised to propanoic acid:

  $CH_3CH_2CH_2OH + 2[O] \rightarrow CH_3CH_2COOH + H_2O$
- the oxidation of an aldehyde. Aldehydes are oxidised to carboxylic acids by the same reagent and under the same conditions as primary alcohols, for example:

  $CH_3CHO + [O] \rightarrow CH_3COOH$
- the hydrolysis of an ester (p. 152)
- the hydrolysis of a nitrile. Nitriles are compounds that contain the C≡N group. When heated under reflux with dilute acid, a nitrile is hydrolysed to a carboxylic acid and an ammonium salt, for example:

  $CH_3CN + H^+ + 2H_2O \rightarrow CH_3COOH + NH_4^+$

  If heated under reflux with aqueous alkali, the salt of the carboxylic acid is formed. Subsequent addition of a strong acid, such as dilute sulfuric acid, protonates the carboxylate ion and the carboxylic acid is formed.

  $CH_3CN + OH^- + H_2O \rightarrow CH_3COO^- + NH_3$ then

  $CH_3COO^- + H^+ \rightarrow CH_3COOH$

  Hydroxynitriles (p. 136) are hydrolysed to hydroxyacids in a similar way, for example:

  $CH_3CH(OH)CN + H^+ + 2H_2O \rightarrow CH_3CH(OH)COOH + NH_4^+$

  2-hydroxypropanenitrile → 2-hydroxypropanoic acid (lactic acid)
- the iodoform reaction of a methyl ketone or a secondary alcohol (p. 140). On addition of a solution of iodine in aqueous sodium hydroxide to a methyl ketone (such as butanone, $CH_3CH_2COCH_3$), the salt of the acid with one *fewer* carbon atoms (propanoic acid, CH_3CH_2COOH) is produced. The free carboxylic acid is formed by adding excess strong acid to the mixture.

  $CH_3CH_2COCH_3 + 3I_2 + 4OH^- \rightarrow CH_3CH_2COO^- + CHI_3 + 3I^- + 3H_2O$ then

  $CH_3CH_2COO^- + H^+ \rightarrow CH_3CH_2COOH$

  Butan-2-ol, $CH_3CH_2CH(OH)CH_3$, would also produce propanoic acid in this reaction.

Vinegar is a 3% solution of ethanoic acid. It is produced from ethanol, which is made by the fermentation of grains and sugars. Ethanol is then oxidised by the oxygen in the air in a reaction catalysed by enzymes in specific bacteria.

Chemical reactions

The reactions of carboxylic acids are illustrated using ethanoic acid as the example.

As an acid

Carboxylic acids are weak acids (p. 90).

Reaction with water

Carboxylic acids react reversibly with water:

$$CH_3COOH + H_2O \rightleftharpoons CH_3COO^- + H_3O^+$$

Aqueous solutions of ethanoic acid have pH ≈ 3.

The –OH group in carboxylic acids is more acidic than the –OH group in alcohols.

There are two reasons for this. First, the C=O group pulls electrons away from the –OH group, making the hydrogen atom more δ^+ and therefore easier to remove as an H^+ ion:

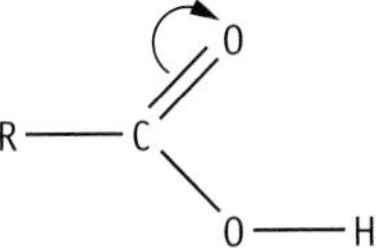

Second, the carboxylate anion has the negative charge shared between two oxygen atoms. The p_z-orbital of the carbon atom (containing one electron), the p_z-orbital of the oxygen double-bonded to it (containing one electron), and the p_z-orbital of the other oxygen (containing its one electron and the one gained by the formation of the O^- ion) all overlap. This can be shown as:

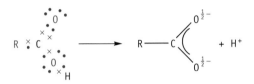

Reaction with bases

Carboxylic acids form salts with bases such as sodium hydroxide:

$$CH_3COOH + NaOH \rightarrow CH_3COONa + H_2O$$

All carboxylic acids are weak acids (see Table 8.2 for their acid dissociation constants). The concentration of a carboxylic acid can be found by titrating it with a strong base, such as sodium hydroxide, using phenolphthalein as the indicator. The amount (moles) of citric acid in lemon juice can be estimated in this way.

Reaction with carbonates and hydrogencarbonates

Carboxylic acids produce carbon dioxide when added to a carbonate:

$$2CH_3COOH(aq) + Na_2CO_3 \rightarrow 2CH_3COONa(aq) + CO_2(g) + H_2O(l)$$

Similarly, with hydrogencarbonate:

$$CH_3COOH(aq) + NaHCO_3(s) \rightarrow CH_3COONa(aq) + CO_2(g) + H_2O(l)$$

Esterification

Carboxylic acids do not react readily with nucleophiles. However, they do react with alcohols in the presence of concentrated sulfuric acid (as a catalyst) in a reversible reaction to form esters:

$$\text{acid} + \text{alcohol} \rightleftharpoons \text{ester} + \text{water}$$

For example:

$$\underset{\text{ethanoic acid}}{CH_3COOH} + \underset{\text{ethanol}}{C_2H_5OH} \overset{H_2SO_4(l)}{\rightleftharpoons} \underset{\text{ethyl ethanoate}}{CH_3COOC_2H_5} + H_2O$$

The catalyst acts by protonating the ethanoic acid, which then loses a water molecule:

$$CH_3COOH + H_2SO_4 \rightarrow HSO_4^- + CH_3COOH_2^+ \rightarrow CH_3C^+O + H_2O$$

The lone pair of electrons on the oxygen of the alcohol bonds to the positive carbon atom in the CH_3C^+O ion and then an H^+ is removed, thus reforming the catalyst:

$$C_2H_5OH + CH_3C^+O \rightarrow CH_3COOC_2H_5 + H^+$$
$$H^+ + HSO_4^- \rightarrow H_2SO_4$$

Methyl butanoate is prepared by the reaction of butanoic acid with methanol in the presence of a few drops of concentrated sulfuric acid:

$$CH_3CH_2CH_2COOH + CH_3OH \rightleftharpoons CH_3CH_2CH_2COOCH_3 + H_2O$$

Methyl butanoate, an ester, smells of pineapples and is used to flavour sweets.

This can be shown experimentally:

- Add a few drops of concentrated sulfuric acid to a mixture of butanoic acid and methanol.
- Warm for several minutes in a water bath.
- Pour the contents into a beaker of dilute sodium hydrogencarbonate solution.

The smell of pineapples can be noticed readily.

e When carrying out this experiment, care must be taken not to get any butanoic acid on your hands. Its odour is extremely unpleasant and lasts for days. Therefore, protective gloves must be worn.

Reduction

Carboxylic acids are reduced by lithium tetrahydridoaluminate(III), $LiAlH_4$, dissolved in dry ether (ethoxyethane). The H^- ion in the AlH_4^- is a powerful nucleophile. It adds on to the δ^+ carbon atom in the –COOH group. A series of reactions takes place and the final product has to be hydrolysed by dilute acid. The carboxylic acid is reduced to a primary alcohol. The overall equation is:

$$CH_3COOH + 4[H] \rightarrow CH_3CH_2OH + H_2O$$

Carboxylic acids are *not* reduced by hydrogen gas in the presence of a catalyst of nickel or platinum, unlike alkenes, which are reduced to alkanes.

$$CH_2{=}CHCOOH \xrightarrow{LiAlH_4} CH_2{=}CHCH_2OH$$

$$CH_2{=}CHCOOH \xrightarrow{H_2/Pt} CH_3CH_2COOH$$

e The symbol [H] is used in organic reactions involving a reducing agent such as $LiAlH_4$, because the full equation is too complex. However, the equation must still balance, which is why 4[H] is written in this equation.

The H^- ion in $LiAlH_4$ attacks δ^+ sites. Therefore, $LiAlH_4$ will not reduce a C=C group as it has a high electron density.

Formation of acid chloride

Ethanoic acid can be converted to ethanoyl chloride by adding solid phosphorus pentachloride to the dry acid:

$$CH_3COOH + PCl_5 \rightarrow CH_3COCl + POCl_3 + HCl$$

This is a similar reaction to that between PCl_5 and an alcohol. In both reactions, the –OH group is replaced by a chlorine atom and clouds of misty fumes of hydrogen chloride are given off.

The conversion can also be performed using thionyl chloride, $SOCl_2$:

$$RCOOH + SOCl_2 \rightarrow RCOCl + SO_2 + HCl$$

The advantage of this reagent is that the inorganic products are gases, so only the acid chloride is left in the reaction flask.

Formation of a halogenoacid

If chlorine is bubbled into a boiling carboxylic acid in the presence of sunlight or ultraviolet (UV) light, a chlorine atom replaces one of the hydrogen atoms in the alkyl chain:

$$CH_3COOH + Cl_2 \rightarrow CH_2ClCOOH + HCl$$

This is an example of free-radical substitution. It is similar to the reaction of methane with chlorine in the presence of UV light.

If propanoic acid is used in place of ethanoic acid, the product is 2-chloropropanoic acid, not 3-chloropropanoic acid:

$$CH_3CH_2COOH + Cl_2 \rightarrow CH_3CHClCOOH + HCl$$

Summary

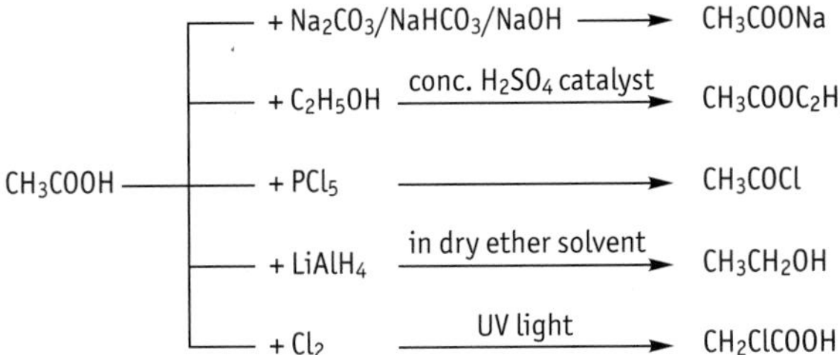

Esters

Esters have the structural formula:

$$R-C(=O)-O-R'$$

The group R can be a hydrogen atom or an organic residue, such as $-CH_3$. R′ must be a residue of an alcohol with a carbon atom attached to the oxygen of the ester linkage.

Esters are named after the alcohol residue, R′, followed by the name of the carboxylate group, RCOO:

- $HCOOCH_3$ is methyl methanoate.
- $CH_3COOCH_2CH_2CH_3$ is called 1-propyl ethanoate.

The prefix '1-' indicates the position of the ester linkage in the propyl chain.

e Note that the formula of an ester is written with the acid-derived group first, then the stem of the alcohol.

Physical properties

Most esters are liquids at room temperature. The names and boiling points of the first four members of the series of esters derived from ethanoic acid are shown in Table 8.3.

***Table 8.3** Esters derived from ethanoic acid*

Name	Formula	Boiling temperature/°C
Methyl ethanoate	CH_3COOCH_3	57
Ethyl ethanoate	$CH_3COOC_2H_5$	77
1-propyl ethanoate	$CH_3COOCH_2CH_2CH_3$	102
2-propyl ethanoate	$CH_3COOCH(CH_3)_2$	93

Solubility

Despite being polar molecules, all esters are insoluble in water. The reason for this is that they cannot form hydrogen bonds with water molecules because they do not have any δ^+ hydrogen atoms and the δ^- oxygen atoms are sterically hindered, preventing close approach by water molecules.

Preparation

Esters can be prepared by warming an alcohol and a carboxylic acid under reflux with a few drops of concentrated sulfuric acid:

$$C_2H_5OH + CH_3COOH \rightleftharpoons CH_3COOC_2H_5 + H_2O$$

A higher yield is obtained if the alcohol is reacted with an acid chloride at room temperature, because the reaction is not reversible:

$$C_2H_5OH + CH_3COCl \rightarrow CH_3COOC_2H_5 + HCl$$

Chemical reactions

Hydrolysis

Esters are hydrolysed when heated under reflux with either aqueous acid or aqueous alkali:

$$CH_3COOC_2H_5 + H_2O \overset{H^+(aq)}{\rightleftharpoons} CH_3COOH + C_2H_5OH$$

$$CH_3COOC_2H_5 + NaOH \rightarrow CH_3COONa + C_2H_5OH$$

Note the difference between these two reactions. In the first, the acid is a catalyst and the reaction is reversible. Therefore, the yield of acid and alcohol is low.

The hydrolysis with aqueous alkali is not reversible, so there is a good yield of the salt of the carboxylic acid and the alcohol. If the organic acid is required, the solution is cooled and excess dilute strong acid, such as hydrochloric or sulfuric, is added:

$$CH_3COONa + HCl \rightarrow CH_3COOH + NaCl$$

Soaps are produced by the hydrolysis of natural esters such as vegetable oils and animal fats. This is achievd by heating the fats/oils with aqueous sodium hydroxide.

Transesterification

There are two types of transesterification reaction.

This reaction is also called **interesterification.**

Reaction with another organic acid

The acid part of the ester is replaced by the acid reactant. The simplest example is the reaction between ethyl ethanoate and methanoic acid in the presence of an acid catalyst. The products are ethyl methanoate and ethanoic acid:

$$CH_3COOC_2H_5 + HCOOH \rightleftharpoons HCOOC_2H_5 + CH_3COOH$$

This type of reaction is used in the manufacture of low-fat margarine (see p. 153) where the incoming acid is saturated and the product acid is unsaturated.

Reaction with another alcohol

The alcohol part of the ester is replaced by the alcohol reactant. A simple example is the reaction between ethyl ethanoate and methanol in the presence of an acid catalyst. The products are methyl ethanoate and ethanol:

$$CH_3COOC_2H_5 + CH_3OH \rightleftharpoons CH_3COOCH_3 + C_2H_5OH$$

This type of reaction is used in the manufacture of biodiesel (see p. 154) where the reactant alcohol is methanol or ethanol and the product alcohol is propane-1,2,3-triol.

Natural esters

Animal fats and vegetable oils are examples of triglycerides. These are esters formed from acids with long hydrocarbon chains (fatty acids) and the alcohol propane-1,2,3-triol (glycerol).

In both animal fats and vegetable oils, the three fatty acids that form the ester links with propane-1,2,3-triol are usually different — some being fully saturated, some monounsaturated and some polyunsaturated. One thing they have in common is that they contain an even number of carbon atoms. Stearic acid, $C_{17}H_{35}COOH$ is a saturated fatty acid present in animal fats; it is never found on the middle carbon of the triglyceride.

The percentages of the three different kinds of fatty acid in a variety of fats and oils are given in Table 8.4.

***Table 8.4** Percentages of the different types of fatty acid present in some fats and oils*

Source	% polyunsaturated	% monounsaturated	% saturated
Soya oil	62	23	15
Rapeseed oil	32	58	10
Sunflower oil	66	23	11
Olive oil	10	76	14
Butter	7	36	57
Spreading margarine	20–40	40–60	15–20

Vegetable oils contain between 80 and 90% unsaturated fat, whereas butter contains approximately 45% unsaturated fat.

Spreading margarine

Vegetable oils can be converted into a semi-liquid form, which is put into tubs and sold as spreading or low-fat margarine. This can be carried out by two methods:

- **Transesterification** — a catalyst (either inorganic or an enzyme) is added with some stearic acid to the vegatable oil. This takes the place of one of the unsaturated fatty acids. By a process of controlled crystallisation, the harder (less unsaturated) triglycerides crystallise first and are separated off. Stearic acid is used because it does not affect the low-density lipoprotein ('bad cholesterol') levels in the blood.
  Advantage: no *trans* fatty acids are produced.
  Disadvantage: some of the stearic acid goes onto the middle carbon position of the triol. This is thought to be slightly harmful to health.
- **Partial hydrogenation** — a controlled amount of hydrogen is added to the oil in the presence of a catalyst (normally nickel). Some of the double bonds are saturated and a harder fat is produced.

Advantage: there is no interference with the order of the fatty acids on the triol.

Disadvantage: the C=C is weakened as it bonds to the active sites on the catalyst. On desorbtion, some of the molecules rotate around the remaining weakened double bonds and some *trans* isomers are formed. These are believed to increase cholesterol levels in the blood.

The altered oils, made by either method, are then mixed with untreated vegetable oils and with lipid-soluble additives such as vitamins, colouring agents and emulsifiers. This mixture is then blended with water-soluble additives such as milk whey, milk proteins and salt. The result is a spreadable margarine.

Omega-3 and omega-6

Omega-3 oils have one carbon–carbon double bond between the third and fourth carbon atoms, counting from the methyl end. If it is polyunsaturated, the other double bonds are between the sixth and seventh, and ninth and tenth, carbon atoms (perversely counting from the other end from the IUPAC system).

One such example is α-linolenic acid, 9*Z*,12*Z*,15*Z*-octadeca-9,12,15-trienoic acid, $C_{17}H_{29}COOH$ or $HOOC(CH_2)_7CH{=}CHCH_2CH{=}CHCH_2CH{=}CHCH_2CH_3$

Omega-3 acids are found in fish oils and, to a lesser extent, in some vegetable and nut oils.

Omega-6 acids have the first double bond between the sixth and seventh carbon atoms from the methyl end, e.g. linoleic acid, 9*Z*,12*Z*-octadeca-9,12-dienoic acid.

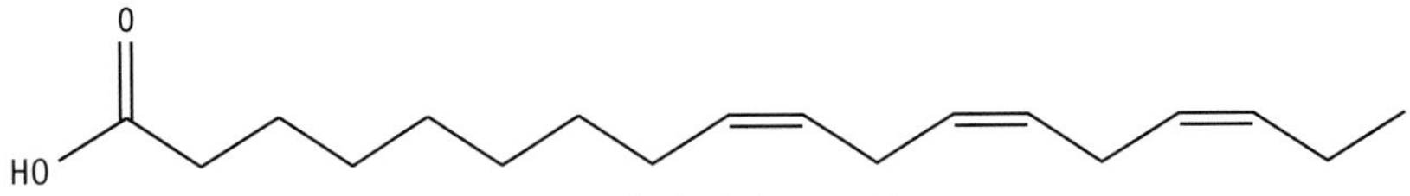

O

HO

α-linolenic (omega-3)

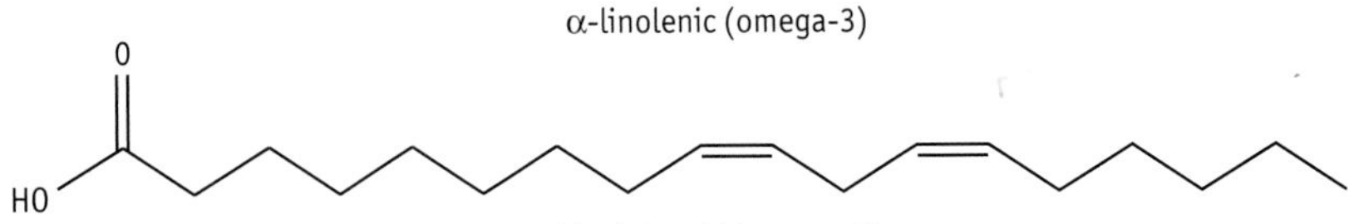

O

HO

Linoleic acid (omega-6)

Biodiesel

Biodiesel is made from natural vegetable oils. One such is rapeseed oil, 1000 kg of which is obtained per hectare of crop. Another source of biodiesel is the seeds from the *Jatropha* tree. This grows on poor soil and so will not use land that could be used for food production. Research is taking place to develop the production of natural oils from algae. The potential yield is thought to be as high as 40 000 kg ha^{-1} and waste domestic water could be used as a growing medium.

Rudolf Diesel first demonstrated his diesel engine using peanut oil, but vegetable oils are not good as fuels for these engines, as they tend to clog the fuel injection nozzles. A more volatile liquid is needed. This is obtained by transesterification. Unlike the process used in making margarine, the natural oil ester is mixed with methanol and a catalyst. Transesterification takes place forming the methyl esters of the fatty acids present in the vegetable oils.

$$\begin{array}{l} RCOOCH_2 \\ | \\ R'COOCH \\ | \\ R''COOCH_2 \end{array} + 3CH_3OH \longrightarrow RCOOCH_3 + R'COOCH_3 + R''COOCH_3 + \begin{array}{l} CH_2OH \\ | \\ CHOH \\ | \\ CH_2OH \end{array}$$

In words, this equation is:

**glycerol esters of three fatty acids + three molecules of methanol →
glycerol + methyl esters of the three fatty acids**

McDonald's is hoping to fuel its entire delivery fleet of lorries with biodiesel made from old frying oil

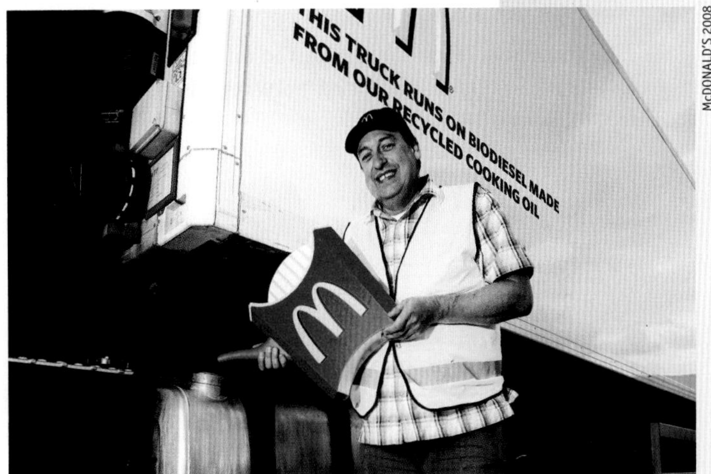

Saponification

When a fat or oil is heated with concentrated sodium hydroxide solution, it is hydrolysed to a mixture of the sodium salts of the fatty acids present in the natural esters in the fat or oil. The equation is often simplified, with the three fatty acids being identical. This is not the case, as no fat has esters derived from, for example, three stearic acid molecules. A more typical equation is:

$$\begin{array}{l} C_{17}H_{35}COOCH_2 \\ | \\ C_{17}H_{31}COOCH \\ | \\ C_{17}H_{35}COOCH_2 \end{array} + 3NaOH \longrightarrow 2C_{17}H_{35}COONa + C_{17}H_{31}COONa + \begin{array}{l} CH_2OH \\ | \\ CHOH \\ | \\ CH_2OH \end{array}$$

Uses of esters

Esters are used as:

- Perfumes and flavourings. Perfumes are complex mixtures of esters and ketones. Simple esters are used as pineapple, pear and other fruit flavours.
- Solvents. Nail varnish contains some ethyl ethanoate, as the solvent of the varnish.
- Anaesthetics. Benzocaine and procaine are esters and are used as local anaesthetics, the latter in dentistry.
- Biofuels. Methyl esters of long-chain acids are used as biofuels.

Polyesters

Polyesters are condensation polymers.

> A condensation polymer is formed when monomers join together with the elimination of a simple molecule, such as water or hydrogen chloride.

The monomers that condense must have two groups, one at each end of the molecule. For example, a dicarboxylic acid has two –COOH groups, a diacid chloride has two –COCl groups and a diol has two –OH groups in the molecule.

A polyester is formed if a dicarboxylic acid, such as benzene-1,4-dicarboxylic acid (terephthalic acid) reacts with a diol, such as ethane-1,2-diol, CH_2OHCH_2OH. One of the –COOH groups in the dicarboxylic acid reacts with one of the –OH groups in the diol. The remaining –OH group in the diol then reacts with one of the –COOH groups in a second diacid molecule. The remaining –COOH group then reacts with the –OH in a second diol molecule and so on, thousands of times.

The reaction also takes place with the dimethyl ester or a diacid chloride. An example is the formation of the polymer Terylene® or PET (**P**oly **E**thylene **T**erephthalate):

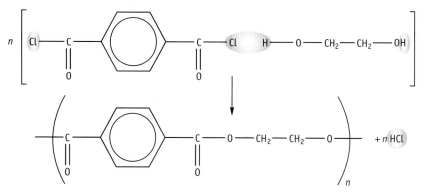

$$n\left[\text{Cl}-\overset{\displaystyle O}{\overset{\|}{C}}-C_6H_4-\overset{\displaystyle O}{\overset{\|}{C}}-\text{Cl}\quad \text{H}-\text{O}-\text{CH}_2-\text{CH}_2-\text{OH}\right]$$

$$\downarrow$$

$$\left(-\overset{\displaystyle O}{\overset{\|}{C}}-C_6H_4-\overset{\displaystyle O}{\overset{\|}{C}}-\text{O}-\text{CH}_2-\text{CH}_2-\text{O}-\right)_n + n\,\text{HCl}$$

Uses of polyesters

Polyesters, such as Terylene, are used in the manufacture of synthetic fibres. Some shirts, sheets, socks and trousers are made from a mix of polyester and a natural fibre such as cotton or wool. The polyester gives the material strength and crease resistance; the natural fibre gives a softer feel and allows the material to absorb some perspiration from the wearer.

Polyesters are also excellent thermal insulators and can be used as fillings for duvets.

PET can be extruded into different shapes and is used to make bottles for fizzy drinks and water. The bottle does not allow the dissolved carbon dioxide to escape as gas and the material is shatterproof, so will not break if dropped.

Polyesters have a wide range of uses

Disposal of polyesters

Polyesters are not biodegradable because enzymes have not evolved that hydrolyse the ester linkages in the artificial fibres. This means that they do not rot in a landfill site. If burnt, toxic fumes are produced if the temperature and amount of air are not closely controlled.

Homopolymeric esters

A **homopolymer** is a polymer made from a single type of monomer. This monomer must have a group at one end that can react with the group at the other end of the molecule. Hydroxyacids are examples.

Polylactic acid

Lactic acid (2-hydropropanoic acid) can be made by the bacterial fermentation of cornstarch. It is dimerised and finally polymerised:

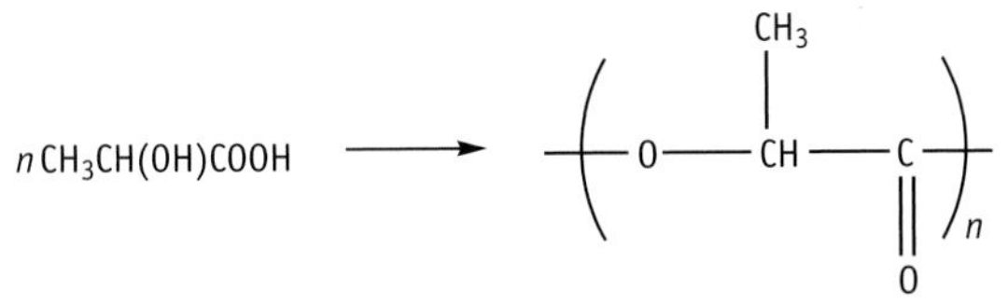

$$n\,CH_3CH(OH)COOH \longrightarrow \left(O - \underset{}{\overset{CH_3}{\overset{|}{C}H}} - \underset{\underset{O}{\|}}{C} \right)_n$$

The polymer is fully biodegradable, but is less good as a fibre than Terylene.

Biopol

3-hydroxybutanoate, $CH_3CH(OH)CH_2COOH$ is produced as an energy store by the microorganism *Alcaligenes*. After extraction, it is polymerised catalytically:

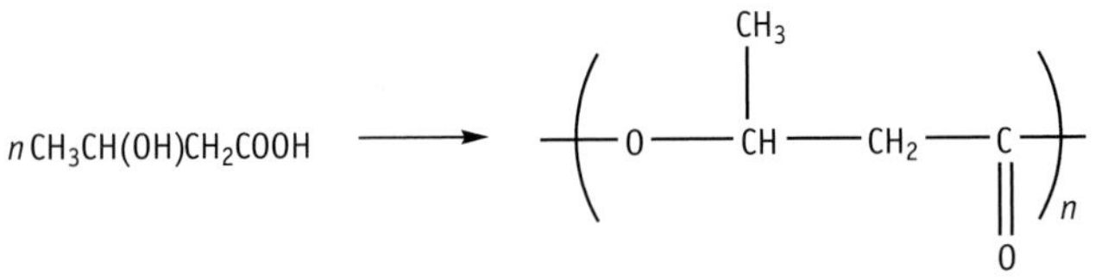

$$n\,CH_3CH(OH)CH_2COOH \longrightarrow \left(O - \overset{CH_3}{\overset{|}{C}H} - CH_2 - \underset{\underset{O}{\|}}{C} \right)_n$$

This polymer is known as Biopol and is fully biodegradable. Greenpeace uses Biopol to make its credit cards. Some bottles containing hair lotions are made of Biopol, as is some disposable plastic cutlery. Its cost and lack of tensile strength makes it, as yet, unsuitable for major use.

Acid chlorides

These are also known as acyl chlorides and have the functional group:

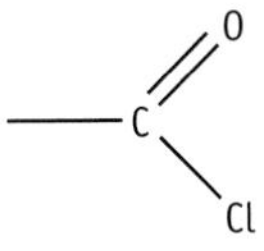

$$-C\begin{matrix} \nearrow O \\ \searrow Cl \end{matrix}$$

Physical properties

- Most acid chlorides are volatile liquids at room temperature.
- They are soluble in several organic solvents, but react with water.

The names, formulae and boiling temperatures of some acid chlorides are given in Table 8.5.

Methanoyl chloride has never been isolated as it decomposes at room temperature into carbon monoxide and hydrogen chloride.

Name	Formula	Boiling temperature/°C
Ethanoyl chloride	CH_3COCl	51
Propanoyl chloride	CH_3CH_2COCl	80
Butanoyl chloride	$CH_3CH_2CH_2COCl$	102
Benzoyl chloride	C_6H_5COCl	197

Table 8.5
Some acid chlorides

Preparation

Acid chlorides are prepared from carboxylic acids. The –OH group is replaced by a chlorine atom. The reagents that can be used are phosphorus pentachloride, phosphorus trichloride or thionyl chloride. Starting from ethanoic acid, the equations are:

$$CH_3COOH + PCl_5 \rightarrow CH_3COCl + POCl_3 + HCl$$
$$3CH_3COOH + PCl_3 \rightarrow 3CH_3COCl + H_3PO_3$$
$$CH_3COOH + SOCl_2 \rightarrow CH_3COCl + SO_2 + HCl$$

The organic product is ethanoyl chloride.

Chemical reactions

In acid chlorides, the carbon atom of the C=O group is very δ^+ as the carbon atom is joined to two highly electronegative atoms; the C–Cl bond is not very strong and Cl^- is a good leaving group. Taken together, this makes the molecule very susceptible to nucleophilic attack, in an addition–elimination reaction. The end result is a **substitution** in which the nucleophile replaces the chlorine atom.

Acid chlorides react much more quickly than carboxylic acids. Because of the hydrogen of the –OH group, the carbon in carboxylic acids is less δ^+ than the carbon in acid chlorides. The C–OH bond is stronger than the C–Cl bond and OH^- is a less good leaving group than Cl^-.

Reaction with water

Ethanoyl chloride reacts vigorously with water, forming ethanoic acid and clouds of hydrogen chloride gas:

$$CH_3COCl + H_2O \rightarrow CH_3COOH + HCl$$

The first step is the addition of water. The oxygen atom in water has a lone pair of electrons. This makes it a nucleophile. The lone pair forms a bond with the δ^+ carbon atom in the C=O group. Simultaneously, the π-bond breaks and the C=O oxygen becomes O^-:

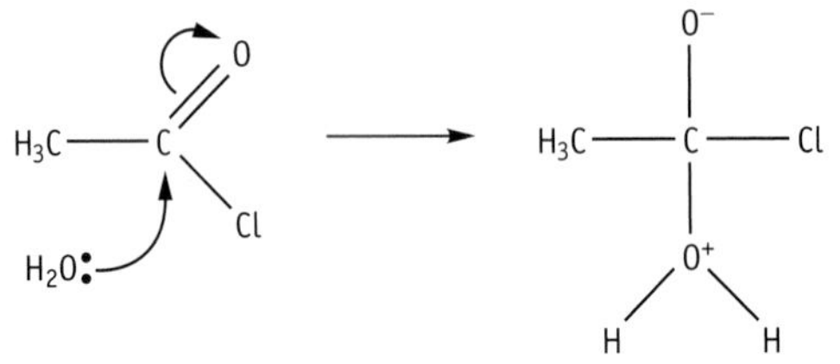

H_3C—C(=O)—Cl + H_2O: → H_3C—C(O^-)(O^+H_2)—Cl

This is followed by the loss of H^+ and Cl^- as gaseous HCl:

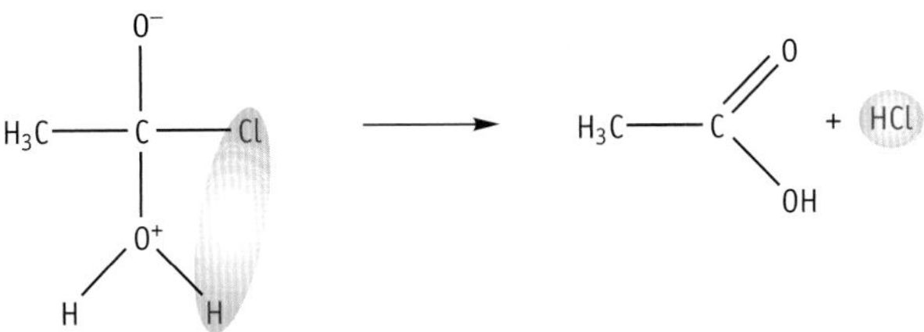

This rapid reaction with water means that when using ethanoyl chloride as a reactant, all reagents and glassware must be dry.

Benzoyl chloride is less reactive than ethanoyl chloride and reacts only slowly with water. Therefore, it can be used with reagents in aqueous solution.

Reaction with alcohols

Ethanoyl chloride reacts rapidly in a non-reversible reaction with alcohols. The products are an ester and misty fumes of hydrogen chloride vapour.

$$\underset{\text{ethanoyl chloride}}{CH_3COCl} + \underset{\text{methanol}}{CH_3OH} \rightarrow \underset{\text{methyl ethanoate}}{CH_3COOCH_3} + HCl$$

The mechanism of this reaction is analogous to that of the reaction with water.

The alcohol must be dry, or the acid chloride will be hydrolysed by the water.

As this is a non-reversible reaction with a good yield, it is a more efficient method of making an ester than reacting the alcohol with a carboxylic acid. The latter reaction is reversible with a maximum possible yield of under 70%.

Reaction with ammonia

The nitrogen atom in ammonia has a lone pair of electrons, which is used in a nucleophilic attack on the C=O group. The product is an amide (p. 267), for example:

$$\underset{\text{ethanoyl chloride}}{CH_3COCl} + NH_3 \rightarrow \underset{\text{ethanamide}}{CH_3CONH_2} + HCl$$

With excess ammonia, the hydrogen chloride forms a white smoke of ammonium chloride.

$$HCl + NH_3 \rightarrow NH_4Cl$$

Reaction with amines

Amines also have a lone pair of electrons on the nitrogen atom. They react with acid chlorides in the same way as ammonia. The product is a substituted amide, for example:

$$\underset{\text{ethanoyl chloride}}{CH_3COCl} + CH_3NH_2 \rightarrow \underset{\textit{N}\text{-methylethanamide}}{CH_3CONHCH_3} + HCl$$

The prefix *N*- indicates that the methyl group is attached to a nitrogen atom.

Reaction with lithium tetrahydridoaluminate(III)

Acid chlorides, like carboxylic acids and esters, are reduced by lithium tetrahydridoaluminate(III) in dry ether solution to form a primary alcohol, for example:

$$CH_3COCl + 4[H] \rightarrow CH_3CH_2OH + HCl$$

Reaction with aromatic compounds

Ethanoyl chloride reacts with benzene and with phenol (see pp. 247 and 252).

Summary

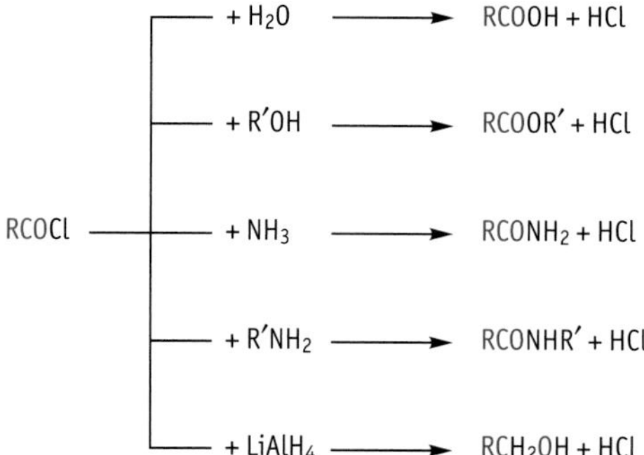

Questions

1 Name the following compounds:

a $(CH_3)_2C(OH)COOH$

b $CH_3CHClCOCl$

c $C_2H_5COOCH(CH_3)_2$

2 Explain why propanoic acid is soluble in water whereas propane is insoluble.

3 If the molar mass of ethanoic acid is measured when dissolved in an organic solvent such as benzene, it is found to be 120 g mol^{-1}. Explain this.

4 Identify the organic product of the reaction of excess phosphorus pentachloride with 2-hydroxypropanoic acid (lactic acid).

5 Identify the organic product obtained on reducing CHOCH=CHCH(OH)COOH with:

a lithium tetrahydridoaluminate(III) in dry ether followed by dilute hydrochloric acid

b hydrogen gas and a platinum catalyst

6 Write equations for the reactions of propanoic acid with:

a aqueous sodium hydroxide

b magnesium

c methanol

State the conditions necessary for the reaction between propanoic acid and methanol.

7 Aspirin can be prepared by the reaction of ethanoyl chloride with:

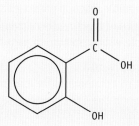

Write the structural formula of aspirin.

8 Describe how you would prepare a pure, high-yield sample of benzoic acid, C_6H_5COOH, from ethyl benzoate, $C_6H_5COOC_2H_5$.

9 Describe how you would prepare a sample of propanoyl chloride from propanoic acid.

10 Write equations for the reaction of propanoyl chloride with:

a ammonia

b water

c 2-aminoethanol, $NH_2CH_2CH_2OH$

11 Explain the meaning of the term **trans-esterification.**

12 In terms of their chemical composition, explain the difference between animal fats and vegetable oils.

13 Explain why it is preferable to use an acid chloride, rather than a carboxylic acid, when making large quantities of an ester.

14 2-hydroxypropanoic acid can be polymerised. Draw the structure of *two* repeat units of this polymer.

15 Explain why, when making a low-fat spreading margarine, transesterification is preferable to hydrogenation.

16 Describe the chemistry of converting a vegetable oil into biodiesel.

17 Give one advantage of growing *Jatropha* trees, rather than soyabeans, for the manufacture of biodiesel.

Spectroscopy and chromatography

Chapter 9

Introduction

Electromagnetic radiation consists of an oscillating electric and magnetic field of a wide range of frequencies (Table 9.1). In a vacuum, light travels at a speed of $3.00 \times 10^8\,\text{m}\,\text{s}^{-1}$ (670 000 000 miles per hour).

Frequency is measured in hertz, Hz ($1\text{Hz} = 1\,\text{s}^{-1}$). The frequency determines the colour and the energy of the light, which is calculated by the expression formulated by Max Planck:

$$E = h\nu$$

where h is Planck's constant and ν is the frequency.

The speed of light, its frequency and the wavelength are combined in the equation:

$$c = \lambda\nu$$

where c is the speed of light and λ is the wavelength.

The frequency can also be expressed as a **wavenumber**:

$$\text{wavenumber} = 1/\lambda$$

where the units of λ are cm.

Type of radiation	Frequency/MHz	Wavenumber/cm^{-1}	Energy/J per photon
X-rays	$> 10^{11}$	$> 3 \times 10^6$	$> 7 \times 10^{-17}$
Ultraviolet	1×10^9	3×10^4	7×10^{-19}
Visible — blue	6×10^8	2.1×10^4	4.2×10^{-19}
Visible — yellow	5×10^8	1.6×10^4	3.2×10^{-19}
Visible — red	4×10^8	1.4×10^4	2.8×10^{-19}
Infrared	$< 3 \times 10^8$	$< 10\,000$	$< 2 \times 10^{-19}$
Microwaves	$< 3 \times 10^4$	< 1	$< 2 \times 10^{-22}$
Radio waves	< 100	$< 3 \times 10^{-3}$	$< 7 \times 10^{-26}$

Table 9.1 *Frequency, wave-number and energy of different types of electromagnetic radiation*

When electromagnetic radiation is passed through a diffraction grating, it is split up into a spectrum according to the frequency of the radiation. Visible light is split by passing it through a prism. Blue light is refracted (bent) more than red light.

A rainbow is caused by droplets of rain splitting up white light into its component colours.

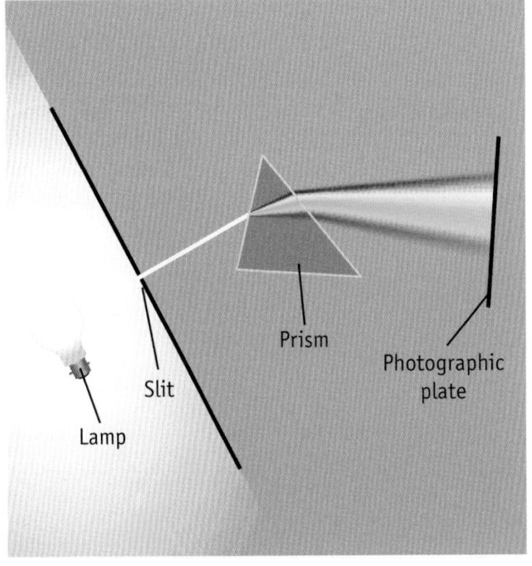

Figure 9.1 *Splitting of visible light by a prism*

A rainbow

Emission spectra

When some inorganic compounds are heated in a Bunsen flame, heat energy is converted into light energy. The heat from the Bunsen flame promotes an electron in a metal to a higher energy level. This is an unstable state. The electron spontaneously drops back to a lower level and radiation is given off. This radiation is in the form of ultraviolet or visible light.

The electron can be promoted to a number of higher levels of different energy, so the light given off is usually of several different frequencies. If this light is observed by eye, the colour is due to all these frequencies. If the light is observed in a spectrometer, bright lines are seen at different frequencies.

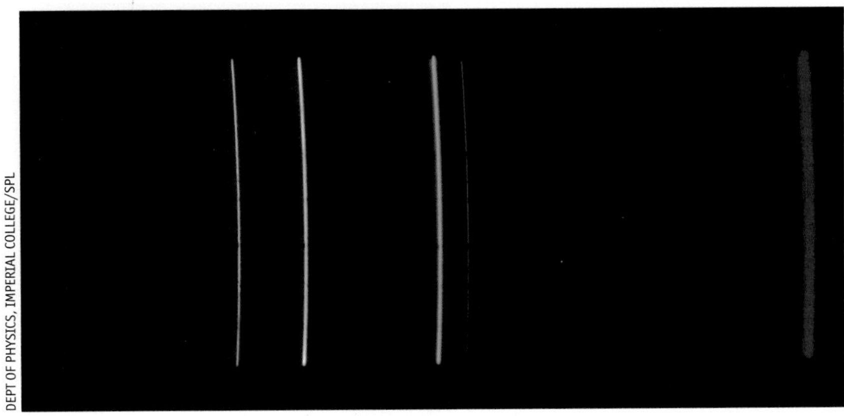

DEPT OF PHYSICS, IMPERIAL COLLEGE/SPL

Spectral lines of cadmium

The energy of a spectral line is the difference between the energy of the electron in the higher state, E_2, and that in the lower state, E_1.

$$\Delta E = E_2 - E_1 = h\nu$$

The colours given by group 1 and group 2 cations can be used to show their presence in a compound.

Table 9.2
Flame colours of group 1 and group 2 cations

Group 1	Flame colour	Group 2	Flame colour
Lithium	Red or Magenta	Calcium	Orange-red
Sodium	Yellow	Strontium	Crimson or red
Potassium	Lilac	Barium	Pale green
Rubidium	Red		

Absorption spectra

When electromagnetic radiation is passed through a substance, some of the radiation is absorbed.

X-rays

X-rays have high energy per photon. When a photon is absorbed, an inner electron is moved to an empty outer orbit.

X-rays are strongly absorbed by atoms with a high atomic number. Bones contain calcium ions and absorb some X-rays, but soft tissue is mainly carbon, oxygen and hydrogen and does not absorb X-rays. Use is made of this when taking, for example, an X-ray picture of a suspected fracture.

Barium has a large atomic number and is opaque to X-rays. Taking X-ray pictures of the human gut after the patient has eaten a meal of insoluble barium sulfate can show up blockages in the digestive system.

The state of the arteries around the heart can be analysed in a procedure known as an angiogram. A solution of a compound containing several iodine atoms is injected into the blood around the heart via a catheter introduced into an artery in the patient's groin. The heart area is then X-rayed and any narrowing of the arteries can be detected.

A heart angiogram

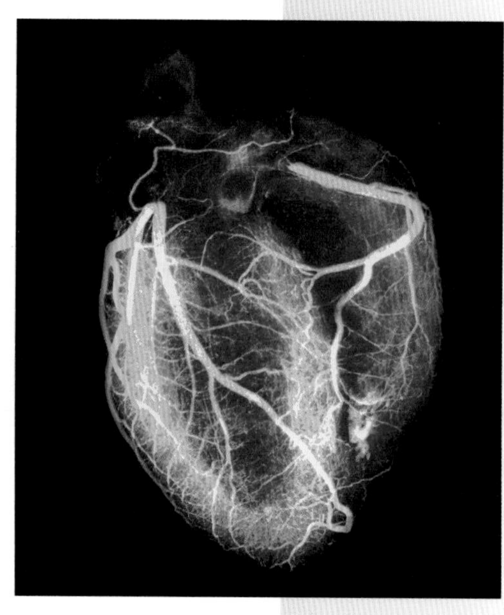

SPL

Ultraviolet (UV) radiation

UV radiation is high energy. When it is absorbed by a molecule, a bond is broken. For example, ozone absorbs UV rays that provide the energy for the reaction:

$$O_3 \rightarrow O_2 + O\bullet$$

It also causes the covalent bond in a chlorine molecule to break. However, radiation of lower energy, such as blue light, also does this.

Visible radiation

When visible light is absorbed by a molecule or ion, either a bond is broken or a bonding electron is moved to a level of higher energy.

The energy of blue light is 4.2×10^{-19} J per photon. This is sufficient to cause a chlorine molecule to break into two radicals:

$$Cl_2 \rightarrow 2Cl\bullet$$

The Cl–Cl bond energy is $242\,kJ\,mol^{-1}$ which equals 4.0×10^{-19} J per molecule. Thus blue light or higher energy radiation (UV) will homolytically split a chlorine molecule and initiate the free-radical chain substitution reaction of chlorine with alkanes.

$$C_2H_6 + Cl_2 \xrightarrow{h\nu} CH_3CH_2CH + HCl$$

Energy per bond $= \dfrac{\text{energy per mole}}{\text{Avogadro constant}} = \dfrac{242 \times 1000\,J\,mol^{-1}}{6.02 \times 10^{23}\,mol^{-1}} = 4.02 \times 10^{-19}\,J$

Complexes of transition metal ions absorb light in the visible region of the spectrum. The light energy of a particular band of frequencies is absorbed and an electron is moved from the lower level of the split *d*-orbitals to the upper level. The colour of the ion is the complementary colour to that absorbed. On collision with another molecule, the energy is released as heat. The electron returns to its lower level and is then able to absorb more light energy (see Unit 5, Chapter 11, p. 218).

Infrared spectra

Absorption of infrared radiation causes a bond to stretch or bend, thus gaining vibrational energy. A particular bend or stretch is only infrared active if it is accompanied by a change in dipole moment. Thus, oxygen and nitrogen, along with all other diatomic elements, do not absorb infrared radiation. This means that they are not greenhouse gases.

Linear carbon dioxide and tetrahedral methane are non-polar, even though they have polar bonds. This is because the individual bond dipoles cancel out due to the molecules being symmetrical. Any asymmetrical bending or stretching will cause the dipole moment to change from zero and so the molecule will then be infrared active. All vibrations in polar molecules such as water and nitric oxide result in changes in their dipole moments and so they absorb infrared radiation. This means that carbon dioxide, methane, water and nitric oxide are greenhouse gases.

Different bonds in a covalent molecule absorb radiation of different frequencies, which are normally measured as wavenumbers. An infrared spectrum usually has a range from $4000\,cm^{-1}$ to $600\,cm^{-1}$.

The frequency of the infrared radiation absorbed depends on the energy required to cause the bond to bend or stretch. Thus particular bonds have specific frequencies at which they absorb. The remainder of the molecule also has some effect, so the absorption due to the stretching of a particular bond occurs over a range of frequencies. For example, the stretching of the C–H bond in alkanes absorbs over the range $2853–2962\,cm^{-1}$, depending on the exact molecular environment of the C–H bond.

The C=O bond absorbs at around 1700 cm^{-1}, but the actual value depends on the other atoms attached to the C=O group. This is shown in Table 9.3.

Type of compound	Wavenumber/cm^{-1}
Aliphatic aldehyde	1720–1740
Aliphatic ketone	1700–1730
Aromatic aldehyde	1690–1715
Aromatic ketone	1680–1700
Carboxylic acid	1700–1725
Ester	1735–1750
Acid chloride	1815–1825
Amide	1640–1680

***Table 9.3** Absorption frequencies of different C=O groups*

Other types of bond absorb at different frequencies, but the actual value again depends on the neighbouring atoms and groups.

Bond	Functional group	Wavenumber/cm^{-1}
O–H	Alcohols (hydrogen bonded)*	3200–3600
	Alcohols (not hydrogen bonded)	3600–3700
	Carboxylic acids	2500–3300
N–H	Amines (hydrogen bonded)*	3300–3500
C–H	Alkanes	2850–3000
	Alkenes and arenes	3000–3100
C–C	Alkanes	1360–1490
C=O	See Table 9.3	
C=C	Aromatic	1450–1650
	Alkenes	1650–1700

*These peaks are very broad due to intermolecular hydrogen bonding.

***Table 9.4** Absorption frequencies of some common groups*

The region below about 1300 cm^{-1} is known as the **fingerprint region**. It shows a complex series of peaks that depends on the exact compound being analysed. Just as human fingerprints can be matched by computer to give a unique identification, so computer analysis of the fingerprint region can be used to identify a pure unknown organic substance.

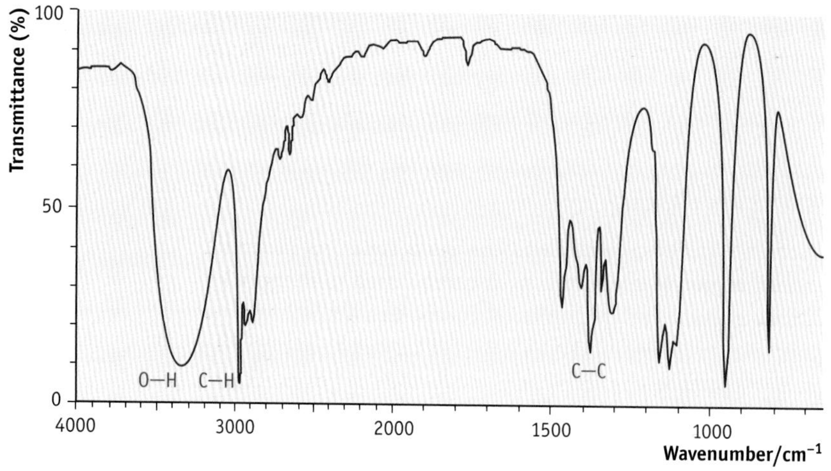

***Figure 9.2** Infrared spectrum of propan-2-ol*

In the infrared spectrum of propan-2-ol shown in Figure 9.2:

- the broad peak at around 3330 cm^{-1} is due to hydrogen-bonded O–H
- the peak at 2970 cm^{-1} is due to the C–H bond
- the peak at 1380 cm^{-1} is due to the C–C bond

Progress of reaction

Infrared spectroscopy can be used to investigate the extent of, or rate of, a chemical reaction.

Extent of reaction

The final step in the preparation of ibuprofen is:

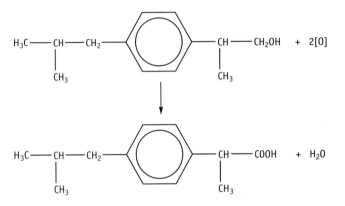

$$H_3C{-}CH(CH_3){-}CH_2{-}C_6H_4{-}CH(CH_3){-}CH_2OH \quad + \quad 2[O]$$

$$\downarrow$$

$$H_3C{-}CH(CH_3){-}CH_2{-}C_6H_4{-}CH(CH_3){-}COOH \quad + \quad H_2O$$

The extent of the reaction can be estimated by measuring the height of the peak at 1720 cm^{-1}, which is not present in the absorption spectrum of the starting primary alcohol.

Rate of reaction

The rate of the nucleophilic substitution reaction between 1-chloropropane and potassium cyanide to produce butanenitrile can be studied using an infrared spectrometer.

$$CH_3CH_2CH_2Cl + KCN \rightarrow CH_3CH_2CH_2C{\equiv}N + KCl$$

Both organic substances absorb at 2974 cm^{-1} and 1460 cm^{-1} (due to the C–H bonds) but only butanenitrile has a strong absorption peak at 2260 cm^{-1}, due to the C≡N bond. The peak height depends on the amount of butanenitrile present in the mixture. Its value is measured at regular intervals and a graph of [butanenitrile] against time plotted. Half-lives can be used to indicate the order of this reaction.

Identification of an organic compound

The infrared spectrum of a pure compound is compared with spectra in a database. The fingerprint region is specific and so can be used to identify an unknown. For example, all aliphatic alcohols absorb around 3600 cm^{-1} (O–H stretch), 2900 cm^{-1} (C–H stretch) and 1400 cm^{-1} (C–H bend), but the absorption patterns of their fingerprint regions differ. Therefore, in the infrared breathalyser, the absorption detected must be in the fingerprint region of ethanol and not at higher frequencies.

Water has a peak due to O–H; most organic substances have peaks due to C–H bonds.

Test of purity

The infrared spectrum of a pure compound should match that in the database. Any stray peaks will be due to impurities. For example, a peak at around 1700 cm^{-1} in the spectrum of an alcohol would probably be due to a carbonyl or acid impurity.

Microwaves

The household microwave oven produces oscillating radiation at a frequency of 2.45 GHz (a wavelength of 12 cm). The rotation of polar molecules, such as water, increases as they become aligned with the oscillating electric field produced by the microwaves. When the molecules collide, this rotational energy is converted into kinetic energy and the water heats up. Microwave heating is more efficient at heating water than it is at heating fats or carbohydrates, because water is more polar.

The molecules in frozen water are fixed in position and so cannot rotate, thus ice does not absorb microwaves. Defrosting in a microwave oven works because the surface of the frozen food becomes covered with a thin film of liquid water. As this heats up, it melts nearby ice, until all the food is defrosted.

Commercial microwaves operate at 0.92 GHz. The pharmaceutical industry uses microwaves to heat up batches of chemicals. Using microwave heating, rather than conventional heating, means that there is less heat loss and less chance of overheating fragile molecules.

The nitrogen and oxygen molecules in air are non-polar and do not absorb microwaves.

Radio waves (NMR)

Most materials are transparent to radio waves, which is why it is possible to have indoor television and radio aerials.

All nuclei in atoms are spinning and in some cases there is a resultant magnetic field. If the nucleus is exposed to a strong external magnetic field, the spinning nucleus aligns itself so that its own magnetic field is parallel with, or anti-parrallel to, the external field. There is a slight energy difference between these two states. Its magnitude depends on the molecular environment of the nucleus and the strength of the external magnetic field.

Nuclear magnetic resonance

If a nucleus has an odd number of protons or an odd number of neutrons and an odd mass number, its spin will produce a slight magnetic field.

The nucleus of a hydrogen atom, ^{1}H, produces a measurable magnetic field and the effect of an applied magnetic field on the nucleus of the proton can be studied.

The ^{13}C isotope produces a magnetic field and so can also be used in NMR spectroscopy. The deuterium isotope, ^{2}H, does not produce a magnetic field when spinning and so does not produce an NMR spectrum.

If an external magnetic field is applied, the magnetic moment caused by the spinning nucleus will either be aligned with the applied field or be opposed to it. The former state is of lower energy. This is shown in Figure 9.3.

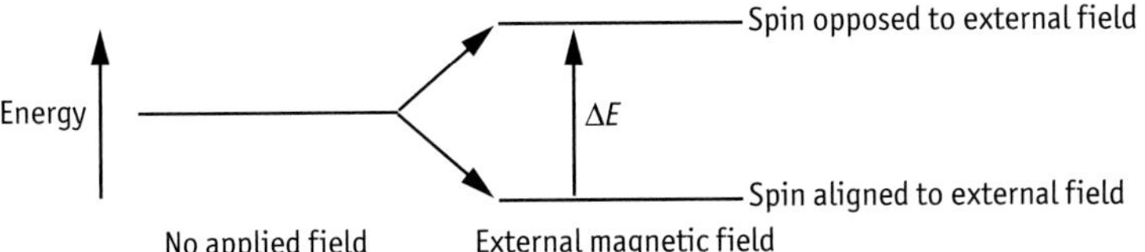

***Figure 9.3** Nuclear-spin energy levels split by an external magnetic field*

Electromagnetic radiation is absorbed as the spin of the nucleus flips from the lower-energy aligned orientation to the higher-energy opposed orientation. The frequency of the radiation that causes this is given by $v = \Delta E/h$. If a magnetic field of 15 kilogauss is applied, the splitting is such that radio waves of about 60 MHz are absorbed. The actual frequency depends on the environment of the nucleus and the strength of the applied magnetic field. Nuclei with spins that are aligned with the applied field will absorb the energy of the radiation and flip over to the spin-opposed state.

After a short time, normally of the order of a second, the nucleus spontaneously changes its spin and reverts to the lower-energy spin-aligned state giving off a particular radio frequency as it drops. The average time for this process is called the **relaxation time**.

Obtaining an NMR spectrum

The sample of the material, dissolved in a suitable solvent, is placed between the poles of a powerful electromagnet. Between the poles of the magnet there is also a radio-frequency coil that is then activated with oscillating radio waves. The absorption of these waves by the sample followed by the re-emission of the radio waves is detected and the results fed to a computer that shows the results as a trace.

The experiment can be carried out by:

- steadily altering the strength of the magnetic field and keeping the frequency of the radio waves constant
- keeping the magnetic field constant and sending pulses of radio waves of gradually increasing frequency through the sample

The extent of the splitting depends on the strength of the magnetic field. Therefore, the ΔE value of the hydrogen nuclei being investigated must be compared with that of a standard. The standard always used is the hydrogen nuclei in TMS, tetramethylsilane, $(CH_3)_4Si$.

The sample is dissolved in a solvent that does not contain any protons. Carbon tetrachloride, CCl_4, or D_2O are the usual choices. A small amount of TMS is added to provide the zero line for comparison and the sample is placed in the spectrometer.

Under these conditions, hydrogen nuclei in different chemical environments have different ΔE values. The extent to which they differ from the ΔE of the TMS protons is called the **chemical shift, δ.**

Low-resolution NMR

Low-resolution NMR investigates the value of the chemical shift and the area under the absorption peak (the peak height). The chemical shift indicates the environment of the hydrogen atom in the molecule and the peak height indicates the number of hydrogen atoms in that environment.

Chemical shift

The chemical shift, relative to the protons in TMS, depends mainly on the chemical environment of the hydrogen atom in the molecule. Some typical δ values are given in Table 9.5.

***Table 9.5** Chemical shifts of common groups containing hydrogen*

Group	δ (ppm)	Group	δ (ppm)
C–CH_3	0.8–1.2	CH_2OH	3.3–4.0
C–CH_2–C	1.1–1.5	C–C(H)(C)–OH	3.2–4.1
C–C(C)(C)–**H**	1.5	–C(**H**)–C(=O)– in aldehydes, ketones, acids, esters and amides	2.0–3.0
C–**H** on benzene ring	6.8–8.2	CHO	9.0–10.0
C=C–**H** in alkenes	4.5–6.5	COOH	10.0–12.0
C=C–CH	1.8–2.0	C–OH	1.0–6.0*

* The value of the shift for hydroxyl hydrogen atoms depends on the solvent and the extent of hydrogen bonding and, therefore, it might be difficult to identify such compounds using NMR spectra alone.

Peak height

The peak height is often indicated by a number above the peak. It can also be estimated by looking at the spectrum. Some spectra have an integration trace superimposed on the NMR spectrum. The NMR spectrum of ethoxyethane, $CH_3CH_2–O–CH_2CH_3$, with the integration trace added is shown in Figure 9.4.

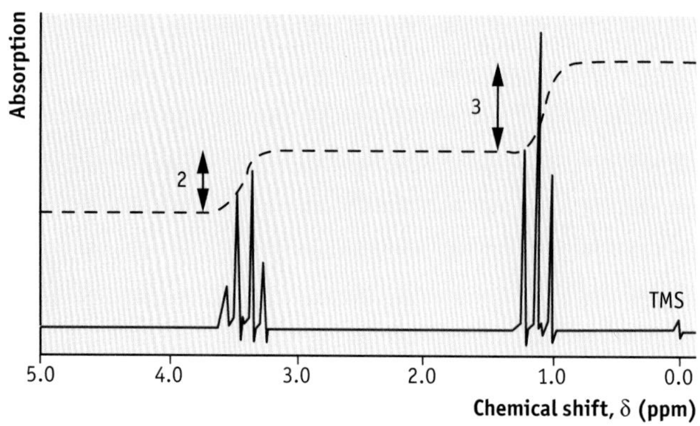

***Figure 9.4** NMR spectrum of ethoxyethane with integrated trace*

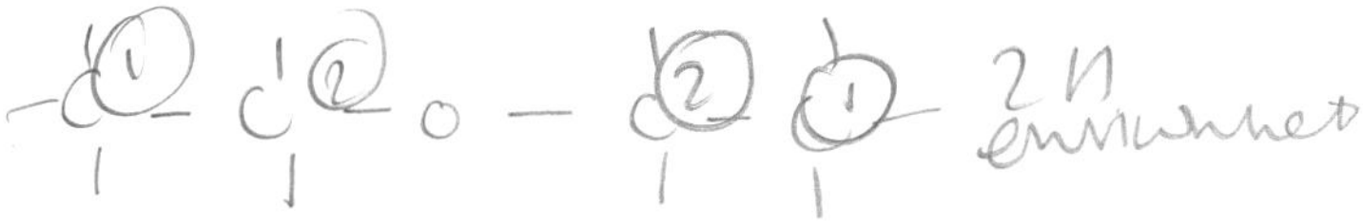

There are two different chemical environments for the hydrogen atoms — the CH_2 hydrogens and the CH_3 hydrogens. There are four CH_2 hydrogen atoms and six CH_3 hydrogen atoms, so the peaks heights are in the ratio 2:3.

The peak at $\delta = 1.0$ is caused by the hydrogen atoms in the CH_3 group; the peak at $\delta = 3.4$ is caused by the CH_2 hydrogen atoms.

The NMR spectrum of propanal is shown below:

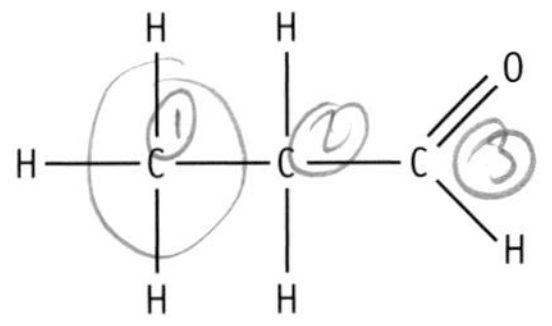

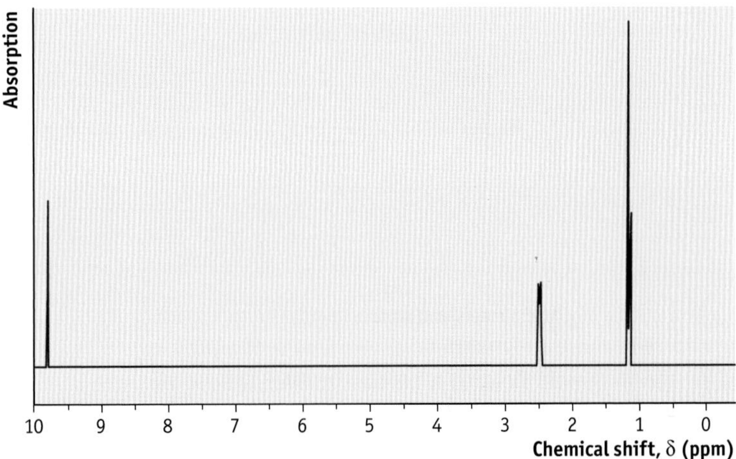

There are three peaks:

- The peak at $\delta = 9.8$ is caused by the hydrogen atom on the CHO group.
- The peak at $\delta = 2.4$ is caused by the two hydrogen atoms in the CH_2 group that is next to the CHO group.
- The peak at $\delta = 1.1$ is caused by the hydrogen atoms in the CH_3 group.

The areas under the peaks have not been calculated.

Worked example

Examine the NMR spectrum of propan-2-ol, $CH_3CH(OH)CH_3$, and identify the peaks.

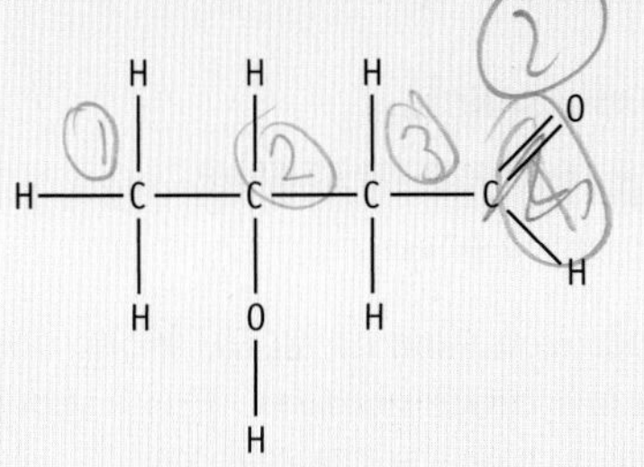

e Do not state that the chemical shift is caused by a group. It is caused by the protons in the nuclei of the hydrogen atoms in that group.

Note the splitting of the peaks in the spectrum of propan-2-ol.

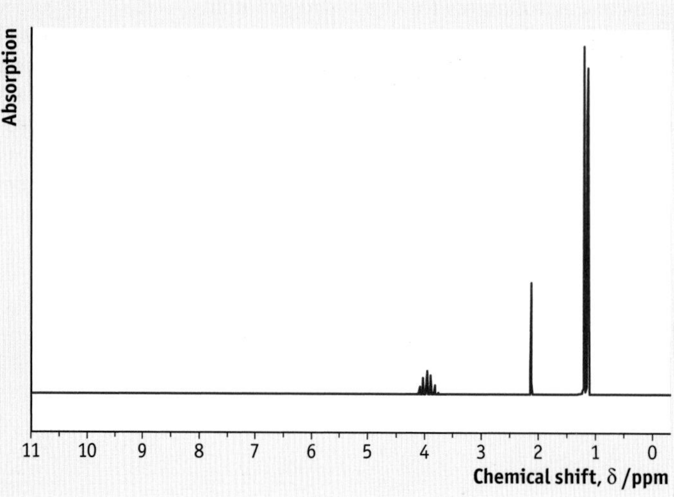

Answer

The tall peak at $\delta = 1.1$ is due to the six hydrogen atoms in the CH_3 groups.
The peak at $\delta = 4.0$ is due to the CH hydrogen on the CH(OH) group.
The peak at $\delta = 2.2$ is, therefore, caused by the hydrogen atom in the OH group.

High-resolution NMR spectra

If the spectrum is investigated with a high-resolution spectrometer, the peaks are seen to be split. Splitting occurs because the magnetic environment of a proton in one group is affected by the magnetic field of a proton in a neighbouring group. If the field of a hydrogen atom is aligned with the applied field it is reinforced and increases the field on the neighbouring hydrogen atom.

Consider the effect of a proton in a CH group on the hydrogen atoms in the neighbouring CH_2 group:

- The CH hydrogen atom can have its spin aligned ↑ or opposed ↓.
- This results in two different fields affecting the CH_2 hydrogen atoms, so their peaks are split into two.
- The CH_2 hydrogen atoms affect the field on the CH hydrogen atom. Here, the situation is more complex because there are two neighbouring hydrogen atoms and there are four ways in which their spins can be aligned — ↑↑, ↑↓, ↓↑ or ↓↓. The effect of ↑↓ is the same as that of ↓↑ so they produce one peak, twice as high as the others.This results in the CH peak splitting into three, in a 1:2:1 ratio.

This process is called spin coupling or **spin–spin splitting**.

If the proton of a hydrogen atom has *n* hydrogen atoms on neighbouring carbon atoms, its peak will be split into $(n + 1)$ sub-peaks.

The only exception to this rule is that the hydrogen atom on an OH group does not normally cause splitting, particularly if it is hydrogen-bonded. This is shown by the high-resolution NMR spectrum of propan-1-ol, $CH_3CH_2CH_2OH$:

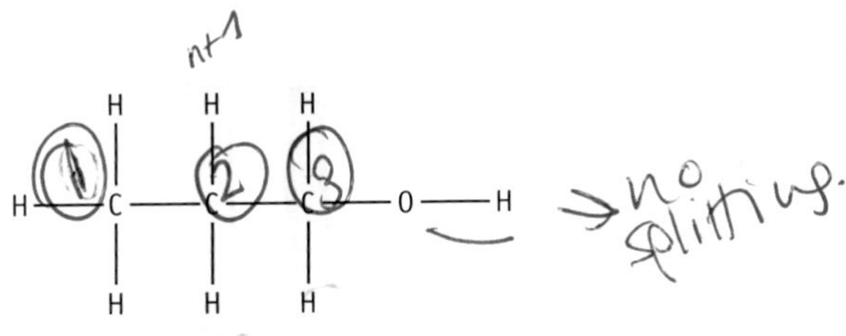

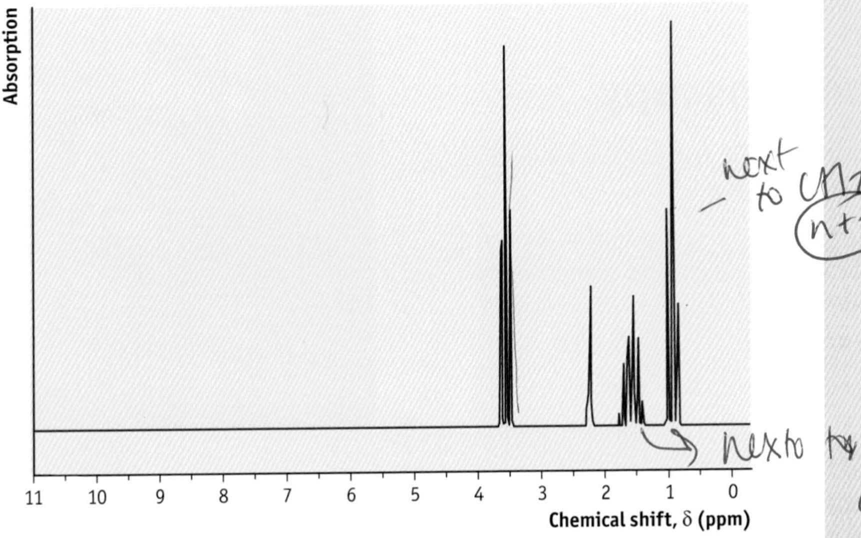

- The peak at 0.9 is due to the hydrogen atoms in the CH_3 group. It is split into three by the neighbouring two CH_2 hydrogen atoms.
- The peak at 1.5 is due to the hydrogen atoms in the middle CH_2 group. It is split into six by the five neighbouring hydrogen atoms.
- The peak at 3.5 is split into three. The peak is the caused by the two hydrogen atoms in the CH_2OH group. It is split into three by the two neighbouring hydrogen atoms.
- The peak at 2.4 is caused by the OH hydrogen. It neither causes splitting nor is split.

Examination of the splitting indicates the number of hydrogen atoms on the neighbouring carbon atoms. If a peak is split into 2, the neighbouring carbon atom has one H atom attached. If the peak is split into 5, there are four neighbouring hydrogen atoms. This means that there is either a CH_3 group and a CH group, or two CH_2 groups, as neighbours.

Look back at Figure 9.4 and try to explain the splitting in the NMR spectrum of ethoxyethane

MRI

Magnetic resonance imaging is used as a diagnostic tool for medical problems in soft tissue and in collagen in bones. The technique is slightly different and requires extremely powerful electromagnets. This is achieved by cooling the metal core of the magnet to a temperature of 4 K with liquid helium. The metal then becomes a superconductor and high electric currents can be made to flow through it without any heat being produced. This results in an extremely powerful magnetic field of 10 kilogauss. Pulses of radio waves at around 40 MHz are then radiated through the patient. The hydrogen nuclei in soft tissue have a different chemical shift to those in harder tissue. Computers can generate pictures of slices through the part of the body being investigated and, therefore, problems can be detected. The patient suffers no ill effects because the procedure is non-invasive.

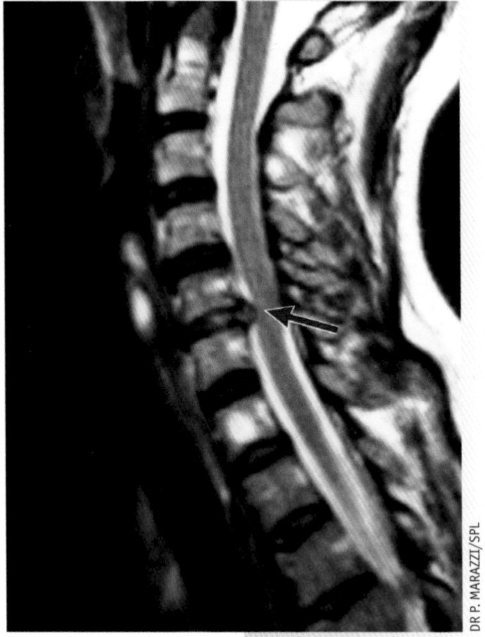

Coloured MRI scan of a patient with a slipped disc (arrowed)

Mass spectroscopy

When a molecule, M, is bombarded by high-energy electrons, it becomes ionised:

$$M(g) + e^- \rightarrow M^+(g) + 2e^-$$

The molecular ion produced might subsequently break up into a smaller positive ion, X^+, and a radical, Y•:

$$M^+(g) \rightarrow X^+(g) + Y\bullet$$

In a **mass spectrometer**, these positive ions are then accelerated by an electric field and deflected by a magnetic field.

The lighter ions are deflected more than the heavier ions and so a spectrum of ions of different masses is produced.

The peak with the largest mass/charge ratio, *m/e*, is caused by the **molecular ion**. The peaks at lower *m/e* values arise from fragments of that ion. A small peak at M + 1, where M is the molecular ion value, may be observed. This is due to the presence of ^{13}C atoms in the molecule. Natural carbon contains 1.1% ^{13}C.

e In some texts, the mass to charge ratio is written as *m/z*, rather than as *m/e*.

Fragmentation of the molecular ion

The fragments give clues to the groups that are present in the molecule. A CH_3 group has a mass of 15 units and so an ion of mass (*M* – 15) indicates that the substance has a CH_3 group:

$$(RCH_3)^+ \rightarrow R^+ + CH_3\bullet$$

The masses of some common fragments that are often lost are given in Table 9.6.

***Table 9.6** Common fragments lost*

m/e units	Group
15	CH_3
29	C_2H_5 or CHO
31	CH_2OH
45	COOH or CH_2CH_2OH
77	C_6H_5

The mass spectrum of a substance with the empirical formula C_2H_4O, which has only one OH group and a highest *m/e* value of 88, also has peaks at *m/e* values 43 and 59. These fragments enable identification of the isomer.

The molecular ion peak at *m/e* = 88 proves that the molecular formula is $C_4H_8O_2$.

The three suggested formulae are:

- $CH_3CH_2CH_2COOH$
- $(CH_3)_2CHCOOH$
- $(CH_3)_2CHC(=O)—CH_2OH$

The m/e value of 43 is 45 units less than 88 and is caused by the loss of COOH. This discounts the third structure, as it doesn't have a COOH group.

The m/e value of 59 is 29 units less than 88 and is caused by the loss of C_2H_5. This discounts the second structure, as it doesn't have a CH_2CH_3 group.

Thus, only the first structure is possible.

Worked example

Study the mass spectrum of propanoic acid, CH_3CH_2COOH.

Identify the species responsible for the peaks at m/e values of 74, 73, 45 and 29 and state how they are formed.

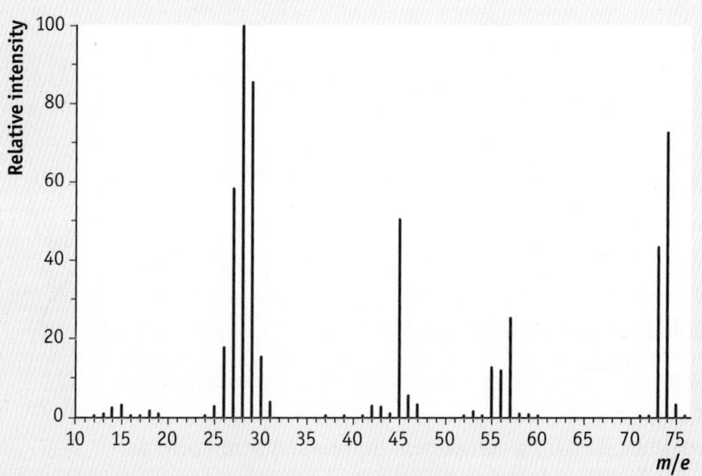

Answer

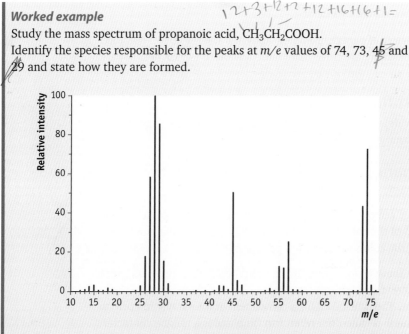

The molar mass of propanoic acid is 74 g mol^{-1} and so the peak at 74 was caused by the molecular ion, $(CH_3CH_2COOH)^+$, produced by the removal of an electron from the gaseous molecule:

$$CH_3CH_2COOH(g) + e^- \rightarrow (CH_3CH_2COOH)^+(g) + 2e^-$$

The peak at 73 is caused by the molecular ion losing one of its hydrogen atoms.

$$(CH_3CH_2COOH)^+ \rightarrow (CH_3CH_2COO)^+ + H\bullet$$

The peak at 45 is 29 less than the molecular peak. It results from the loss of a C_2H_5 group and so is caused by the $(COOH)^+$ ion:

$$(CH_3CH_2COOH)^+ \rightarrow (COOH)^+ + C_2H_5\bullet$$

The peak at 29 is due to the $(C_2H_5)^+$ ion:

$$(CH_3CH_2COOH)^+ \rightarrow (C_2H_5)^+ + COOH\bullet$$

e Do not forget to put the positive charge on the formulae of the ions that produce the peaks in a mass spectrum.

Chromatography

Chromatography of all types works on the same principle. The sample is dissolved in a solvent and washed through a stationary phase by a mobile phase called the eluent. The competition between the sample molecules adsorbed by the stationary phase and dissolved by the eluent results in separation. Substances that are more soluble in the eluent and less adsorbed by the stationary phase move faster through the apparatus.

The speed at which a specific substance passes through depends on the nature of the stationary phase and of the eluent. For a given stationary phase and eluent, this is measured by the R_f value of the substance.

$$R_f \text{ value} = \frac{\text{distance moved by substance}}{\text{distance moved by eluent}}$$

All students will be aware of paper chromatography, but the resolution obtained by this method is poor. There are three main more advanced techniques that are used.

Thin-layer chromatography, TLC

This is described in detail in Unit 5 Chapter 13, p. 275. Knowledge of this technique will be examined in Unit Test 5 only.

High-performance liquid chromatography (HPLC)

This technique used to be called high-pressure liquid chromatography.

This technique uses a column packed with a solid of uniform particle size — the stationary phase. The sample to be separated is dissolved in a suitable solvent and added to the top of the column. The liquid eluent — the mobile phase — is then forced through the column under high pressure.

The principle is the same as for other chromatographic techniques. Different substances have different strengths of interaction between the stationary phase and the mobile phase. The time taken for a component in the sample mixture to pass through the column is called the **retention time** and is a unique characteristic of the substance, the composition of the eluent, the nature of the stationary phase and the pressure. This means that different components will pass through one after the other with gaps between each.

The use of high pressure increases the speed at which the eluent passes through the column and so reduces the extent to which the band of a component spreads out due to diffusion. This means that HPLC has a much higher resolution than paper or thin-layer chromatography. The column can be connected to an infrared spectrometer, which can be used to identify each component in the mixture.

The early columns used a polar stationary phase and a non-polar eluent. However, this has been superseded by a non-polar stationary phase and a polar eluent. This technique is called reversed phase high-performance liquid chromatography, RPHPLC. It is used in the pharmaceutical industry to separate a drug from a mixture and to check the purity of its products.

Gas–liquid chromatography, GLC

Gas–liquid chromatography involves a sample being injected into the top of a chromatography column in a thermostatically controlled oven. The sample evaporates and is forced through the column by the flow of an inert, gaseous mobile phase. The column itself contains a liquid stationary phase that is adsorbed onto the surface of an inert solid.

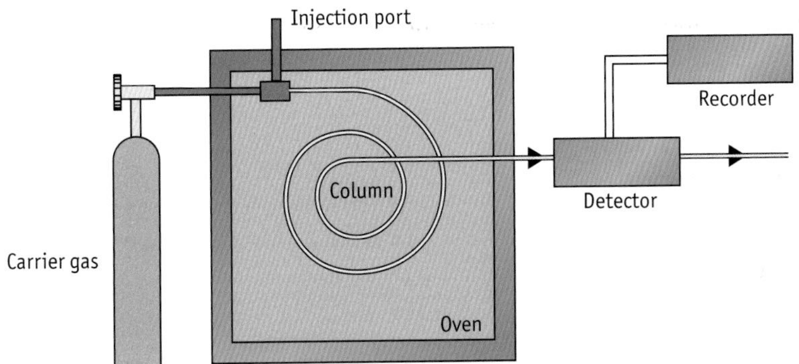

Figure 9.5
A representation of GLC

The carrier gas can be hydrogen, argon, oxygen, nitrogen or air. The rate at which a substance passes through the column depends on the extent to which it interacts with the liquid stationary phase.

A detector measures the thermal conductivity of the gas exiting the column. When one of the components of the original mixture is mixed with the carrier gas, the thermal conductivity changes.

This method is suitable for separating mixtures of volatile liquids.

The identity of the components in the mixture can be found by removing samples leaving the column and measuring their infrared spectra. These are then compared with spectra in a database.

Questions

1 State the effect on a molecule such as water of it absorbing

a infrared radiation

b microwaves

2 Ba^{2+} ions are poisonous. Barium carbonate and barium sulfate are both insoluble in water. Explain why it would be dangerous to use barium carbonate, rather than barium sulfate, when taking X-rays of the human digestive tract.

3 Describe the mechanism of the reaction of chlorine with methane in the presence of white light.

4 Explain why water vapour is a powerful greenhouse gas.

5 Examine the infrared spectra, A and B, below. One spectrum is that of propanal; the other is that of propanoic acid.

Identify the peaks in A at 1715 cm^{-1} and 1421 cm^{-1} and the peaks in B at 2986 cm^{-1}, 1716 cm^{-1} and 1416 cm^{-1}. Hence, decide which spectrum is that of propanoic acid.

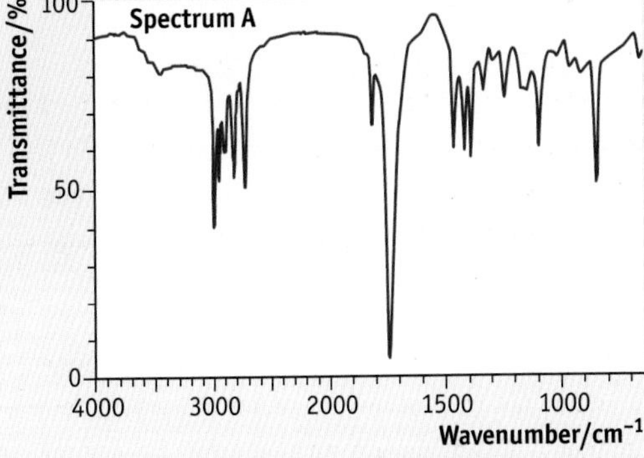

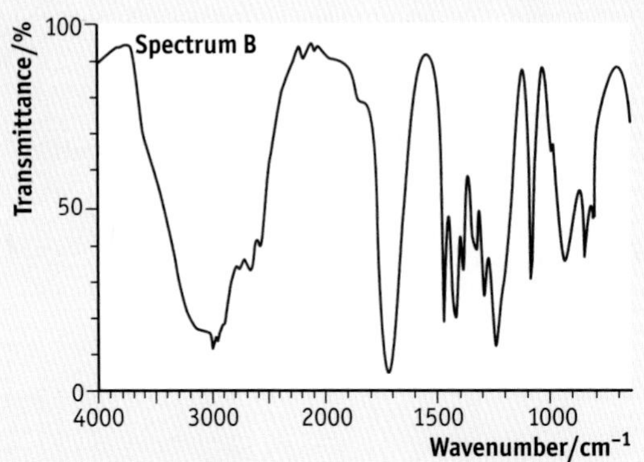

6 Explain what is wrong with the statement:

In a mass spectrometer, the molecular ion, $(CH_3CH_2OH)^+$, splits up as shown in the equation:

$$(CH_3CH_2OH)^+ \rightarrow (CH_3CH_2)^+ + OH^+$$

7 The mass spectrum below is that of propanal.

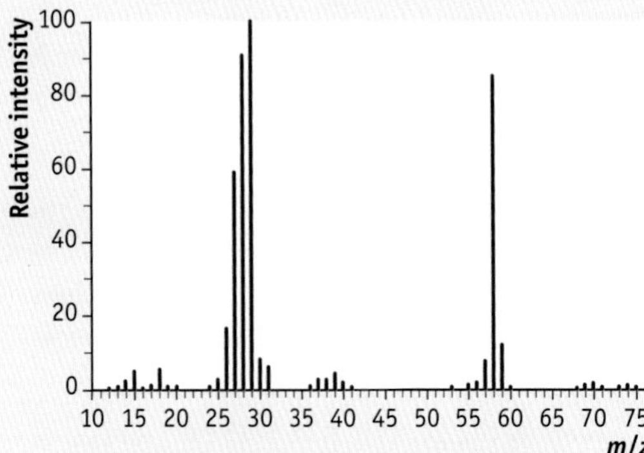

a Identify the peaks at m/e values of 58, 29 and 15 and write equations to show the formation of the species responsible for these peaks.

b Why is there a peak at $m/e = 59$?

8 Bromine has two isotopes, ^{79}Br and ^{81}Br, in approximately equal proportions. Sketch the mass spectrum of bromine (Br_2) in the range m/e 155 to 165.

9 A particular peak in a high-resolution NMR spectrum is split into 6 peaks. Give the formulae of the neighbouring groups. Is this consistent with the compound being butan-1-ol?

10 How many peaks are there in the low-resolution NMR spectrum of butanone, $CH_3CH_2COCH_3$? What is the ratio of their heights?

11 How could low-resolution NMR spectra be used to distinguish between the isomers propan-1-ol and propan-2-ol?

12 The NMR spectrum of ethanol is shown below.

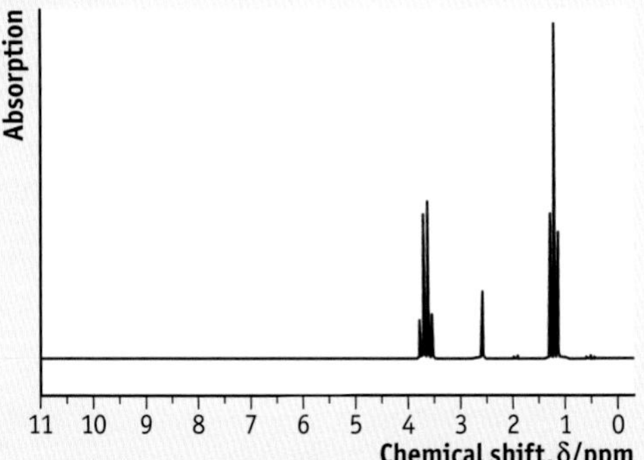

Identify the hydrogen atoms in the structure of ethanol that cause each set of peaks.

13 Explain why, in a high-resolution NMR spectrum, the hydrogen atoms in a CH_3 group cause the peak of a neighbouring hydrogen atom to split into four.

14 How would the NMR spectra of C_2H_5OH and C_2H_5OD differ (where D represents a deuterium, ^2_1H, atom)?

15 Search the internet and write short notes on MRI.

Unit 5

Transition metals, arenes and organic nitrogen chemistry

Electrochemistry and redox equilibria

Chapter 10

Introduction

This topic assumes knowledge of the redox chemistry covered in the AS course. There are some important concepts that must be revisited before embarking on the new A2 work.

The definitions of oxidation and reduction are:

> Oxidation is loss of electrons by an atom, ion or molecule or the increase in oxidation number of an element
>
> Reduction is gain of electrons by an atom, ion or molecule or the decrease in oxidation number of an element.

e OIL RIG — oxidation is loss; reduction is gain.

Oxidation numbers

Oxidation numbers can be worked out using a series of rules:

- The oxidation number of an uncombined element is zero.
- The oxidation number of the element in a monatomic ion is the charge on the ion.
- The sum of the oxidation numbers of the atoms in a neutral compound is zero.
- The sum of the oxidation numbers in a polyatomic ion equals the charge on the ion.
- All group 1 metals have an oxidation number of +1 in their compounds, and all group 2 metals have an oxidation number of +2 in their compounds.
- Fluorine always has the oxidation number –1 in its compounds.
- Hydrogen has the oxidation number +1 in its compounds, apart from when it is combined with a metal, when the oxidation number is –1.
- Oxygen has the oxidation number –2 in its compounds, apart from in peroxides and superoxides or when it is combined with fluorine.

Half-equations

Ionic half-equations always have electrons on either the left-hand side or the right-hand side.

When zinc is added to dilute hydrochloric acid, the hydrogen ions are reduced to hydrogen, because their oxidation number decreases from +1 to zero. The half-equation is:

$$2H^+(aq) + 2e^- \rightleftharpoons H_2(g)$$

As this is a reduction reaction, the electrons are on the left-hand side of the half-equation.

e Half-equations must balance for charge as well as for numbers of atoms.

The zinc atoms are oxidised to zinc ions, because their oxidation number increases from zero to +2:

$$Zn(s) \rightleftharpoons Zn^{2+}(aq) + 2e^-$$

As this is oxidation, the electrons are on the right-hand side of the equation.

Ⓔ State symbols must be used in half-equations.

Overall equations

When half-equations are combined to give the overall equation, the stoichiometry must be such that the numbers of electrons cancel. To do this, one or both half-equations must be multiplied by integers so that the number of electrons is the same in both. The two half-equations are then added together to get the overall equation.

The total change in oxidation number of the species being oxidised is equal to the total change in oxidation number of the species being reduced.

Ⓔ You should make sure that both reactants are on the left-hand side of the overall equation.

Worked example

a Write the half-equation for the oxidation of Sn^{2+} ions to Sn^{4+} ions in aqueous solution.

b Write the half-equation for the reduction of MnO_4^- ions to Mn^{2+} ions in aqueous acidic solution.

c Hence, write the overall equation for the oxidation of tin(II) ions by manganate(VII) ions in aqueous acidic solution.

Answer

a This is oxidation, so the electrons are on the right-hand side of the half-equation. The oxidation number of tin changes by 2, so there are two electrons in the half-equation:

$$Sn^{2+}(aq) \rightarrow Sn^{4+}(aq) + 2e^-$$

b This is reduction, so the electrons are on the left-hand side. Five electrons are needed because the oxidation number of manganese changes by 5 (from +7 to +2):

$$MnO_4^-(aq) + 8H^+(aq) + 5e^- \rightarrow Mn^{2+}(aq) + 4H_2O(l)$$

c Multiply the first equation by 5 and the second equation by 2 to obtain the same number of electrons in each equation. Then add the two equations and cancel the electrons. The overall equation is:

$$5Sn^{2+}(aq) + 2MnO_4^-(aq) + 16H^+(aq) \rightarrow 5Sn^{4+}(aq) + 2Mn^{2+}(aq) + 8H_2O(l)$$

Ⓔ Tin changes oxidation number by $5 \times 2 = 10$ and manganese changes oxidation number by $2 \times 5 = 10$ as well.

Redox as change in oxidation number

In any redox reaction the oxidation number of one element in a species increases and that of another element in another species decreases. The total of the two changes in oxidation number are equal.

Worked example

An acidified solution containing dichromate(VI) ions, $Cr_2O_7^{2-}$, oxidises iodide ions to iodine. The dichromate(VI) ions are reduced to Cr^{3+} ions. Write the overall ionic equation for the reaction.

Answer

Each chromium atom changes in oxidation number from +6 to +3. There are two chromium atoms per dichromate(VI) ion, so the total change of oxidation number is 6. The oxidation number of the iodide ions changes from −1 to 0, so there must be $6I^-$ on the left-hand side of the equation. The equation is:

$$Cr_2O_7^{2-}(aq) + 6I^-(aq) + 14H^+(aq) \rightarrow 2Cr^{3+}(aq) + 3I_2(s) + 7H_2O(l)$$

14 H^+ ions are needed on the left-hand side to form $7H_2O$ with the oxygen atoms from the $Cr_2O_7^{2-}$ ions and also to make the equation balance for charge (6+ on each side).

Redox as electron transfer

When a piece of zinc is placed in a solution of copper(II) sulfate, a reaction takes place in which the zinc is oxidised to zinc ions and the copper ions are reduced to copper metal.

Reaction between zinc and copper(II) sulfate

The two half-equations are:

$$Zn(s) \rightleftharpoons Zn^{2+}(aq) + 2e^-$$

$$Cu^{2+}(aq) + 2e^- \rightleftharpoons Cu(s)$$

In the reaction shown in the photograph, the copper ions collide with zinc atoms in the surface of the metal and remove two electrons from each zinc atom. However, this reaction can also occur without the copper ions and the zinc atoms coming into contact.

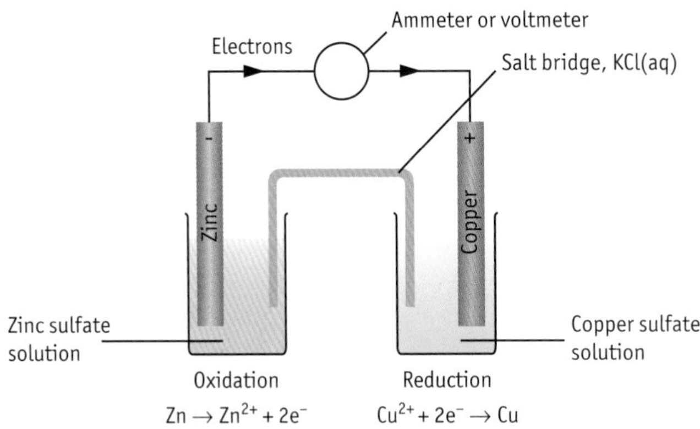

Figure 10.1 *An electrochemical cell of Zn/Zn²⁺ and Cu/Cu²⁺*

The zinc rod is in equilibrium with the solution of zinc ions:

$$Zn(s) \rightleftharpoons Zn^{2+}(aq) + 2e^-$$

The position of equilibrium is to the right-hand side, so electrons are produced making the zinc rod negatively charged relative to the solution.

Oxidation takes place at the zinc rod, which is called the anode.

e Oxidation and anode begin with a vowel.

The copper ions are in equilibrium with the copper rod:

$$Cu^{2+}(aq) + 2e^- \rightleftharpoons Cu(s)$$

The copper rod has given some of its delocalised electrons to the copper ions, so the rod becomes positively charged relative to the solution.

Reduction (of copper ions) takes place at the copper rod, which is called the cathode.

e Reduction and cathode begin with a consonant.

This sets up a potential difference (emf) between the two metal rods. A current will flow when the rods are connected by a wire and the solutions are connected by a **salt bridge**, which contains a concentrated solution of an inert electrolyte, such as potassium chloride or potassium nitrate. The current is carried in the salt bridge by the movement of the ions. The anions (Cl^- or NO_3^-) move towards the anode and the cations (K^+) move towards the cathode. The current is carried in the wire by the flow of electrons from the zinc rod to the copper rod.

e The ions in the salt bridge must *not* react with the ions in the cells.

e At first sight, it appears that positive ions are being attracted to the positive cathode. In fact, the solution around the cathode has had Cu^{2+} ions removed from it and so the solution has a net negative charge. It is this that attracts cations. (Remember that cations always go to the cathode.)

The convention for diagrams such as that in Figure 10.1 is to put the electrode at which oxidation is taking place on the left.

Electrode potentials

Standard electrode potential, $E^\ominus$

Standard electrode potential is also called the standard reduction potential, because the equation is normally written as a reduction half-equation with the electrons on the left.

A zinc rod dipped into a solution of zinc ions generates a potential relative to the solution. However, an electric potential cannot be measured. To avoid this problem, the electrode potential difference is measured against another electrode. By convention, the standard electrode potential of hydrogen is defined as zero.

The term *standard* means that:

- All solutions are at a concentration of $1.0\,mol\,dm^{-3}$.
- All gases are at a pressure of 1.0 atm.
- The system is at a stated temperature, usually 298 K (25°C).

For zinc, the standard reduction potential is for a piece of zinc dipping into a $1.0\,mol\,dm^{-3}$ solution of zinc ions. It is written either as:

$$Zn^{2+}(aq) + 2e^- \rightleftharpoons Zn(s) \qquad E^\ominus = -0.76\ V$$

or as:

$$Zn^{2+}(aq)/Zn(s) \qquad E^\ominus = -0.76\ V$$

The oxidised form of the couple is written on the left, followed by a slash and then the reduced form.

If the substance is a gas, such as hydrogen or chlorine, the electrode consists of a platinum plate dipping into a $1.0\,mol\,dm^{-3}$ solution of ions of the element with the gaseous element, at 1.0 atm pressure, bubbling over the surface of the platinum.

If the reduction involves two cations — for example, M^{3+} being reduced to M^{2+} — the electrode consists of a platinum rod dipping into a solution containing M^{3+} ions and M^{2+} ions, both at a concentration of 1.0 mol dm^{-3}. An example is the reduction of Fe^{3+} to Fe^{2+}.

For manganate(VII) ions/ manganese(II) ions in acid solution, a platinum electrode dipping into a solution containing MnO_4^-, Mn^{2+} and H^+ ions, all at a concentration of 1.0 mol dm^{-3} is used.

Types of electrode

Standard hydrogen electrode (SHE)

A **standard hydrogen electrode** consists of hydrogen gas at 1.0 atm pressure bubbling over a platinum plate which is dipping into a solution that is 1.0 mol dm^{-3} in H^+ ions (such as 1.0 mol dm^{-3} HCl), at a temperature of 298 K.

$$H^+(aq) + e^- \rightleftharpoons \tfrac{1}{2}H_2(g) \qquad E^\ominus = 0\ V$$

or

$$H^+(aq)/H_2(g), Pt \qquad E^\ominus = 0\ V$$

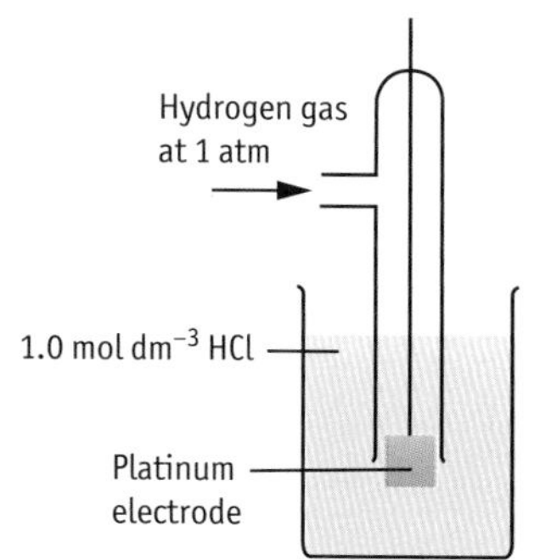

Figure 10.2
A standard hydrogen electrode

Calomel electrode

A standard hydrogen electrode is not easy to use, so a secondary standard is normally used as the reference electrode. This is a **calomel electrode**, which consists of mercury in contact with a saturated solution of mercury(I) chloride. This has a reduction potential of +0.27 V.

For a Zn/Zn^{2+} electrode joined to a calomel electrode:

$$E^\ominus_{cell} = E^\ominus(calomel) - E^\ominus(Zn^{2+}/Zn)$$

where $E^\ominus(Zn^{2+}/Zn)$ is the standard *reduction* potential of zinc ions to zinc and $E^\ominus_{cell} = 1.03$ V.

$$E^\ominus(Zn^{2+}/Zn) = E^\ominus(calomel) - E^\ominus_{cell} = +0.27 - 1.03 = -0.76\ V$$

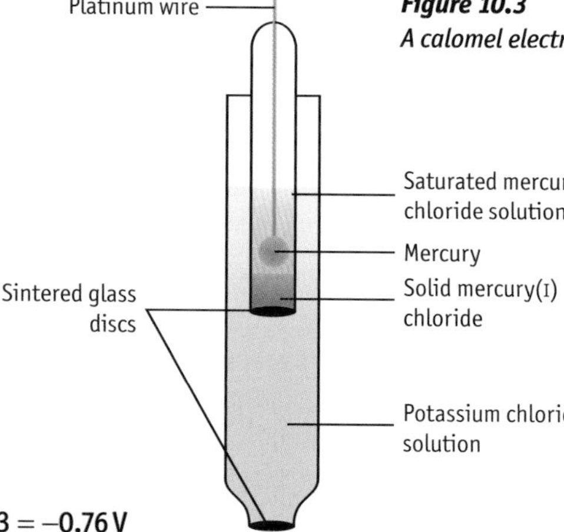

Figure 10.3
A calomel electrode

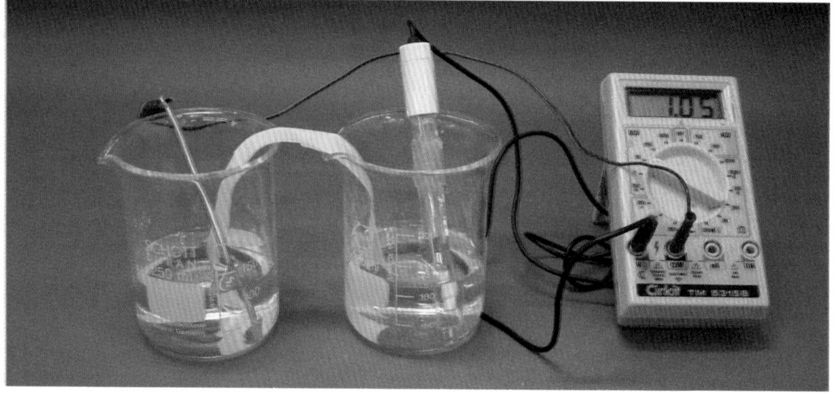

Using a calomel electrode

ZAHOOR UL-HAQ

Glass electrode

A pH meter utilises a half-cell that is based on the reduction of hydrogen ions:

$$H^+(aq) + e^- \rightleftharpoons \frac{1}{2}H_2(g)$$

which is linked to a reference electrode, such as a calomel electrode.

The potential of the H^+/H_2 half-cell depends on the concentration of H^+ ions. This means that the voltage produced is a measure of $[H^+]$ and therefore of the pH of the solution. The hydrogen half-cell, which acts as an H^+/H_2 system, is called a **glass electrode**.

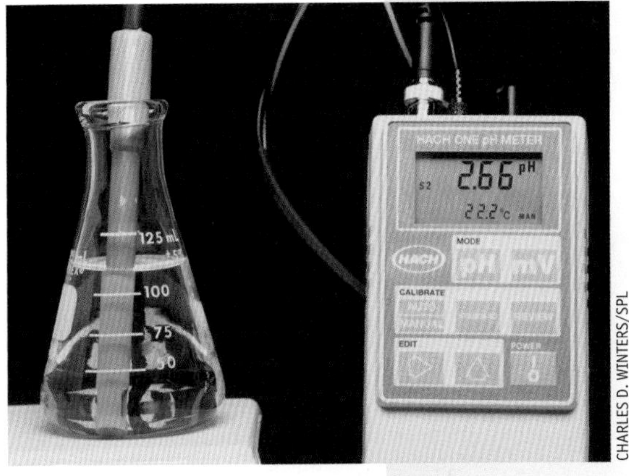

Using a pH meter to measure the pH of methanoic acid

Measurement of electrode potentials

Standard electrode potential of a metal

The standard electrode potential of a metal is measured using the system illustrated in Figure 10.4.

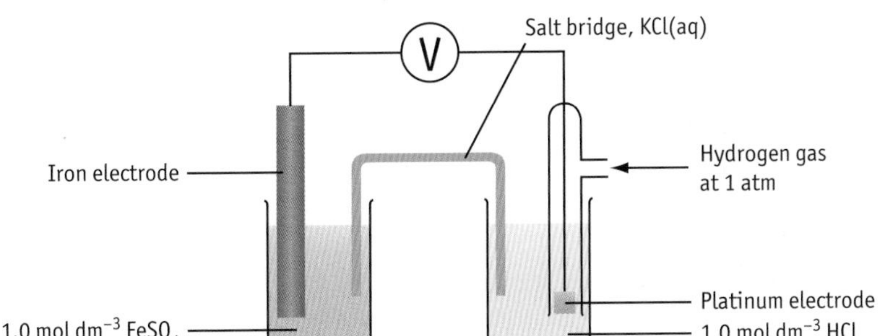

Figure 10.4 *Measurement of the standard electrode potential of iron*

The metal is dipping into a $1.0\,mol\,dm^{-3}$ solution of its ions with a salt bridge to a standard hydrogen electrode, SHE, or a calomel electrode. The two are connected externally by a high-resistance voltmeter or a potentiometer. Using the apparatus shown in Figure 10.4, the voltmeter would read 0.44 V.

$$E^\ominus_{cell} = E^\ominus(\text{standard hydrogen electrode}) - E^\ominus(Fe^{2+}(aq)/Fe(s))$$

So the standard electrode potential of iron is:

$$E^\ominus(Fe^{2+}(aq)/Fe(s)) = E^\ominus(SHE) - E^\ominus_{cell} = 0 - (+0.44) = -0.44\,V$$

This information can be written as a half-equation:

$$Fe^{2+}(aq) + 2e^- \rightleftharpoons Fe(s) \qquad E^\ominus = -0.44\,V$$

or as:

$$E^\ominus(Fe^{2+}(aq)/Fe(s)) = -0.44\,V$$

Copper and other metals below hydrogen in the electrochemical series have positive standard electrode potentials.

If a calomel electrode ($E^\ominus = +0.27\,V$) is used in place of a standard hydrogen electrode, the measured cell potential, $E^\ominus_{cell}$, is +0.71 V.

e A reference electrode has to be used, as a potential cannot be measured. There has to be a potential difference. A salt bridge is used to complete the electrical circuit.

The standard electrode potential is then calculated as:

$$E^{\ominus}(Fe^{2+}(aq)/Fe(s)) = +0.27\,V - E^{\ominus}_{cell} = +0.27 - (+0.71) = -0.44\,V$$

Standard electrode potential of a gas

If the standard electrode potential of a gas such as chlorine is required, the iron electrode compartment in Figure 10.4 is replaced by a compartment with a platinum plate dipping into a 1.0 mol dm^{-3} solution of sodium chloride, with chlorine gas, at 1.0 atm pressure, bubbling over the platinum.

$$\tfrac{1}{2}Cl_2(g) + e^- \rightleftharpoons Cl^-(aq) \qquad E^{\ominus} = +1.36\,V$$

Standard electrode potential of an ion pair

Manganate(VII) is a powerful oxidising agent in acid solution. The half-equation is:

$$MnO_4^-(aq) + 8H^+(aq) + 5e^- \rightleftharpoons Mn^{2+}(aq) + 4H_2O(l)$$

One electrode compartment is made of a piece of platinum dipping into a solution that is 1.00 mol dm^{-3} in MnO_4^-, H^+ *and* Mn^{2+} ions. This is connected to a standard hydrogen or calomel electrode, as in Figure 10.5.

$$MnO_4^-(aq),\ H^+(aq)/Mn^{2+}(aq),\ Pt \qquad E^{\ominus} = +1.52\,V$$

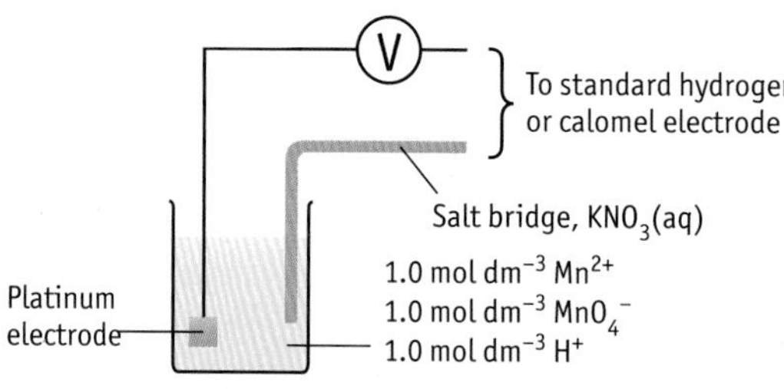

***Figure 10.5** Measurement of the standard electrode potential of MnO_4^-/Mn^{2+}*

e A potential can never be measured, only a potential difference. This is why the $Fe^{2+}(aq)/Fe(s)$ cell is coupled to a standard hydrogen (or calomel) electrode.

Comparative values of $E^{\ominus}$

Standard electrode potentials can be listed either in alphabetical order or in numerical order. Alphabetical order has the advantage of easy use in a long list of data. Numerical order lists the electrode potentials with the most negative first, then in order of increasing value, finishing with the most positive. The numerical-order method means that the weakest oxidising agent is the left-hand species at the top of the list ($Li^+(aq)$ in Table 10.1) and the strongest reducing agent is the right-hand species at the top of the list (Li(s) in Table 10.1). Similarly, the strongest oxidising agent is the left-hand species at the bottom of the list ($F_2(g)$ in Table 10.1) and the weakest reducing agent is the right-hand species at the bottom of the list ($F^-(aq)$ in Table 10.1).

Non-standard conditions

If the conditions are not standard, the value of the electrode potential will alter. The direction of change can be predicted using Le Chatelier's principle.

Change in concentration

Consider the redox half-equation:

$$Cr_2O_7^{2-}(aq) + 14H^+(aq) + 6e^- \rightleftharpoons 2Cr^{3+}(aq) + 7H_2O(l) \qquad E^{\ominus} = +1.33\,V$$

If the concentrations of dichromate(VI) ions and hydrogen ions are increased above 1.0 mol dm^{-3}, the position of equilibrium is driven to the right. This causes

Alphabetical list

Electrode reaction	$E^\ominus$/V
$Ag^+(aq) + e^- \rightleftharpoons Ag(s)$	+0.80
$Al^{3+}(aq) + 3e^- \rightleftharpoons Al(s)$	−1.66
$Ba^{2+}(aq) + 2e^- \rightleftharpoons Ba(s)$	−2.90
$\frac{1}{2}Br_2(l) + e^- \rightleftharpoons Br^-(aq)$	+1.07
$Ca^{2+}(aq) + 2e^- \rightleftharpoons Ca(s)$	−2.87
$\frac{1}{2}Cl_2(g) + e^- \rightleftharpoons Cl^-(aq)$	+1.36
$HOCl(aq) + H^+(aq) + e^- \rightleftharpoons \frac{1}{2}Cl_2(g) + H_2O(l)$	+1.64
$Cr^{3+}(aq) + e^- \rightleftharpoons Cr^{2+}(aq)$	−0.41
$Cr_2O_7^{2-}(aq) + 14H^+(aq) + 6e^- \rightleftharpoons 2Cr^{3+}(aq) + 7H_2O(l)$	+1.33
$Cu^+(aq) + e^- \rightleftharpoons Cu(s)$	+0.52
$Cu^{2+}(aq) + 2e^- \rightleftharpoons Cu(s)$	+0.34
$Cu^{2+}(aq) + e^- \rightleftharpoons Cu^+(aq)$	+0.15
$\frac{1}{2}F_2(g) + e^- \rightleftharpoons F^-(aq)$	+2.87
$Fe^{2+}(aq) + 2e^- \rightleftharpoons Fe(s)$	−0.44
$Fe^{3+}(aq) + e^- \rightleftharpoons Fe^{2+}(aq)$	+0.77
$H^+(aq) + e^- \rightleftharpoons \frac{1}{2}H_2(g)$	0.00
$\frac{1}{2}I_2(s) + e^- \rightleftharpoons I^-(aq)$	+0.54
$Li^+(aq) + e^- \rightleftharpoons Li(s)$	−3.04
$MnO_4^-(aq) + 8H^+(aq) + 5e^- \rightleftharpoons Mn^{2+}(aq) + 4H_2O(l)$	+1.52
$MnO_4^-(aq) + e^- \rightleftharpoons MnO_4^{2-}(aq)$	+0.56
$MnO_4^{2-}(aq) + 2H_2O(l) + 2e^- \rightleftharpoons MnO_2(s) + 4OH^-(aq)$	+0.59
$\frac{1}{2}H_2O_2(aq) + H^+(aq) + e^- \rightleftharpoons H_2O(l)$	+1.77
$\frac{1}{2}O_2(g) + 2H^+(aq) + 2e^- \rightleftharpoons H_2O(l)$	+1.23
$\frac{1}{2}O_2(g) + H_2O(l) + e^- \rightleftharpoons 2OH^-(l)$	+0.40
$O_2(g) + 2H^+(aq) + 2e^- \rightleftharpoons H_2O_2(aq)$	+0.68
$Pb^{4+}(aq) + 2e^- \rightleftharpoons Pb^{2+}(aq)$	+1.69
$Sn^{2+}(aq) + 2e^- \rightleftharpoons Sn(s)$	−0.14
$Sn^{4+}(aq) + 2e^- \rightleftharpoons Sn^{2+}(aq)$	+0.15
$S(s) + 2H^+(aq) + 2e^- \rightleftharpoons H_2S(g)$	+0.14
$Zn^{2+}(aq) + 2e^- \rightleftharpoons Zn(s)$	−0.76

Numerical list

Electrode reaction $E^\ominus$/V	
$Li^+(aq) + e^- \rightleftharpoons Li(s)$	−3.04
$Ba^{2+}(aq) + 2e^- \rightleftharpoons Ba(s)$	−2.90
$Ca^{2+}(aq) + 2e^- \rightleftharpoons Ca(s)$	−2.87
$Al^{3+}(aq) + 3e^- \rightleftharpoons Al(s)$	−1.66
$Zn^{2+}(aq) + 2e^- \rightleftharpoons Zn(s)$	−0.76
$Fe^{2+}(aq) + 2e^- \rightleftharpoons Fe(s)$	−0.44
$Cr^{3+}(aq) + e^- \rightleftharpoons Cr^{2+}(aq)$	−0.41
$Sn^{2+}(aq) + 2e^- \rightleftharpoons Sn(s)$	−0.14
$H^+(aq) + e^- \rightleftharpoons \frac{1}{2}H_2(g)$	0.00
$S(s) + 2H^+(aq) + 2e^- \rightleftharpoons H_2S(g)$	+0.14
$Cu^{2+}(aq) + e^- \rightleftharpoons Cu^+(aq)$	+0.15
$Sn^{4+}(aq) + 2e^- \rightleftharpoons Sn^{2+}(aq)$	+0.15
$Cu^{2+}(aq) + 2e^- \rightleftharpoons Cu(s)$	+0.34
$\frac{1}{2}O_2(g) + H_2O(l) + e^- \rightleftharpoons 2OH^-(l)$	+0.40
$Cu^+(aq) + e^- \rightleftharpoons Cu(s)$	+0.52
$\frac{1}{2}I_2(s) + e^- \rightleftharpoons I^-(aq)$	+0.54
$MnO_4^-(aq) + e^- \rightleftharpoons MnO_4^{2-}(aq)$	+0.56
$MnO_4^{2-}(aq) + 2H_2O(l) + 2e^- \rightleftharpoons MnO_2(s) + 4OH^-(aq)$	+0.59
$O_2(g) + 2H^+(aq) + 2e^- \rightleftharpoons H_2O_2(aq)$	+0.68
$Fe^{3+}(aq) + e^- \rightleftharpoons Fe^{2+}(aq)$	+0.77
$Ag^+(aq) + e^- \rightleftharpoons Ag(s)$	+0.80
$\frac{1}{2}Br_2(l) + e^- \rightleftharpoons Br^-(aq)$	+1.07
$\frac{1}{2}O_2(g) + 2H^+(aq) + 2e^- \rightleftharpoons H_2O(l)$	+1.23
$Cr_2O_7^{2-}(aq) + 14H^+(aq) + 6e^- \rightleftharpoons 2Cr^{3+}(aq) + 7H_2O(l)$	+1.33
$\frac{1}{2}Cl_2(g) + e^- \rightleftharpoons Cl^-(aq)$	+1.36
$MnO_4^-(aq) + 8H^+(aq) + 5e^- \rightleftharpoons Mn^{2+}(aq) + 4H_2O(l)$	+1.52
$HOCl(aq) + H^+(aq) + e^- \rightleftharpoons \frac{1}{2}Cl_2(g) + H_2O(l)$	+1.64
$Pb^{4+}(aq) + 2e^- \rightleftharpoons Pb^{2+}(aq)$	+1.69
$\frac{1}{2}H_2O_2(aq) + H^+(aq) + e^- \rightleftharpoons H_2O(l)$	+1.77
$\frac{1}{2}F_2(g) + e^- \rightleftharpoons F^-(aq)$	+2.87

increasing strength as an oxidising agent

***Table 10.1** Standard reduction potential values at 298 K*

the value of the electrode potential, E, to be higher than the standard value: $E^\ominus > +1.33\,V$.

Consider the redox equilibrium:

$$\frac{1}{2}Cl_2(g) + e^- \rightleftharpoons Cl^-(aq) \quad E^\ominus = +1.36\,V$$

If the concentration of chloride ions is increased, the equilibrium position will shift to the left. This causes the value of the electrode potential, E, to be lower than the standard value: $E^\ominus < +1.36\,V$.

This can have a serious effect on the spontaneity of a reaction.

Change in pressure

A change in pressure affects gaseous reactants only. Oxygen acting as an oxidising agent has the redox half-equation:

$$O_2(g) + 2H_2O(l) + 4e^- \rightleftharpoons 4OH^-(aq) \quad E^\ominus = +0.40\,V$$

This assumes that the pressure of oxygen is 1.0 atm. In air, this is not the case. The partial pressure of oxygen in air is about 0.2 atm. This drives the redox equilibrium to the left, making $E < +0.40\,\text{V}$. Therefore, oxygen in the air is a less good oxidising agent than pure oxygen. Conversely, a scuba diver's air tank has oxygen at a partial pressure much greater than 1 atm. So, if damp air is pumped into the tank, internal rusting will occur more than it does with iron in normal air.

Change in temperature

The effect of a change in temperature depends on whether the redox half-equation is exothermic or endothermic. If it is exothermic, an increased temperature drives the position of equilibrium to the left (in the endothermic direction). This makes the value of the electrode potential less positive (or more negative).

Chlorine is reduced exothermically in aqueous solution. Therefore, an increase in temperature makes the new electrode potential less than +1.36 V.

Altering a reduction potential half-equation

Changing direction

If the electrode equation is reversed, its sign must also be reversed, for example:

$$Zn^{2+}(aq) + 2e^- \rightleftharpoons Zn(s) \qquad E^\ominus = -0.76\,\text{V}$$
$$Zn(s) \rightleftharpoons Zn^{2+}(aq) + 2e^- \qquad E^\ominus = -(-0.76) = +0.76\,\text{V}$$
$$\tfrac{1}{2}Cl_2(g) + e^- \rightleftharpoons Cl^-(aq) \qquad E^\ominus = +1.36\,\text{V}$$
$$Cl^-(aq) \rightleftharpoons \tfrac{1}{2}Cl_2(g) + e^- \qquad E^\ominus = -(+1.36) = -1.36\,\text{V}$$

ⓔ This is similar to enthalpy calculations — if the equation of a reaction is reversed, the sign of ΔH has to be changed.

Multiplying by integers

The units of E are volts *not* volts per mole, so multiplying a redox half-equation has *no* effect on the value of E:

$$\tfrac{1}{2}Cl_2(g) + e^- \rightleftharpoons Cl^-(aq) \qquad E^\ominus = +1.36\,\text{V}$$
$$Cl_2(g) + 2e^- \rightleftharpoons 2Cl^-(aq) \qquad E^\ominus = +1.36\,\text{V}$$

ⓔ This is *unlike* ΔH calculations, where the units are kJ mol^{-1} and the value depends on the number of moles in the equation as written.

Calculation of $E^\ominus_{cell}$ for a reaction

The standard cell potential, $E^\ominus_{cell}$, can be calculated from standard electrode potential data. Since these data are normally given as reduction potentials, the two reactants must be identified. One reactant will be on the left-hand side of one half-equation and the other on the right-hand side of the second half-equation.

The half-equation with the reactant on the right-hand side must be reversed. This alters the sign of its $E^\ominus$ value. If necessary, the two half-equations are multiplied by integers to give the same number of electrons in each equation, but this does not alter the $E^\ominus$ values. The overall equation is obtained by adding these half-equations together. The $E^\ominus$ value of the reversed half-equation is then added to the $E^\ominus$ value of the unchanged half-equation to give the value of $E^\ominus_{cell}$.

Worked example 1

Use the data below to calculate the value of $E^\ominus_{cell}$ for the reaction in which Fe^{3+} ions oxidise I^- ions in aqueous solution. Write the overall equation.

$$Fe^{3+}(aq) + e^- \rightleftharpoons Fe^{2+}(aq) \qquad E^\ominus = +0.77\,\text{V}$$
$$\tfrac{1}{2}I_2(s) + e^- \rightleftharpoons I^-(aq) \qquad E^\ominus = +0.54\,\text{V}$$

Answer

The reactants are Fe^{3+} and I^-, so the second equation has to be reversed and the sign of its $E^\ominus$ value changed.

Both equations have one electron, so no multiplying is needed.

$$Fe^{3+}(aq) + e^- \rightleftharpoons Fe^{2+}(aq) \qquad E^\ominus = +0.77\,V$$
$$I^-(aq) \rightleftharpoons \tfrac{1}{2}I_2(s) + e^- \qquad E^\ominus = -(+0.54) = -0.54\,V$$

These two equations and their $E^\ominus$ values are then added together:

$$Fe^{3+}(aq) + I^-(aq) \rightarrow Fe^{2+}(aq) + \tfrac{1}{2}I_2(s) \qquad E^\ominus_{cell} = +0.77 + (-0.54) = +0.23\,V$$

Worked example 2

Use the data below to calculate the value of $E^\ominus_{cell}$ for the reaction of dichromate(VI) ions in acidified potassium dichromate(VI) oxidising chloride ions in hydrochloric acid.

$$Cr_2O_7^{2-}(aq) + 14H^+(aq) + 6e^- \rightleftharpoons 2Cr^{3+}(aq) + 7H_2O(l) \qquad E^\ominus = +1.33\,V$$
$$\tfrac{1}{2}Cl_2(g) + e^- \rightleftharpoons Cl^-(aq) \qquad E^\ominus = +1.36\,V$$

Answer

The reactants are $Cr_2O_7^{2-}$ and Cl^-, so the second equation has to be reversed and the sign of its $E^\ominus$ value changed.

To get the overall equation, the number of electrons must be the same in both half-equations, so the second equation has to be multiplied by 6. This does not alter the $E^\ominus$ value. The half-equations are then added together.

$$Cr_2O_7^{2-}(aq) + 14H^+(aq) + 6e^- \rightleftharpoons 2Cr^{3+}(aq) + 7H_2O(l) \qquad E^\ominus = +1.33\,V$$
$$6Cl^-(aq) \rightleftharpoons 3Cl_2(g) + 6e^- \qquad E^\ominus = -1.36\,V$$

$$Cr_2O_7^{2-}(aq) + 14H^+(aq) + 6Cl^-(aq) \rightleftharpoons 2Cr^{3+}(aq) + 7H_2O(l) + 3Cl_2(g)$$
$$E^\ominus_{cell} = 1.33 + (-1.36) = -0.03\,V$$

This is the safest way to calculate the value of $E^\ominus$ of a redox reaction. It also generates the overall equation.

There are other methods, but they pose particular difficulties:

- The 'anticlockwise rule' works only if the data are presented in increasing numerical order, with the most negative (least positive) $E^\ominus$ value on top. Each half equation then occurs in an anticlockwise direction. For example, the direction of change for the combination of the I_2/I^- and Fe^{3+}/Fe^{2+} can be predicted using this rule.

$$I_2(aq) + 2e^- \rightleftharpoons 2I^-(aq) \qquad E^\ominus = +0.54\,V$$
$$Fe^{3+}(aq) + e^- \rightleftharpoons Fe^{2+}(aq) \qquad E^\ominus = +0.77\,V$$

This shows that the top equation goes backwards and the bottom equation goes forwards. Fe^{3+} ions will oxidise I^- ions. As the top reaction is being reversed, its sign must also be reversed. So the value of $E^\ominus_{cell}$ is:

$$E^\ominus_{cell} = +0.77 - (+0.54) = +0.23\,V$$

- The rule '$E^\ominus_{cell} = E^\ominus_{oxidising\ agent} - E^\ominus_{reducing\ agent}$' can be misremembered and also does not give the overall equation. In addition, it requires correct identification of the oxidising agent and the reducing agent. The reducing agent is found on the right of the reduction potential half-equation, and $E_{\text{reducing agent}}$ is the value of the reduction potential.

In worked example 2 above, the oxidising agent is dichromate(VI) and the reducing agent is the Cl^- ions.

Thus $E^\ominus_{\text{cell}} = E^\ominus_{\text{oxidising agent}} - E^\ominus_{\text{reducing agent}} = +1.33 - (+1.36) = -0.03\,\text{V}$

Relationship between $E^\ominus$, ΔS_{total} and K

The value of the total entropy change, ΔS_{total}, is directly proportional to the standard cell potential of a reaction. In turn, $\ln K$ is directly proportional to ΔS_{total}, where K is the equilibrium constant for the reaction.

This means that a positive $E^\ominus$ results in a positive ΔS_{total} and hence a value of K greater than 1. The more positive the value of $E^\ominus$, the larger is the value of the equilibrium constant.

A negative value of $E^\ominus$ results in a negative ΔS_{total} and a value of K that is less than 1.

e The actual relationship is complex and need not be learnt for A-level chemistry.

$$\Delta S_{\text{total}} = nE^\ominus F$$

where n is the number of electrons transferred in the overall equation, and F is the Faraday constant ($96\,500\,\text{C}\,\text{mol}^{-1}$)

$$\ln K = \Delta S_{\text{total}}/R = nE^\ominus F/R$$

where R is the gas constant ($8.31\,\text{J}\,\text{K}^{-1}\,\text{mol}^{-1}$)

Feasibility of a redox reaction

A reaction is thermodynamically spontaneous (feasible) if the value of ΔS_{total} is positive. Since the total entropy change is directly proportional to the value of $E^\ominus_{\text{cell}}$, a redox reaction that has a positive value of $E^\ominus_{\text{cell}}$ will be thermodynamically spontaneous. Likewise, if the cell potential is negative, then the reaction is not thermodynamically spontaneous and the position of equilibrium will lie to the left.

A redox reaction is thermodynamically feasible if the value of the cell potential is positive.

Some books state that unless $E^\ominus_{\text{cell}}$ is more than 0.3 V, the reaction will not proceed very far. This is an oversimplification. As was shown above, the relationship between $E^\ominus_{\text{cell}}$ and ΔS_{total} and hence $\ln K$ also depends on the number of electrons transferred in the reaction.

The relationship between ΔS_{total}, K and the extent of the reaction is shown in Table 3.2. In worked example 1 above (pp. 188–189), the cell potential for the oxidation of iodide ions by iron(III) ions is +0.23 V. The value of $\Delta S^\ominus_{\text{total}} = +74\,\text{J}\,\text{K}^{-1}\,\text{mol}^{-1}$ and the numerical value of $K = 7.4 \times 10^3$.

This value shows that the reaction very much favours the products.

The oxidation of chloride ions by manganate(VII) ions in acid solution

$$MnO_4^-(aq) + 5Cl^-(aq) + 8H^+(aq) \rightarrow Mn^{2+}(aq) + 2\tfrac{1}{2}Cl_2(g) + 4H_2O(l)$$

has $E^\ominus_{cell} = +0.16\,V$, but as there are five electrons transferred, the numerical value of $\Delta S_{total} = +518$ and $K = 1.6 \times 10^{16}$.

This shows that a standard solution of manganate(VII) ions in acid solution almost completely oxidises a standard solution of chloride ions, despite $E^\ominus_{cell}$ being only +0.16 V.

Remember:
- A positive cell potential means that the reaction is thermodynamically feasible.
- A negative cell potential means that the reaction is not thermodynamically feasible.

Actuality of reaction

A reaction may be thermodynamically feasible ($E^\ominus_{cell}$ positive) but might not take place. The reasons for this can be kinetic or connected with non-standard conditions.

Kinetic reasons

The sign of $E^\ominus_{cell}$ enables the prediction of whether a reaction is thermodynamically feasible. However, thermodynamic feasibility is no guarantee that the reaction will take place under standard conditions. The reaction may have such a high activation energy that it is too slow to be observed at room temperature. For example, $E^\ominus_{cell}$ for the reaction $H_2(g) + \frac{1}{2}O_2(g) \rightarrow H_2O(l)$ is +1.23 V, but a mixture of hydrogen and oxygen will not react unless heated or unless a catalyst is present (see fuel cells on p. 196).

The reaction between persulfate ions, $S_2O_8^{2-}$, and iodide ions, I^-, is thermodynamically feasible as $E^\ominus_{cell} = +1.47\,V$. However, it does not occur unless a catalyst of iron ions (either Fe^{2+} or Fe^{3+}) is added. The activation energy is high because of the need for two negative ions to collide in the uncatalysed route.

The reactants are kinetically stable with respect to the products if the activation energy of the reaction is too high.

The reactants are thermodynamically stable with respect to the products if ΔS_{total} is negative or if $E^\ominus_{cell}$ is negative.

Non-standard conditions

The data provided in the data booklet are always *standard* electrode potentials. If the concentration of a reactant or product is not 1.0 mol dm^{-3}, the value of E_{cell} will differ from that of $E^\ominus_{cell}$.

This might result in a reaction taking place that is predicted to be unfeasible.

Worked example

Consider the redox reaction:

$Cu^{2+}(aq) + 2I^-(aq) \rightarrow CuI(s) + \frac{1}{2}I_2(s)$

Explain why this reaction will take place, given that:

$Cu^{2+}(aq) + e^- \rightleftharpoons Cu^+(aq)$ $\quad E^\ominus = +0.15\,V$

$\frac{1}{2}I_2(s) + e^- \rightleftharpoons I^-(aq)$ $\quad E^\ominus = +0.54\,V$

Answer

The reactants are Cu^{2+} and I^- ions. The feasibility is predicted by reversing the sign of the $E^\ominus$ value of the second equation and adding the two $E^\ominus$ values together.

$E^\ominus_{cell} = +0.15 + (-0.54) = -0.39\,V$

The negative value predicts that the reaction will not take place under standard conditions where the concentrations of Cu^+ and I^- ions are both 1.0 mol dm^{-3}. However, this is not the case in this reaction because copper(I) iodide is precipitated and so $[Cu^+]$ is almost zero. This drives the equilibrium of the Cu^{2+}/Cu^+ reaction to the right, increasing its electrode potential and making the value of the non-standard cell potential positive. The reaction is now feasible and, as the activation energy is low, the reaction takes place rapidly.

Disproportionation reactions

If an element exists in three different oxidation states (which could include the zero state of the uncombined element), disproportionation becomes a possibility. The feasibility of such a reaction can be predicted from standard electrode potential data.

A disproportionation reaction is a special kind of redox reaction.

Worked example

Predict whether or not copper(I) ions will disproportionate into copper metal and copper(II) ions. If so, write the overall equation.

$Cu^{2+}(aq) + e^- \rightleftharpoons Cu^+(aq) \quad E^\ominus = +0.15\,V$

$Cu^+(aq) + e^- \rightleftharpoons Cu(s) \quad E^\ominus = +0.52\,V$

Answer

As this is a disproportionation reaction, a single species is both reduced and oxidised at the same time. In this example that species is the Cu^+ ion, so the first equation must be reversed and added to the second equation to give the overall equation:

$Cu^+(aq) \rightleftharpoons Cu^{2+}(aq) + e^- \quad E^\ominus = -0.15\,V$

$Cu^+(aq) + e^- \rightleftharpoons Cu(s) \quad E^\ominus = +0.52\,V$

$2Cu^+(aq) \rightleftharpoons Cu^{2+}(aq) + Cu(s) \quad E^\ominus_{cell} = -0.15 + (+0.52) = +0.37V$

The value of $E^\ominus_{cell}$ is positive, so the disproportionation reaction is feasible.

This is a disproportionation reaction because copper in the +1 state is simultaneously oxidised to copper in the +2 state and reduced to copper in the zero state.

Evaluation of experimental results compared with calculated $E^\ominus_{cell}$ values

Vanadium forms compounds in the +2, +3, +4 and +5 oxidation states. The colour of the ions depends on the oxidation state of vanadium.

Oxidation state	Formula of the ion	Colour
+2	V^{2+}	Lavender
+3	V^{3+}	Green
+4	VO^{2+}	Blue
+5	VO_2^+ VO_3^-	Yellow Colourless

***Table 10.2** Oxidation states of vanadium and colours of the ions*

Solutions of vanadium in the +2,+3, +4 and +5 oxidation states

Ammonium vanadate(V), NH_4VO_3, is a colourless solid, but when added to water, the solution is yellow. This is because of the reaction:

$$VO_3^- + H_2O \rightleftharpoons VO_2^+ + 2OH^-$$

Addition of acid drives the equilibrium to the right and the yellow colour becomes more intense.

This is not a redox reaction because vanadium is in the +5 oxidation state in both VO_3^- and VO_2^+.

The standard reduction potentials for the redox changes of vanadium are shown in Table 10.3.

Reduction half-equation	Change in oxidation state	$E^\ominus$/V
$VO_2^+ + 2H^+ + e^- \rightleftharpoons VO^{2+} + H_2O$	+5 to +4	+1.00
$VO^{2+} + 2H^+ + e^- \rightleftharpoons V^{3+} + H_2O$	+4 to +3	+0.34
$V^{3+} + e^- \rightleftharpoons V^{2+}$	+3 to +2	−0.26

***Table 10.3** Standard reduction potentials for vanadium*

Worked example

Which of $Fe^{2+}(aq)$ and $Sn^{2+}(aq)$ will reduce vanadium(V) ions to vanadium(IV) and which to vanadium (III)?

$Fe^{3+}(aq) + e^- \rightleftharpoons Fe^{2+}(aq)$ $E^\ominus = +0.77\,V$

$Sn^{4+}(aq) + 2e- \rightleftharpoons Sn^{2+}(aq)$ $E^\ominus = +0.15\,V$

Answer

(1) Iron(II) ions as the reducing agent

(a) Reduction of vanadium(V) to vanadium (IV):

$VO_2^+ + 2H^+ + e^- \rightleftharpoons VO^{2+} + H_2O$ $E^\ominus = +1.00\,V$

$Fe^{2+} \rightleftharpoons Fe^{3+} + e^-$ $E^\ominus = -0.77\,V$

Overall:

$VO_2^+ + 2H^+ + Fe^{2+} \rightleftharpoons VO^{2+} + H_2O + Fe^{3+}$ $E^\ominus_{cell} = +1.00 + (-0.77) = +0.23\,V$

$E^\ominus_{cell}$ for the reaction is positive so the reduction is feasible.

(b) Subsequent reduction to vanadium(III)

$VO^{2+} + 2H^+ + e^- \rightleftharpoons V^{3+} + H_2O$ $E^\ominus = +0.34\,V$

$Fe^{2+} \rightleftharpoons Fe^{3+} + e^-$ $E^\ominus = -0.77\,V$

Overall:

$VO^{2+} + 2H^+ + Fe^{2+} \rightleftharpoons V^{3+} + H_2O + Fe^{3+}$ $E^\ominus_{cell} = +0.34 + (-0.77) = -0.43\,V$

$E^\ominus_{cell}$ for the reaction is negative and so the reduction is not feasible.

The conclusion is that iron(II) ions will reduce vanadium(V) to vanadium(IV) but not to vanadium(III).

(2) Tin(II) ions as the reducing agent

(a) Reduction of vanadium(V) to vanadium (IV):

$VO_2^+ + 2H^+ + e^- \rightleftharpoons VO^{2+} + H_2O$ $E^\ominus = +1.00\,V$

$Sn^{2+} \rightleftharpoons Sn^{4+} + 2e^-$ $E^\ominus = -0.15\,V$

Overall:

$2VO_2^+ + 4H^+ + Sn^{2+} \rightleftharpoons 2VO^{2+} + 2H_2O + Sn^{4+}$ $E^\ominus_{cell} = +1.00 + (-0.15) = +0.85\,V$

$E^\ominus_{cell}$ for the reaction is positive so the reduction is feasible.

(b) Subsequent reduction to vanadium(III)

$VO^{2+} + 2H^+ + e^- \rightleftharpoons V^{3+} + H_2O$ $E^\ominus = +0.34\,V$

$Sn^{2+} \rightleftharpoons Sn^{4+} + 2e^-$ $E^\ominus = -0.15\,V$

Overall:

$2VO^{2+} + 4H^+ + Sn^{2+} \rightleftharpoons 2V^{3+} + 2H_2O + Sn^{4+}$ $E^\ominus_{cell} = +0.34 + (-0.15) = +0.19\,V$

$E^\ominus_{cell}$ for the reaction is positive and so the reduction is feasible.

The conclusion is that tin(II) ions will reduce vanadium(V) first to vanadium(IV) and then to vanadium(III).

The value of $E^\ominus_{cell}$ for the reduction of vanadium(III) to vanadium(II) by tin(II) is −0.41 V and so the reaction does not take place.

The predictions made in the worked example above can be checked experimentally:

- Mix together equal volumes of dilute hydrochloric acid and iron(II) sulfate solution.
- Add a few drops of ammonium vanadate(V) solution
- Observe the final colour. (Warm if necessary.)

Repeat the experiment using a solution of tin(II) chloride instead of iron(II) sulfate.

Check that the final colours obtained agree with the predictions.

$E^\ominus_{cell}$ values predict that sulfur dioxide should reduce vanadium(V) to vanadium(III). But, as the activation energy for reduction to V^{3+} is too high, so the reaction stops at vanadium(IV).

Reduction to vanadium(II)

This can be achieved by warming a solution of powdered zinc with ammonium vanadate(V) in the presence of 50% hydrochloric acid solution. The experiment is carried out in a conical flask fitted with a Bunsen valve in order to exclude air. The VO_2^+ ions are steadily reduced and the final colour is lavender, due to V^{2+} ions.

The solution containing yellow VO_2^+ ions first turns green. This is caused by a mixture of yellow VO_2^+ ions and blue VO^{2+} ions. The solution turns blue when all the VO_2^+ ions have been reduced to blue VO^{2+} ions, green again as V^{3+} ions are formed and finally lavender when reduction to V^{2+} is complete.

Practical aspects of electrochemistry

Electricity from chemical reactions

There are three ways in which useful amounts of electricity can be produced from chemical reactions. Chemical energy is converted directly into electrical energy. The process is not 100% efficient as there are some heat losses.

Disposable batteries

Standard AA batteries consist of a zinc anode and a cathode of a carbon rod packed round with granules of manganese(IV) oxide. The electrolyte is a paste of ammonium chloride. The zinc loses electrons to form zinc ions. At the cathode, manganese(IV) oxide is reduced to a manganese(III) compound.

An AA alkaline battery

Alkaline batteries are similar, except that the electrolyte is sodium hydroxide. Here, the anode reaction is:

$$Zn(s) + 2OH^-(aq) \rightarrow Zn(OH)_2(s) + 2e^-$$

The cathode reaction is

$$2MnO_2(s) + 2H_2O(l) + 2e^- \rightarrow 2MnO(OH)(s) + 2OH^-(aq)$$

Since the concentration of ions remains constant, the voltage does not fall until all the zinc has been used, at which point the battery becomes flat.

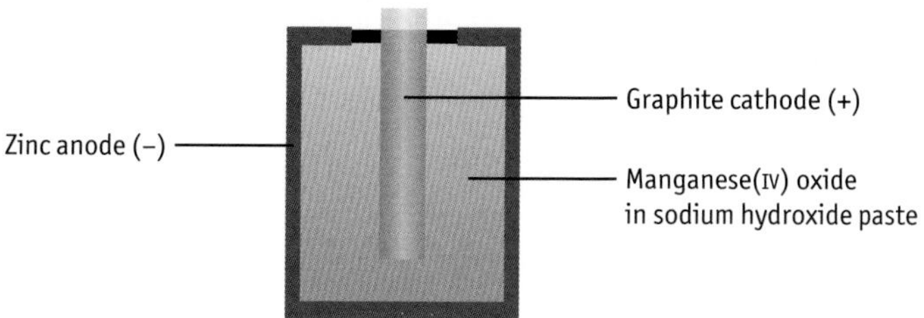

Figure 10.6 *An alkaline battery*

Mercury batteries can be used in watches, cameras and heart pacemakers. The anode is zinc, the cathode is steel and the electrolyte is a paste of mercury(II) oxide in alkali.

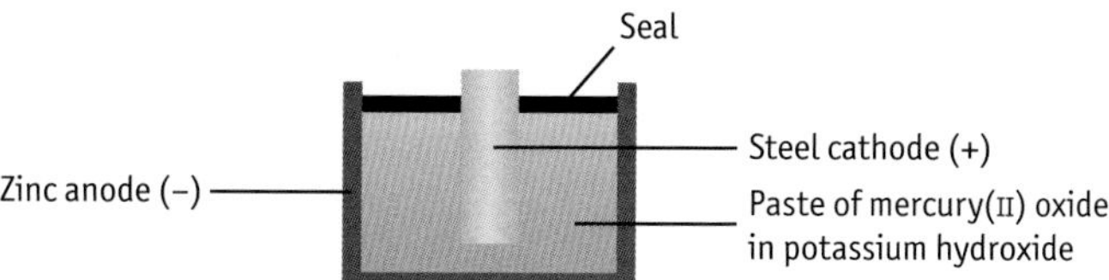

Figure 10.7 *A mercury button battery*

Other button batteries are made of lithium and manganese(IV) oxide, or zinc and silver oxide with suitable electrolytes.

Rechargeable batteries

The lead–acid battery is used in cars. Each cell consists of two lead plates. The cathode is coated with solid lead(IV) oxide. The electrolyte is a fairly concentrated solution of sulfuric acid.

The anode reaction is:

$$Pb(s) + SO_4^{2-}(aq) \rightarrow PbSO_4(s) + 2e^-$$

The cathode reaction is:

$$PbO_2(s) + 4H^+(aq) + SO_4^{2-}(aq) + 2e^- \rightarrow PbSO_4(s) + 2H_2O(l)$$

The overall discharging reaction is:

$$Pb(s) + PbO_2(s) + 4H^+(aq) + 2SO_4^{2-}(aq) \rightarrow 2PbSO_4(s) + 2H_2O(l)$$

The potential of each cell is +2.0 V. Normally, six cells are arranged in series creating a battery with a potential of 12 V.

A car battery

When all the lead(IV) oxide has been reduced, the battery is flat.

The discharging reaction is reversible. If an external potential greater than 12 V is applied, the reaction is driven backwards and the plates restored to their original composition.

The overall charging reaction is:

$$2PbSO_4(s) + 2H_2O(l) \rightarrow Pb(s) + PbO_2(s) + 4H^+(aq) + 2SO_4^{2-}(aq)$$

Hybrid cars have two engines. One is a conventional gasoline engine and the other is an electric engine powered by rechargeable batteries. When the driver applies the brakes, the kinetic energy of the car is converted to electric energy which recharges the batteries.

Cars powered by batteries alone have a limited range, the average being 50 miles.

Many digital cameras contain rechargeable lithium cells that use a solid polymer electrolyte. There are several types with different materials for the anode and cathode. One type of battery has a lithium anode and a titanium(IV) sulfide cathode. The reactions are:

$$Li \rightleftharpoons Li^+ + e^-$$

$$TiS_2 + e^- \rightleftharpoons TiS_2^-$$

The Li^+ and the TiS_2^- ions then form solid $LiTiS_2$.

The main principles of storage (rechargeable) cells are that:

- The chemical reactions at both electrodes must be able to be reversed when an electrical potential is applied.
- The oxidised and reduced forms of the anode and cathode must be solid.

A zero-emission London bus

TOPFOTO

Fuel cells

Manned spacecraft are powered by hydrogen–oxygen fuel cells. These cells are also being developed for commercial use. Some London buses have such a fuel cell. The electricity it produces powers the electric motor of the bus. The buses are advertised as 'zero-emission' buses because, when they are operating only water, and no carbon dioxide, is produced.

The principle behind a hydrogen–oxygen fuel cell is that hydrogen gas, in an alkaline solution of potassium hydroxide, is oxidised at the anode and oxygen is reduced at the cathode.

The electrodes act both as electrical conductors and as catalysts for the reactions. They are made from metals such as platinum, nickel or rhodium and must be very porous to allow the gases to pass through and come into contact with the electrolyte.

The two *reduction* half-equations are:

$$2H_2O(l) + 2e^- \rightleftharpoons H_2(g) + 2OH^-(aq) \qquad E^\ominus = -0.83V$$
$$\tfrac{1}{2}O_2(g) + H_2O(l) + 2e^- \rightleftharpoons 2OH^-(aq) \qquad E^\ominus = +0.40V$$

The overall equation and the cell potential are obtained by reversing the first half-equation and adding it to the second half-equation. This gives:

$$H_2(g) + \tfrac{1}{2}O_2(g) \rightleftharpoons H_2O(g) \quad E^\ominus_{cell} = +0.40 - (-0.83) = +1.23\,V$$

Steam can be seen from the exhaust pipe near the roof on zero-emission buses. This is the water that has been produced by the reaction of the hydrogen fuel with oxygen.

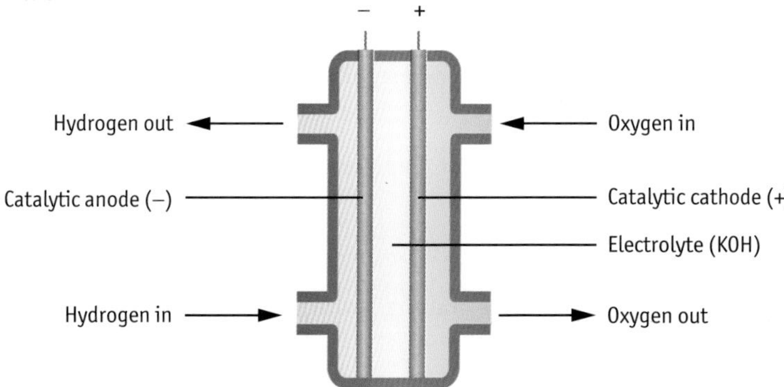

Figure 10.8 *A hydrogen fuel cell*

The statement that there are no emissions (implying no CO_2) is misleading. The hydrogen could be produced by electrolysis, which consumes 96 million coulombs of electricity per tonne of hydrogen. This is equivalent to a current of 26 000 amp h^{-1}. This electricity will almost certainly have been produced by burning fossil fuels such as coal, oil or gas. Therefore, although the bus does not produce carbon dioxide, it is created at the power station or at the plant making the hydrogen from methane. Carbon dioxide is produced in approximately the same amount as if the bus had been powered by a diesel engine. The only way that a hydrogen–oxygen fuel cell could be rightly described as having 'zero emissions' is if the electricity had been produced by nuclear power or some form of renewable energy.

Another source of hydrogen is the endothermic reaction of methane with steam in a two-stage process:

$$CH_4 + 2H_2O \rightarrow CO_2 + 4H_2$$

Some prototype cars have been designed to run on fuel cells that use hydrogen. There are two, as yet, insurmountable problems:

- the storage of hydrogen gas in the car
- the problem of refuelling the vehicle at a garage forecourt

These pose dangers since gaseous hydrogen under pressure is highly flammable.

Research is being carried out into designing fuel cells that will use other fuels (e.g. ethanol), rather than hydrogen. The two half-equations for ethanol are:

$$C_2H_5OH(l) + 12OH^-(aq) \rightleftharpoons 2CO_2(g) + 9H_2O(l) + 12e^-$$
$$O_2(g) + 2H_2O(l) + 4e^- \rightleftharpoons 4OH^-(aq)$$

Ethanol could be made by fermenting sugar or cereals, such as corn, wheat or rice. The carbon dioxide photosynthesised by the crop would be converted into ethanol by the fermentation process and then back to carbon dioxide in the fuel cell. This would have a smaller carbon footprint than the fossil fuel. However, it is not a truly carbon-neutral process, because fossil fuels would have to be used during the growth of the crop and in the manufacture of ethanol. Another major drawback to the production of bioethanol is that it uses agricultural land. This means that food crops are not grown on that land, which results in a reduction in world food production. China and India are developing economically and are increasing their demand for meat, which requires a considerable amount of grain to produce. This, combined with the drive to produce bioethanol, has pushed up the price of sugar and grain, and resulted in starvation in poor countries.

Methanol is another possibility for a fuel cell. However, methanol is made from carbon monoxide and hydrogen:

$$CO + 2H_2 \rightarrow CH_3OH$$

The carbon monoxide comes from coal or methane and the hydrogen itself has a significant carbon footprint.

Breathalysers

Driving after consuming alcohol is a dangerous activity, not only to the occupants of the car but also to other road-users and pedestrians. There is a legal limit to the amount of alcohol that car drivers may have in their bloodstreams. The problem is how to test this amount accurately.

The earliest method to detect if drivers had had too much alcohol was to make them attempt to walk along a straight chalk line or to touch their noses with their eyes shut. Neither method is likely to stand up in a court of law, so scientists designed non-subjective tests for levels of alcohol in the bloodstream.

All methods depend on the equilibrium between ethanol dissolved in the blood and that present in the air breathed out. The first method consisted of a tube containing potassium dichromate(VI) absorbed onto silica gel. The driver had to blow through the tube and fill a plastic bag attached to the end of the tube. Ethanol in the breath reduced the dichromate(VI) ions, causing a colour change from orange to green. The amount of alcohol in the breath and hence in the bloodstream was measured by the extent of 'greening'. The equation for the reaction is:

$$C_2H_5OH + 2Cr_2O_7^{2-} + 16H^+ \rightarrow 2CO_2 + 11H_2O + 4Cr^{3+}$$

2×-2	$4 \times +6$	$2 \times +4$	$4 \times +3$

e The oxidation numbers of each of the two carbon atoms change by 6 (12 in total); those of each of the chromium atoms change by 3. Therefore, four chromium atoms are needed.

Although this method works, it is not particularly accurate. It also had the disadvantage that the tube containing the potassium dichromate(VI) could not be reused. As a result, a machine capable of measuring the alcohol content of expired air was designed. There are two types:

- Infrared analyser — the driver breathes into a box and the infrared absorption in the fingerprint region of ethanol is measured. This is more accurate than the dichromate method. Its main drawback is that the peak height does not depend linearly on the concentration of ethanol.
- Ethanol fuel cell — this is the most accurate method. A fixed volume of breath is pushed through a fuel cell. The ethanol in the breath is oxidised at the catalytic anode and oxygen in the air is reduced at the cathode. The total quantity of electricity produced depends linearly on the amount of ethanol in the breath sample. The equations for the reactions are

$$C_2H_5OH(l) + 12OH^-(aq) \rightleftharpoons 2CO_2(g) + 9H_2O(l) + 12e^-$$
$$O_2(g) + 2H_2O(l) + 4e^- \rightleftharpoons 4OH^-(aq)$$

The spectrometer cannot use the O–H stretching absorption because water also absorbs at this frequency. The C–H and C–C bands cannot be used because other organic compounds that may be present in breath absorb at these frequencies.

e It is not necessary to learn these equations. What matters is that the ethanol is oxidised and either the dichromate or the oxygen in the air is reduced.

Oxidising and reducing agents

Oxidising agents

An oxidising agent is a species that removes electrons from another species, thus oxidising it. It is itself reduced by the gain of electrons.

Some oxidising agents, the species produced when they react and their standard reduction potentials are given in Table 10.4.

***Table 10.4** Common oxidising agents listed in order of decreasing power*

Oxidising agent	Oxidising species	Reduced species	$E^\ominus$/V
Ozone	O_3 in $H^+(aq)$	O_2, H_2O	+2.07
Persulfate ions	$S_2O_8^{2-}$	SO_4^{2-}	+2.01
Hydrogen peroxide	H_2O_2 in $H^+(aq)$	H_2O	+1.77
Chloric(I) acid	$HOCl$ in $H^+(aq)$	Cl_2, H_2O	+1.64
Manganate(VII) ions	MnO_4^- in $H^+(aq)$	Mn^{2+}, H_2O	+1.52
Lead(IV) oxide	PbO_2 in $H^+(aq)$	Pb^{2+}, H_2O	+1.47
Chlorine	Cl_2	Cl^-	+1.36
Dichromate(VI) ions	$Cr_2O_7^{2-}$ in $H^+(aq)$	Cr^{3+}, H_2O	+1.33
Manganese(IV) oxide	MnO_2 in $H^+(aq)$	Mn^{2+}, H_2O	+1.23
Oxygen	O_2 in $H^+(aq)$	H_2O	+1.23
Iodate(V) ions	IO_3^- in $H^+(aq)$	I_2, H_2O	+1.19
Bromine	Br_2	Br^-	+1.07
Iron(III) ions	Fe^{3+}	Fe^{2+}	+0.77
Iodine	I_2	I^-	+0.54
Tetrathionate ions	$S_4O_6^{2-}$	$S_2O_3^{2-}$	+0.09

e The oxidation number of an element in the oxidising agent decreases (becomes less positive or more negative). The oxidation number of an element in the species being oxidised increases (becomes more positive or less negative).

All oxidising agents should oxidise the reduced form of any species below them in the table. For example, lead(IV) oxide will oxidise chloride ions to chlorine. A laboratory preparation of chlorine is to warm concentrated hydrochloric acid with lead(IV) oxide.

Worked example

Show that iodate(V) ions in acid solution should oxidise iodide ions and write the overall equation for the reaction.

Answer

The two reduction half-equations are:

$$IO_3^-(aq) + 6H^+(aq) + 5e^- \rightleftharpoons \frac{1}{2}I_2(s) + 3H_2O(l) \quad E^\ominus = +1.19\,V$$

$$\frac{1}{2}I_2(s) + e^- \rightleftharpoons I^-(aq) \quad E^\ominus = +0.54\,V$$

Iodate(V) ions and iodide ions are the reactants. The overall equation is obtained by reversing the second equation, multiplying it by 5 and then adding it to the first equation:

$$IO_3^-(aq) + 6H^+(aq) + 5I^-(aq) \rightarrow 3I_2(s) + 3H_2O(l) \quad E^\ominus_{cell} = +1.19 + (-0.54) = +0.65\,V$$

$E^\ominus_{cell}$ is positive, so the redox reaction is feasible.

Estimation of the concentration of a solution of an oxidising agent

As can be seen from Table 10.4, all the oxidising agents (apart from tetrathionate ions) in the list should oxidise iodide ions to iodine.

The method is to add an excess of iodide ions to a known volume of the solution of the oxidising agent. The liberated iodine is then titrated against standard sodium thiosulfate solution.

The feasibility of the redox reactions involved can be worked out using reduction potential data.

Worked example

Write the equation for the oxidation of thiosulfate ions by iodine. Calculate the standard cell potential and comment on the feasibility of the reaction.

Answer

The two reduction half-equations are:

$$\frac{1}{2}I_2(s) + e^- \rightleftharpoons I^-(aq) \quad E^\ominus = +0.54\,V$$

$$\frac{1}{2}S_4O_6^{2-} + e^- \rightleftharpoons S_2O_3^{2-} \quad E^\ominus = +0.09\,V$$

The overall equation is obtained by reversing the second equation and adding it to the first. The overall equation is multiplied by 2, to remove the halves.

$$I_2(s) + 2S_2O_3^{2-}(aq) \rightarrow 2I^-(aq) + S_4O_6^{2-}(aq) \quad E^\ominus_{cell} = +0.54 + (-0.09) = +0.45\,V$$

The value of the standard cell potential is positive, so the reaction is thermodynamically feasible.

The reaction in the worked example above is fundamental to the method for estimating the concentrations of solutions of oxidising agents.

The standard method is as follows:

- A known volume, usually 25.0 cm^3, of the solution of the oxidising agent is pipetted into a conical flask.
- A similar volume of dilute sulfuric acid is added, followed by excess solid potassium iodide. The mixture is swirled to ensure that all the oxidising agent reacts.

- A burette is filled with a standard solution of sodium thiosulfate.
- Sodium thiosulfate solution is added steadily to the conical flask from the burette. As this is done, the colour of the iodine in the solution fades.
- When the solution in the flask is a pale straw colour, a few drops of starch solution are added. This reacts reversibly with the iodine to form a dark blue-black substance.
- Sodium thiosulfate is then added dropwise, until the solution is decolorised.
- The experiment is repeated until two consistent titres are obtained.

If the starch is added too soon an insoluble starch–iodine complex is formed, which makes the titre inaccurate.

The reactions are:

Oxidising agent + Excess iodide ions $\rightarrow$ Iodine

$2[O] + 2I^- \rightarrow I_2$

$I_2 + 2S_2O_3^{2-} \rightarrow 2I^- + S_4O_6^{2-}$

If the oxidising agent produces x mol of I_2, this then reacts with $2x$ mol of sodium thiosulfate.

Worked example

Brass is an alloy of copper and zinc, the proportions of which are varied to create a range of brasses with different properties. The percentage of copper in a small brass screw was determined as follows:

- The screw was weighed. Its mass was found to be 2.19 g.
- Some concentrated nitric acid was added to the screw. This was carried out in a fume cupboard. The copper and the zinc both reacted to form soluble nitrates, nitrogen dioxide and water.
- The solution was neutralised. It, and the washings, were transferred to a $250\,cm^3$ standard flask and the volume made up to $250\,cm^3$ with distilled water.
- $25.0\,cm^3$ samples were measured by pipette into conical flasks and excess solid potassium iodide was added.
- The zinc ions did not react. Copper ions reacted according to the equation;
  $2Cu^{2+} + 4I^- \rightarrow 2CuI + I_2$
- The liberated iodine was then titrated against a $0.106\,mol\,dm^{-3}$ solution of sodium thiosulfate, adding starch as the indicator near the end point.
  $I_2 + 2Na_2S_2O_3 \rightarrow 2NaI + Na_2S_4O_6$

The mean titre was $23.75\,cm^3$.

Calculate the percentage of copper in the brass screw.

Answer

amount (in moles) of sodium thiosulfate $= 0.106\,mol\,dm^{-3} \times 0.02375\,dm^3$
$= 0.0025175\,mol$

2 mol $Na_2S_2O_3^{2-}$ react with 1 mol I_2, which comes from 2 mol Cu^{2+}, so, amount of Cu^{2+} in pipetted sample = amount of $S_2O_3^{2-}$ in titre = 0.0025175 mol

amount of Cu^{2+} in $250\,cm^3 = 10 \times 0.0025175 = 0.025175\,mol$

mass of copper in brass screw $= 63.5\,g\,mol^{-1} \times 0.025175\,mol = 1.599\,g$

% of copper in brass screw $= 1.599 \times 100/2.19 = 73.0\%$

Reducing agents

A reducing agent is a species that gives electrons to another species, thus reducing it. It is itself oxidised by loss of electrons.

Some reducing agents, the species produced when they react and their standard *reduction* potentials are given in Table 10.5.

Reducing agent	Oxidised form	$E^\ominus$/V
$(COO)_2^{2-}$	$2CO_2$	−0.49
H_2S	S	+0.14
Sn^{2+}	Sn^{4+}	+0.15
I^-	$\frac{1}{2}I_2$	+0.54
H_2O_2	O_2	+0.68
Fe^{2+}	Fe^{3+}	+0.77
Br^-	$\frac{1}{2}Br_2$	+1.07
Pb^{2+}	PbO_2 (Pb^{4+})	+1.47

Decreasing strength as a reducing agent

***Table 10.5** Common reducing agents*

Reducing agents become oxidised, so the standard reduction potential of +0.54 V in Table 10.5 is the potential for the half-reaction:

$$\frac{1}{2}I_2(s) + e^- \rightleftharpoons I^-(aq) \qquad E^\ominus = +0.54\,V$$

so

$$I^-(aq) \rightleftharpoons \frac{1}{2}I_2(s) + e^- \qquad E^\ominus = -0.54\,V$$

The reducing agent is on the *right-hand* side of the standard reduction potential half-equation.

A reducing agent will reduce the oxidised form of a species that has a more positive reduction potential. For example, Fe^{2+} reduces bromine to bromide ions: $E^\ominus$ for $Br_2/Br^- = +1.07$ V.

e The oxidation number of an element in the reducing agent increases (becomes more positive or less negative). The oxidation number of an element in the species being reduced decreases (becomes less positive or more negative).

Estimation of the concentration of a reducing agent

As can be seen from Table 10.5 and the $E^\ominus$ value of the MnO_4^-, H^+/Mn^{2+} system, all the reducing agents in the table will reduce manganate(VII) ions in acidic solution. This is the basis of the determination of the concentration of a reducing agent. The method is:

- A sample of known volume, usually 25.0 cm^3, of the reducing agent is pipetted into a conical flask.
- Approximately 25 cm^3 of dilute sulfuric acid is added.
- A burette is filled with a standard solution of potassium manganate(VII).
- Potassium manganate(VII) solution is added steadily, with swirling, until the purple colour disappears slowly.

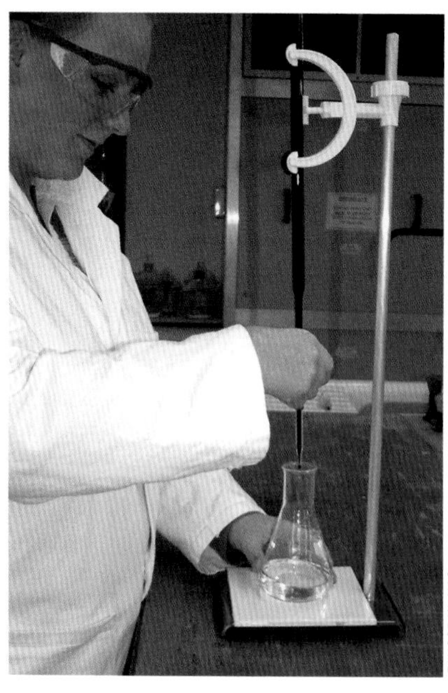

ZAHOOR UL-HAQ

Titration of Fe(II) with manganate(VII)

- The potassium manganate(VII) is then added dropwise, until the solution becomes slightly pink.
- The titration is repeated until at least two consistent titres are obtained.

Worked example

Show that hydrogen peroxide should reduce manganate(VII) ions in acid solution. Write the overall equation for the reaction.

Answer

$O_2(g) + 2H^+(aq) + 2e^- \rightleftharpoons H_2O_2(aq)$ $E^\ominus = +0.68\,V$

$MnO_4^-(aq) + 8H^+(aq) + 5e^- \rightleftharpoons Mn^{2+}(aq) + 4H_2O(l)$ $E^\ominus = +1.52\,V$

The reactants are H_2O_2 and MnO_4^-, so the first equation has to be reversed. To achieve the same number of electrons in each equation, multiply the first equation by 5 and the second equation by 2. This gives ten electrons in each equation, which will cancel when the equations are added.

$5H_2O_2(aq) \rightleftharpoons 5O_2(g) + 10H^+(aq) + 10e^-$ $E^\ominus = -0.68\,V$

$2MnO_4^-(aq) + 16H^+(aq) + 10e^- \rightleftharpoons 2Mn^{2+}(aq) + 8H_2O(l)$ $E^\ominus = +1.52\,V$

$5H_2O_2(aq) + 2MnO_4^-(aq) + 6H^+(aq) \rightleftharpoons 5O_2(g) + 2Mn^{2+}(aq) + 8H_2O(l)$

$E^\ominus_{reaction} = +0.84\,V$

The value of $E^\ominus_{reaction}$ is positive, so hydrogen peroxide should reduce manganate(VII) ions in acid solution.

There is no need to add an indicator because the manganate(VII) solution is so intensely coloured. The titration is stopped when the smallest excess of MnO_4^- ions cause a faint pink colour to be visible.

Worked example

Iron tablets contain hydrated iron(II) sulfate mixed with a filler. Some tablets were crushed and 7.17 g of the powdered tablets were dissolved in water. The solution was made up to 250 cm^3. 25.0 cm^3 of this solution was pipetted into a conical flask and 25 cm^3 of dilute sulfuric acid was added from a measuring cylinder. This mixture was then titrated with a 0.0205 mol dm^{-3} solution of potassium manganate(VII) until a faint pink colour remained. The titration was repeated and the mean titre was found to be 23.40 cm^3. Calculate the percentage of iron in the iron tablets.

Answer

amount (moles) of MnO_4^- in titre = $0.0205\,mol\,dm^{-3} \times 0.02340\,dm^3$
= 0.0004797 mol

The equation for the reaction is:

$5Fe^{2+}(aq) + MnO_4^-(aq) + 8H^+(aq) \rightarrow 5Fe^{3+}(aq) + Mn^{2+}(aq) + 8H_2O(l)$

amount (moles) of Fe^{2+} in 25cm^3 of solution = 0.0004797 mol $MnO_4^- \times 5/1$
= 0.002399 mol

amount (moles) of Fe^{2+} in 250 cm^3 of solution = $10 \times 0.002399 = 0.02399$ mol

mass of iron in sample = $55.8\,g\,mol^{-1} \times 0.02399\,mol = 1.338\,g$

$$\text{\% of iron in tablet} = \frac{\text{mass of iron} \times 100}{\text{mass of powdered tablet}} = \frac{1.338 \times 100}{7.17} = 18.7\%$$

Uncertainty of measurements

The accuracy of a pipette depends on a number of factors, such as its quality, and the temperature and viscosity of the solution being used. Most school pipettes are accurate to at least one decimal place, as are burettes. At worst, the error in a burette is $\pm 0.05\,cm^3$ for each reading and so the maximum error in a titre is $\pm 0.1\,cm^3$.

The concentrations of standard solutions are chosen so that the titre will be between $20\,cm^3$ and $30\,cm^3$. Suppose the titre is $24.00\,cm^3$. The percentage error in its use is:

$$\frac{\text{error} \times 100}{\text{titre}} = \frac{0.1 \times 100}{24.00} = 0.4\%$$

If the titre had been $12.00\,cm^3$, the error would still have been $\pm 0.1\,cm^3$, but the percentage error would be $0.1 \times 100/12.00 = 0.8\%$.

A chemical balance that weighs to two decimal places has an error of ± 0.01 g for each reading (including the zero reading if the balance is tared). Thus the error in the mass is 0.02 g.

In the worked example on p. 203 about iron tablets:

- the percentage error due to weighing is $0.02 \times 100/7.17 = 0.28\%$
- the titration has a percentage error of $0.1 \times 100/23.4 = 0.43\%$

This gives an uncertainty of almost 0.75% in the iron content of the iron tablets.

Worked example

A sample of mass 1.30 g of a *d*-block metal sulfate, $M_2(SO_4)_3$ was dissolved in water. Excess potassium iodide was added and the iodine liberated was titrated against standard sodium thiosulfate solution. The titre showed that 0.00325 mol of the transition metal sulfate was present in the 1.30 g.

a Calculate the molar mass of the $M_2(SO_4)_3$.

b Hence calculate the atomic mass of the element M. Suggest its identity.

c Assume that the error in each weighing is ± 0.01 g, calculate the percentage error in the mass of $M_2(SO_4)_3$ taken.

d Assuming that there are no other significant errors, calculate the maximum and minimum values of the molar mass and hence of the atomic mass of the element M.

e What does this indicate about the reliability of the identification of M?

Answer

a molar mass = mass/moles = $\frac{1.30\,g}{0.00325\,mol^{-1}} = 400\,g\,mol^{-1}$

b mass of the sulfate group = $3 \times (32.1 + 4 \times 16.0) = 288.3$
mass due to 2 atoms of M = $400 - 288.3 = 117.7$
relative atomic mass of M = $1/2 \times 117.7 = 55.85$
so, the element M is iron, as the r.a.m. of iron in the periodic table is 55.8

c total weighing error is ± 0.02 g
percentage error in weighing the solid = $\frac{0.02 \times 100}{1.32} = 1.52\%$

d $\pm 1.52\%$ in $400\,g\,mol^{-1}$ is $\pm \frac{1.52 \times 400}{100}$

$= \pm 6.1\,g\,mol^{-1}$ in the molar mass

so, the maximum molar mass is $406.1\,g\,mol^{-1}$

minimum molar mass is $393.9\,g\,mol^{-1}$

maximum relative atomic mass is $\frac{1}{2} \times (406.1 - 288.3) = 58.9$

minimum relative atomic mass is $\frac{1}{2} \times (393.9 - 288.3) = 52.8$

e These values suggest that the metal M could be manganese ($A_r = 54.9$), iron ($A_r = 55.8$) or chromium ($A_r = 58.9$) and so the validity of the identification of M as iron is poor.

Questions

Refer to Table 10.1 on p. 187 for standard reduction potential values.

1 Give the oxidation number of oxygen in:

a O^{2-} **b** O_2^{2-} **c** O_2^- **d** OF_2

2 Give the oxidation number of sulfur in:

a H_2S **b** SO_4^{2-} **c** S_2Cl_2 **d** H_2SO_3

3 Write the half-equations for:

a the reduction of $Cr_2O_7^{2-}$ ions in acid solution to Cr^{3+} ions and water

b the reduction of chlorine to chloride ions

c the oxidation of iodide ions to iodine

d the oxidation of Fe^{2+} ions to Fe^{3+} ions

4 Use your answers to question 3 to write overall equations for:

a the reaction between acidified dichromate(VI) ions and iodide ions

b the reaction between chlorine and Fe^{2+} ions

5 Draw a labelled diagram of the apparatus that you would use to measure the standard reduction potential of acidified potassium manganate(VII).

6 a Calculate the standard cell potential and hence the thermodynamic feasibility of a reaction in which manganate(VII) ions in acid solution oxidise ethanedioate ions to carbon dioxide, under standard conditions at 25°C.

$$2CO_2(g) + 2e^- \rightleftharpoons C_2O_4^{2-}(aq) \quad E^\ominus = -0.49\,V$$

$$MnO_4^-(aq) + 8H^+(aq) + 5e^- \rightleftharpoons Mn^{2+}(aq) + 4H_2O(l) \quad E^\ominus = +1.52\,V$$

b Suggest why, in practice, the reaction does not occur under standard conditions at 25°C.

7 Explain how the potential of the standard hydrogen electrode would alter if the pressure of hydrogen gas were to be increased.

8 Refer to Table 10.1 and select:

a the strongest oxidising agent

b the weakest oxidising agent

c the strongest reducing agent

d the weakest reducing agent

9 a Calculate the cell potential for each of the following reactions and suggest whether or not the reaction is feasible:

(i) a reaction between lead(IV) ions and chloride ions

(ii) a reaction between tin(IV) ions and Fe^{2+} ions

(iii) a reaction between iodide ions and Fe^{3+} ions

b Write equations for any of the reactions in (i)–(iii) that are feasible.

10 Both potassium manganate(VII) and potassium dichromate(VI) are used in acid solution as oxidising agents. Use the data in Table 10.1 to explain why hydrochloric acid must *not* be used in the oxidation of an organic substance by potassium manganate(VII) but can be used when potassium dichromate(VI) is the oxidising agent.

11 **a** Define the term **disproportionation**.

b Explain why an element must have at least three different oxidation states to be able to undergo a disproportionation reaction.

c Use the E° values in Table 10.1 to predict whether manganate(VI) ions, MnO_4^{2-}, will disproportionate into manganate(VII) ions, MnO_4^-, and manganese(IV) oxide, MnO_2.

d Predict the effect of increasing the pH of the solution on the feasibility of this disproportionation.

12 Calculate E°_{cell} for the oxidation by oxygen of iron in aqueous solution. Write the overall equation for this reaction.

13 Explain why iron corrodes to a lesser extent in oxygenated water if the pH of the water is increased by the addition of alkali.

14 An empty crucible and its lid were weighed. Some metal carbonate, MCO_3 was added and the crucible and lid were reweighed. The crucible was then heated until there was no further loss in mass. The carbonate decomposed according to the equation:

$$MCO_3 \rightarrow MO + CO_2$$

The mass of the carbonate was 1.51 g; the mass of the oxide was 0.93 g.

a Calculate the mass of carbon dioxide produced,

b Calculate the amount (in moles) of carbon dioxide and hence the molar mass of the metal carbonate.

c Calculate the relative atomic mass of the metal M and suggest its identity.

d Assume that there is a possible error of ±0.01 g in each weighing, calculate the percentage error in the molar mass of MCO_3.

e Calculate the possible maximum and minimum values of the molar mass of MCO_3 and hence of the relative atomic mass of M.

f Comment on the reliability of the identity of M that you suggested in part c.

g How might the experiment be improved to give a more reliable identification of the metal?

15 **a** Predict whether ferrate(VI) ions, FeO_4^{2-}, will be reduced to Fe^{3+} or to Fe^{2+} ions by hydrogen peroxide in an acidic solution.

$$FeO_4^{2-}(aq) + 8H^+(aq) + 3e^- \rightleftharpoons Fe^{3+}(aq) + 4H_2O(l) \quad E^\circ = +2.20\,V$$

$$Fe^{3+}(aq) + e^- \rightleftharpoons Fe^{2+}(aq) \quad E^\circ = +0.77\,V$$

$$O_2(g) + 2H^+(aq) + 2e^- \rightleftharpoons H_2O_2(aq) \quad E^\circ = +0.68\,V$$

b Write the overall equation for the reaction that takes place.

16 25.00 cm^3 of a solution of potassium iodate(V) was pipetted into a conical flask. 25 cm^3 of dilute sulfuric acid and 2 g of potassium iodide (an excess) were added. The iodine liberated was titrated with 0.104 mol dm^{-3} sodium thiosulfate solution using starch as an indicator. The mean titre was 23.20 cm^3. Calculate the concentration of the potassium iodate(V) solution.

17 Describe how the concentration of a solution of sodium chlorate(I), NaClO, could be found experimentally.

18 What are the advantages and disadvantages of running a motor vehicle using a fuel cell that is powered by hydrogen?

19 Discuss the advantages and disadvantages of a breathalyser that uses potassium dichromate and a breathlyser that uses a fuel cell.

20 Explain why the strong O–H bond stretching vibration in ethanol cannot be used in infrared breathalysers.

21 Iron(II) sulfate contains water of crystallisation and its formula can be written as $FeSO_4.xH_2O$. Describe an experiment that would enable you to find the value of *x*. The method must involve a titration. Your answer should include what you would do, what measurements you would make and show how you would use your measurements to calculate the value of *x*. (Heating the hydrated solid would decompose the iron(II) sulfate, so this is not an acceptable method.)

22 Search the internet to find details of the composition of rechargeable lithium cells that are used in cameras and camcorders.

Transition metals and the *d*-block elements

Chapter 11

Introduction

The *d*-block elements lie between the *s*-block metals and the *p*-block non-metals in the periodic table. They are shown in Table 11.1.

Table 11.1
The d-block elements

Sc Scandium	Ti Titanium	V Vanadium	Cr Chromium	Mn Manganese	Fe Iron	Co Cobalt	Ni Nickel	Cu Copper	Zn Zinc
Y Yttrium	Zr Zirconium	Nb Niobium	Mo Molybdenum	Tc Technetium	Ru Ruthenium	Rh Rhodium	Pd Palladium	Ag Silver	Cd Cadmium
La Lanthanum	Hf Hafnium	Ta Tantalum	W Tungsten	Re Rhenium	Os Osmium	Ir Iridium	Pt Platinum	Au Gold	Hg Mercury

Electron configuration

Neutral atoms

Argon has the electron configuration $1s^2\ 2s^2\ 2p^6\ 3s^2\ 3p^6$. The next lowest energy level is the 4*s*-orbital, so it fills before the 3*d*. Therefore, potassium has the electron configuration [Ar] $3d^0\ 4s^1$ and calcium [Ar] $3d^0\ 4s^2$.

After calcium, the *d*-block starts and the 3*d*-orbitals begin to be filled. The general configuration for a *d*-block element is [Ar] $3d^x\ 4s^2$ where *x* is the number of the column along the *d*-block. Scandium is the first *d*-block element, so its configuration is [Ar] $3d^1\ 4s^2$, vanadium is the third, so its configuration is [Ar] $3d^3\ 4s^2$ and iron is the sixth so its configuration is [Ar] $3d^6\ 4s^2$.

In the fourth period, Sc to Zn, there are two exceptions to this rule. There is a slight gain in stability in having a full or half-full set of *d*-orbitals. Thus chromium, the fourth *d*-block element, has the electron configuration [Ar] $3d^5\ 4s^1$ *not* [Ar] $3d^4\ 4s^2$; copper has the configuration [Ar] $3d^{10}\ 4s^1$ *not* [Ar] $3d^9\ 4s^2$.

e The stability of a full or half-full set of orbitals only applies to *p*- and *d*-orbitals. The reason for the stability is a quantum-mechanical factor called **exchange energy.** This stabilisation is because of the different numbers of ways in which electrons in the same energy level with *parallel* spins can be considered two at a time. There are three ways of pairing up the three electrons in a p^3 configuration ($p_x^1 + p_y^1$, $p_x^1 + p_z^1$ and $p_y^1 + p_z^1$), but only one for a p^2 configuration ($p_x^1 + p_y^1$).

Positive ions

When a *d*-block element in period 4 loses electrons and forms a positive ion, the outer *s*-electrons are always lost before any *d*-electrons. Losing all the 4*s*-electrons makes the ion smaller than if it had lost its 3*d*-electrons. This means that the lattice energy and the hydration energy are more exothermic. Therefore, in a reaction the overall ΔH is energetically more favourable than if the 3*d*-electrons had been lost.

The electron configurations of *d*-block elements in period 4 and their 2+ and 3+ ions are shown in Table 11.2.

***Table 11.2** Period 4 d-block elements*

Element	Atom	Electron configuration M^{2+} ion	M^{3+} ion
Sc	[Ar] $3d^1\,4s^2$	Does not exist	[Ar] $3d^0\,4s^0$
Ti	[Ar] $3d^2\,4s^2$	[Ar] $3d^2\,4s^0$	[Ar] $3d^1\,4s^0$
V	[Ar] $3d^3\,4s^2$	[Ar] $3d^3\,4s^0$	[Ar] $3d^2\,4s^0$
Cr	[Ar] $3d^5\,4s^1$	[Ar] $3d^4\,4s^0$	[Ar] $3d^3\,4s^0$
Mn	[Ar] $3d^5\,4s^2$	[Ar] $3d^5\,4s^0$	[Ar] $3d^4\,4s^0$
Fe	[Ar] $3d^6\,4s^2$	[Ar] $3d^6\,4s^0$	[Ar] $3d^5\,4s^0$
Co	[Ar] $3d^7\,4s^2$	[Ar] $3d^7\,4s^0$	[Ar] $3d^6\,4s^0$
Ni	[Ar] $3d^8\,4s^2$	[Ar] $3d^8\,4s^0$	[Ar] $3d^7\,4s^0$
Cu	[Ar] $3d^{10}\,4s^1$	[Ar] $3d^9\,4s^0$	Does not exist
Zn	[Ar] $3d^{10}\,4s^2$	[Ar] $3d^{10}\,4s^0$	Does not exist

Transition metals

The *d*-block elements are those between the *s*-block and the *p*-block. All *d*-block elements have an outer electron configuration of nd^x, $(n + 1)s^y$, where x is any number from 1 to 10 and y is 0, 1 or 2.

Transition metals are defined differently:

A transition metal has one or more unpaired d-electrons in one of its ions.

- Scandium forms Sc^{3+} as its only ion. This has no *d*-electrons and so scandium is not a transition metal.
- Titanium forms a Ti^{3+} ion, which has one unpaired *d*-electron, and Ti^{2+}, which has two unpaired electrons. Therefore, titanium is a transition metal.
- Iron has four unpaired electrons in a Fe^{2+} ion (two of the six *d*-electrons are paired in one of the *d*-orbitals) and five in a Fe^{3+} ion. Therefore, iron is a transition metal.
- Zinc does not form a Zn^{3+} ion. Zn^{2+} has ten *d*-electrons that are all paired in five full *d*-orbitals. Therefore, zinc is not a transition metal.

The arrangements of electrons in Fe^{2+} and Zn^{2+} ions are shown in Figure 11.1.

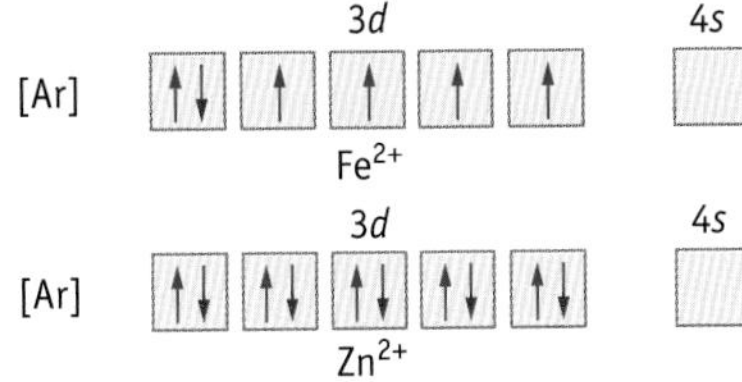

***Figure 11.1** Electron configurations of Fe^{2+} and Zn^{2+} ions*

Ionisation energies

First ionisation energies

In period 3, the first ionisation energies follow a general upward trend from sodium across to argon. This is caused by an increase in the nuclear charge without an increase in the number of shielding electrons in the inner orbits. The effective nuclear charge increases considerably, and this causes a large increase in the first ionisation energies. The first ionisation energy of sodium is 494 kJ mol^{-1} and that of argon is 1520 kJ mol^{-1}.

The pattern is different across the *d*-block. The outer electron that is removed in the reaction:

$$M(g) \rightarrow M^+(g) + e^-$$

is a 4s-electron, and it is shielded by the inner 3*d*-electrons (as well as by the 1*s*-, 2*s*-, 2*p*-, 3*s*- and 3*p*-electrons). Although the nuclear charge increases across the block from scandium to zinc, the number of inner shielding 3*d*-electrons increases as well. This means that the first ionisation energies of the *d*-block elements in period 4 are fairly similar. This is shown in Figure 11.2.

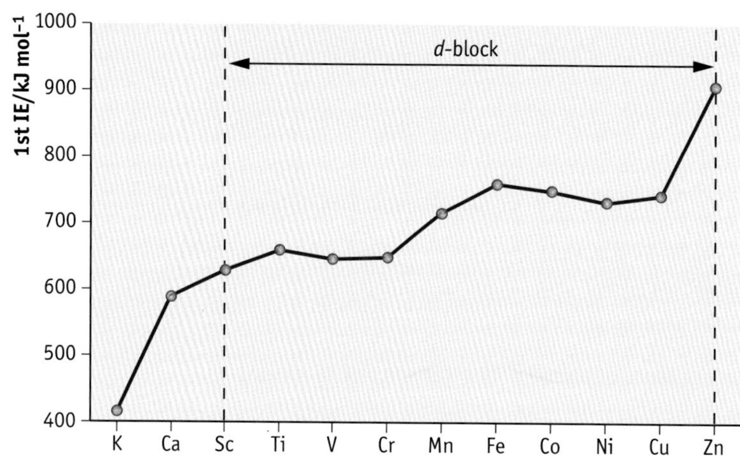

Figure 11.2
First ionisation energies of period 4 elements from potassium to zinc

Successive ionisation energies

There is normally a big jump between successive ionisation energies as a new quantum shell loses an electron. For example, there is a big jump between the second and third ionisation energies of magnesium (electron structure: $1s^2\ 2s^2\ 2p^6\ 3s^2$) because the third electron has to come from the second orbit. This electron is subjected to a much stronger pull from the nucleus as it is much less shielded.

The energy levels of the 3*d*- and the 4*s*-electrons are very similar in *d*-block elements and so the big jump comes after all the 4*s*- and 3*d*-electrons have been removed. There is no big jump between removing the 4*s*-electrons and starting to remove the 3*d*-electrons. This is shown in Table 11.3 and in Figure 11.3.

It can be seen that the difference in energy between the 4*s*- and 3*d*-electrons is quite small, as there is not much of a jump as the first 3*d*-electron is removed.

The numbers in bold in Table 11.3 refer to the first 3*p*-electron being removed.

It can be seen clearly that there is a much larger difference in energy between 3*p*- and 3*d*-electrons, as there is a large jump when the first 3*p*-electron is being removed.

	Titanium ([Ar] $3d^2\,4s^2$)	Vanadium ([Ar] $3d^3\,4s^2$)	Chromium ([Ar] $3d^5\,4s^1$)	Manganese ([Ar] $3d^5\,4s^2$)
1st and 2nd	649	720	937*	794
2nd and 3rd	1410*	1500*	1400	1740*
3rd and 4th	1450	1730	1780	1940
4th and 5th	**5450**	1680	2300	2170
5th and 6th		**6120**	1630	2390
6th and 7th			**7900**	1750
7th and 8th				**7300**

The * refers to the first 3*d*-electron being removed.

Table 11.3 *Difference in successive ionisation energies in kJ mol*$^{-1}$

This is shown graphically in Figure 11.3.

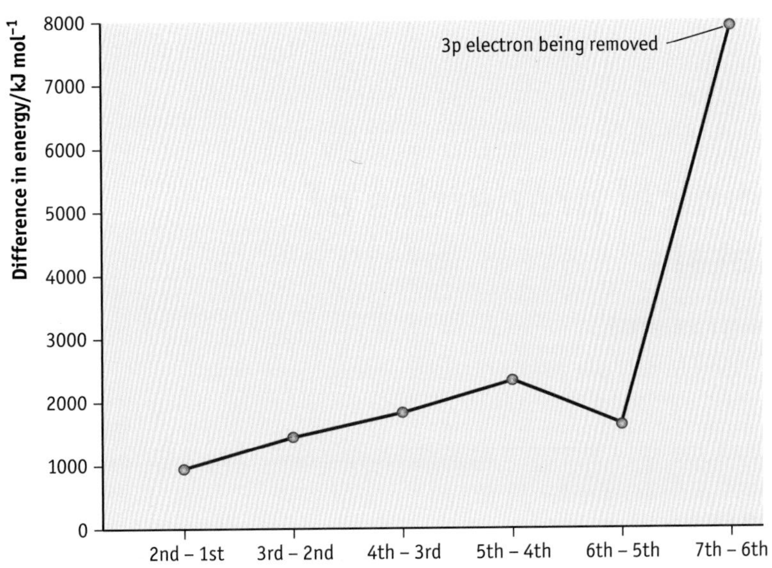

Figure 11.3 *The difference in successive ionisation energies for chromium*

Melting temperature and hardness

Metallic bonding can be described as the attraction between the delocalised electrons and the positive ions formed when the metal atoms lose their valence electrons into the cloud of delocalised electrons. This bonding is moderately strong and depends on the ionic radius and the number of electrons delocalised. The strength of the bond determines the melting temperature of the metal and its tensile strength. For example, potassium has a large ionic radius and only one delocalised electron. Its melting temperature is 64°C and it is so soft that it can be cut with a knife.

In the *d*-block metals, both the *d*-electrons and *s*-electrons are used in bonding. Therefore, the melting temperatures of *d*-block metals are significantly higher than those of *s*-block metals. The metals are also much harder and stronger.

Another effect of the stronger metallic bonding is that the enthalpy of atomisation of the metal is increased considerably. For example, ΔH_a of calcium is +193 kJ mol^{-1} whereas ΔH_a of iron is +418 kJ mol^{-1}.

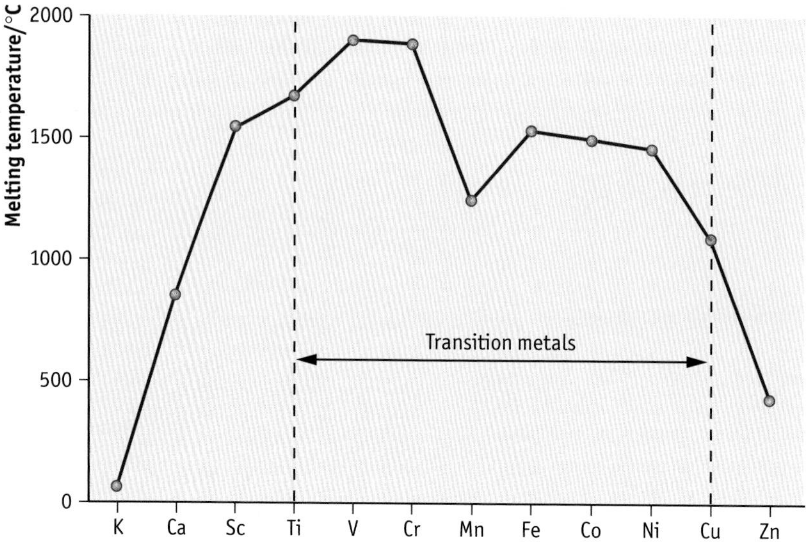

***Figure 11.4** Melting temperatures of the period 4 elements from potassium to zinc*

Common chemical properties of transition metals

Variable oxidation states

The metals in the *s*-block exist in only one oxidation state in their compounds — for example, sodium is always +1 and magnesium is always +2.

The non-transition *d*-block metals also have a single oxidation state. Zinc is always +2 in its compounds and scandium is always +3.

The transition metals exist in several different oxidation states. For example, manganese has stable compounds such as $MnSO_4$ (+2), MnO_2 (+4), K_2MnO_4 (+6) and $KMnO_4$ (+7). It also forms compounds in the +3 and +5 states.

In Table 11.4 oxidation states in red are the stable states; those in blue are less stable.

Some redox reactions of vanadium were covered in Chapter 10, p. 193. Other redox reactions of transition metals are listed on p. 227 in this chapter.

***Table 11.4** Oxidation states of the d-block metals*

Sc	Ti	V	Cr	Mn	Fe	Co	Ni	Cu	Zn
				+7					
			+6	+6	+6				
		+5		+5					
	+4	+4	+4	+4					
+3	+3	+3	+3	+3	+3	+3	+3		
	+2	+2	+2	+2	+2	+2	+2	+2	+2
								+1	

Bonding

In most compounds, in the +2 and +3 oxidation states, the transition metals are ionically bonded. For example, iron(II) sulfate, $FeSO_4$, is ionic, as is chromium(III) sulfate, $Cr_2(SO_4)_3$. Anhydrous chlorides, bromides and iodides are usually covalent, but are ionic when hydrated.

When the transition metal is in an oxidation state of +4 or higher, it is covalently bonded, often in an anion. For example, the manganese atom in an MnO_4^- ion is covalently bonded to the four oxygen atoms by one single and three double bonds:

Never state that there are Mn^{7+} ions. An oxidation state of +7 is different from an Mn^{7+} ion.

Different cation charges

The extra energy required to remove a third electron from an Fe^{2+} ion (the third ionisation energy) to form Fe^{3+} is only +2960 kJ mol^{-1}. This can be recovered either from lattice energy, if it forms a solid, or from hydration energy of the cation, if it goes into solution.

This is not the case with calcium, where removal of a third electron would have to be from an inner shell. The third ionisation energy of calcium is +4940 kJ mol^{-1}, which is too large to be recovered from lattice or hydration energy.

Number of covalent bonds

For each covalent bond to be formed, the element must have an unpaired electron in its valence shell. For transition metals, the valence shell consists of the occupied 4s- and 3*d*-orbitals and the unoccupied 4*p*-orbitals. All these are at a similar energy level and so can be used for bonding. For example, the valence electron configuration of manganese is:

When it forms seven covalent bonds, as in the MnO_4^- ion, it has to have seven unpaired electrons. This is achieved by promoting an electron from the 4*s*-orbital into an empty 4*p*-orbital. These seven unpaired electrons are then used to form seven covalent bonds:

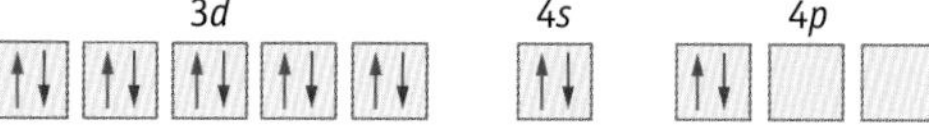

↑ represents an Mn electron; ↓ represents an O electron

As the energy levels of the 3*d*-, 4*s*- and 4*p*-orbitals are similar, little energy is required for promotion and this is recouped from the bond energy released. This means that the *overall* energy change through losing electrons followed by hydration of the ions (or placing ions into a lattice) and the *overall* energy change for promotion of electrons followed by covalent bond formation are both exothermic and, therefore, likely to be thermodynamically feasible.

Zinc has a stable d^{10} configuration. The energy needed to promote any of these electrons would not be regained from the energy released by covalent bond formation. Therefore, zinc does not show variable oxidation states.

Formation of complex ions

Aqua ions

When *d*-block cations are dissolved in water they become hydrated. The oxygen atom in a water molecule has a lone pair of electrons that forms a bond with an empty 3*d*- or 4*p*-orbital in the metal ion.

The exact nature of the bonding is not understood. One theory is that it is an electrostatic attraction between the δ^- oxygen atoms and the positive metal ion. The accepted A-level theory is that a dative covalent bond forms with the oxygen atom as the donor atom. A more accurate theory is that the lone pair of electrons from all the water molecules and the valence electrons of the metal ion form molecular orbitals using the 3*d*-, 4*s*- and 4*p*-orbitals of the *d*-block ion.

e Molecular orbital theory is beyond the scope of A-level chemistry. The most appropriate theory for you to learn is the formation of dative bonds.

An example of an aqua ion is the hydrated chromium(III) ion, which has the formula $[Cr(H_2O)_6]^{3+}$.

- The water molecules are called **ligands**.
- One of the lone pairs of electrons on the oxygen atom of each water molecule forms a dative covalent bond with an empty orbital in the Fe^{3+} ion.
- Six dative bonds form, so the hydrated ion has the **coordination number** 6.
- The ion, with its water molecules bonded to the central metal ion, is called a **complex ion**.

The coordination number is the number of near neighbouring atoms that are bonded to the central ion.

In the example of the $[Cr(H_2O)_6]^{3+}$ ion, the Cr^{3+} ion has six oxygen atoms (one from each of the six water ligands) as near neighbours.

The shape of the complex ion can be predicted using valence-shell electron-pair repulsion (VSEPR) theory. There are six dative bonds, each containing a pair of electrons. These six pairs of bonding electrons repel each other to the position of minimum repulsion, which is also the position of maximum separation. The shape is, therefore, octahedral:

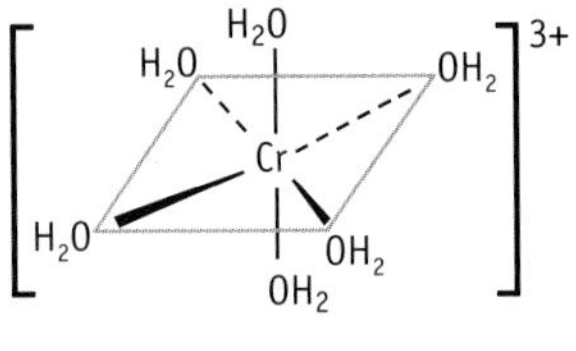

e When drawing hydrated ions make sure that the dative bonds all start from the O of the H_2O and not from the H.

All complex ions with coordination number 6 are octahedral.

In the solid state, hydrated copper(II) ions have four water molecules arranged in a plane around the Cu^{2+} ion. In aqueous solution, two more water molecules are weakly bonded at right angles, forming a distorted octahedron. The two non-planar water molecules are further from the copper ion than are the four planar water molecules. Because of this, the formula of the hydrated ion in solution is sometimes written $[Cu(H_2O)_4]^{2+}$ rather than $[Cu(H_2O)_6]^{2+}$.

Other complex ions

Monodentate ligands

There are many molecules and anions that can form complex ions with transition metal cations:

- Ammonia, NH_3, and organic amines such as ethylamine, $C_2H_5NH_2$, form complexes. An example is the complex of Cu^{2+} and ammonia, which has the formula $[Cu(NH_3)_4(H_2O)_2]^{2+}$.
- Anions such as Cl^- and CN^- also form complexes:
  - $[Fe(CN)_6]^{4-}$ is a complex between an Fe^{2+} ion and six CN^- ions.
  - $[CrCl_4]^-$ is a complex between a Cr^{3+} ion and four Cl^- ions.

It is energetically unfavourable to fit six large ligands around a small cation. The Cl^- ion is much larger than a Cr^{3+} ion and the complex formed has a coordination number of 4. The shape of the ion is tetrahedral, with a bond angle of 109.5°.

The $[CuCl_2]^-$ complex ion has two Cl^- ligands around a central Cu^+ ion. This complex ion is linear with a 180° bond angle.

These examples of complexing species, and also water, are **monodentate ligands**, in which the ion or molecule uses *one* lone pair of electrons to form a dative bond with the *d*-block ion.

Platinum complexes as chemotherapy drugs

A platinum(II) ion, Pt^{2+}, can form a complex with two ammonia molecules and two chloride ions. This is a neutral complex, platin, with formula $[Pt(NH_3)_2Cl_2]$. Its shape is planar and the two chlorine ions can be arranged *cis* or *trans* to each other.

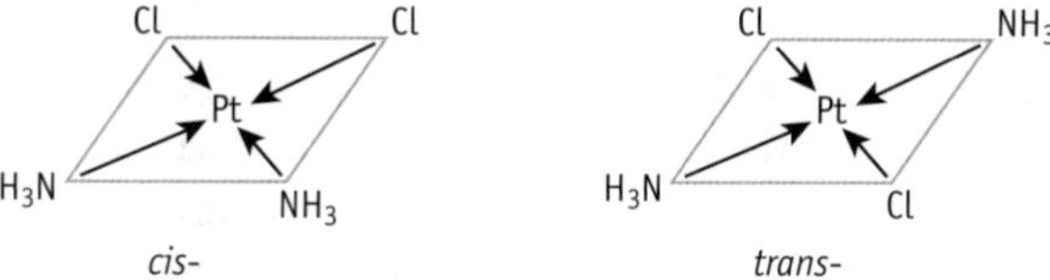

The platinum complex bonds to adjacent guanine molecules in one strand of DNA in cancer cells by a ligand exchange reaction. Each of the chloride ligands is replaced by one of the nitrogen atoms in guanine. This prevents replication of the DNA. The damaged cancer cell is then destroyed by the body's immune system. Unfortunately, *cis*-platin has a number of side effects including kidney damage and the inhibition of hair growth. The latter results in patients undergoing chemotherapy losing their hair temporarily.

A newer anti-cancer drug is the *E*-isomer of a complex that contains two platinum ions. This complex bonds to two guanine molecules in different strands of the DNA in a cancer cell, completely blocking its replication.

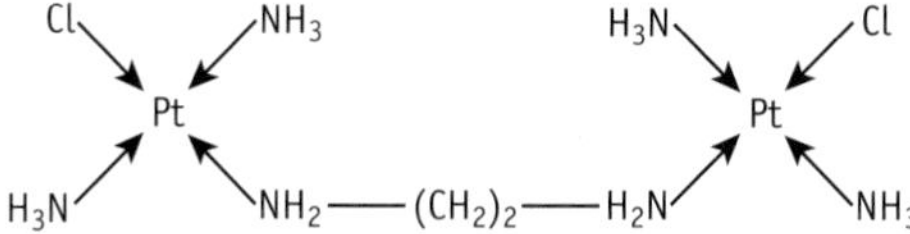

Bi- and poly-dentate ligands

Some more complex molecules or ions have lone pairs in different places and can form two dative bonds with the central metal ion. The geometry has to be correct for this to happen, as a ring is formed. Such ligands are called **bidentate ligands.** Two examples are 1,2-diaminoethane, $NH_2CH_2CH_2NH_2$, and ethanedioate ions, $^-OOCCOO^-$.

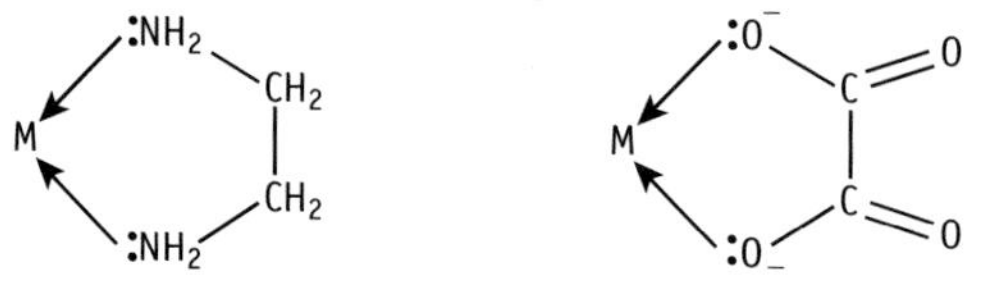

The ring involving the bidentate ligand contains five atoms. Five-membered rings, such as in glucose, are stable stereo structures with no bond strain.

Other species can form five or six dative bonds with the central metal ion. These are called **polydentate ligands**. One of the best reagents for forming complexes of this type is the disodium salt of EDTA. EDTA loses two further H^+ ions to give an ion with two amine groups and four carboxylate groups:

EDTA stands for ethylenediamine-tetraacetic acid

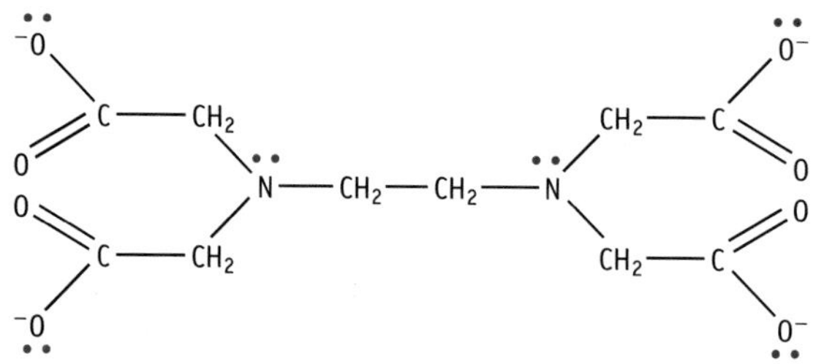

Therefore, EDTA has six sites containing a lone pair (marked in red) and so can form six dative bonds.

In the haemoglobin molecule, the iron ions are complexed with an organic species that can form five bonds to the metal. The sixth position is taken up either by an oxygen molecule (arterial blood) or by a water molecule (venous blood).

Some common ligands are listed in Table 11.5.

***Table 11.5** Common ligands*

Ligand type	Name	Formula
Neutral	Water	H_2O
	Ammonia	NH_3
	Methylamine	CH_3NH_2
Negative ions	Fluoride	F^-
	Chloride	Cl^-
	Cyanide	CN^-*
	Thiocyanate	SCN^-
	Hydroxide	OH^-
	Sulfate	SO_4^{2-}
Bidentate	1,2-diaminoethane	$NH_2CH_2CH_2NH_2$
	Ethanedioate	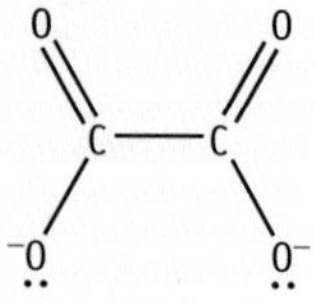

* The cyanide ion normally bonds through the carbon atom.

Chromium in the +3 oxidation state forms many complexes. For example, when excess sodium hydroxide is added to a suspension of chromium(III) hydroxide, a green solution of the hexahydroxy complex is formed:

$$[Cr(H_2O)_3(OH)_3] + 3OH^- \rightarrow [Cr(OH)_6]^{3-} + 3H_2O$$

Catalytic activity

Transition metals and their compounds are good catalysts. This is particularly true of the elements at the right-hand side of the *d*-block.

Heterogeneous catalysts

A heterogeneous catalyst is in a different phase from the reactants. Many industrial processes use transition metals or their compounds as heterogeneous catalysts. For example, in the Haber process, iron in the solid state is used to catalyse the reaction between hydrogen and nitrogen gases.

Metal catalysts work by providing active sites onto which the reactant molecules can bond (adsorb). The sequence of reaction is:

Step 1 (fast): gaseous reactants + active site → adsorbed reactants

Step 2 (slow): adsorbed reactants → adsorbed product

Step 3 (fast): adsorbed product → gaseous product + empty active site

The cycle is then repeated.

A transition metal can act as a catalyst because its energetically available *d*-orbitals can accept electrons from a reactant molecule or its *d*-electrons can form a bond with a reactant molecule. This can be illustrated by the catalytic hydrogenation of an alkene. The alkene bonds to an active site by its π-electrons becoming involved with an empty *d*-orbital in the catalyst. The σ-bond in the hydrogen molecule breaks and each hydrogen atom forms a bond through a *d*-electron on an atom in the catalyst. The two hydrogen atoms then bond with the partially broken π-bond in the alkene and the alkane formed is released from the surface of the catalyst.

Some heterogeneous catalysts work because of the variable oxidation state of the transition metal. For example, vanadium(V) oxide is used as the catalyst in the oxidation of sulfur dioxide to sulfur trioxide in the manufacture of sulfuric acid. The sulfur dioxide is oxidised by the vanadium(V) oxide, which is reduced to vanadium(IV) oxide:

$$SO_2(g) + V_2O_5(s) \rightarrow SO_3(g) + 2VO_2(s)$$

The oxygen then oxidises the vanadium(IV) oxide back to vanadium(V) oxide:

$$\tfrac{1}{2}O_2(g) + 2VO_2(s) \rightarrow V_2O_5(s)$$

The overall equation is:

$$SO_2(g) + \tfrac{1}{2}O_2(g) \rightarrow SO_3(g)$$

Homogeneous catalysts

Homogeneous catalysts are in the same phase as the reactants. They always work via an intermediate compound or ion. For example, Fe^{2+} ions catalyse the oxidation of iodide ions by persulfate ions in aqueous solution:

$$2I^-(aq) + S_2O_8^{2-}(aq) \rightarrow I_2(s) + 2SO_4^{2-}(aq)$$

The two reactants are negative ions and so repel each other, making the reaction very slow. When positively charged Fe^{2+} ions are added, they are oxidised by the negatively charged persulfate ions:

$$S_2O_8^{2-}(aq) + 2Fe^{2+}(aq) \rightarrow 2SO_4^{2-}(aq) + 2Fe^{3+}(aq)$$

The Fe^{3+} ions are then reduced by iodide ions, regenerating the Fe^{2+} catalyst.

$$2Fe^{3+}(aq) + 2I^-(aq) \rightarrow 2Fe^{2+}(aq) + I_2(s)$$

Common physical properties of transition metal ions

Coloured ions

The *d*-orbitals in a transition metal ion are all at the same energy level, but they point in different directions. Three of the orbitals point between the *x*-, *y*- and *z*-axes and two point along these axes.

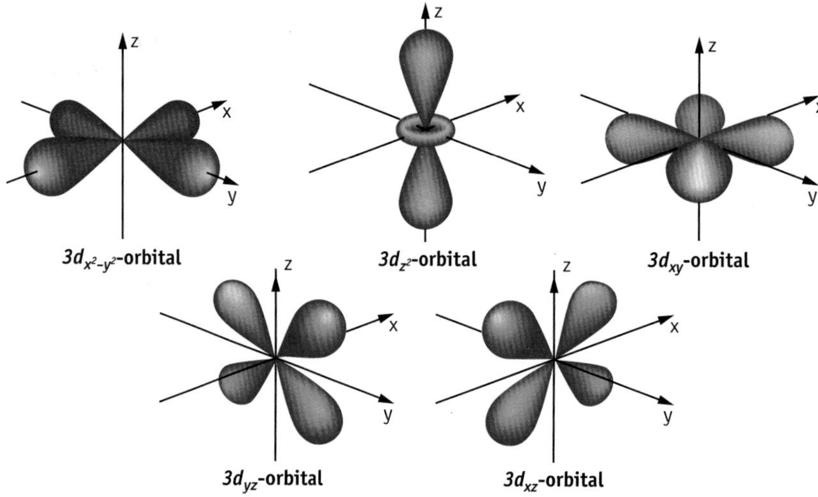

Figure 11.5
Shapes of the d-orbitals

When six ligands approach the ion, they do so along the *x*-, *y*- and *z*-axes. This causes greater repulsion with the two *d*-orbitals that point along the *x*-, *y*- and *z*-axes than with the three that point between the axes. The energy levels of the *d*-orbitals are split into a lower group of three and an upper group of two (Figure 11.6).

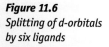

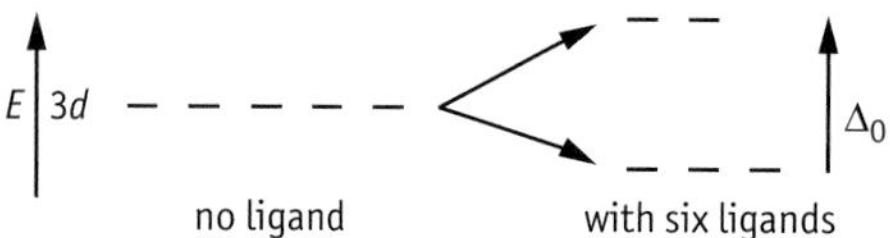

Figure 11.6
Splitting of d-orbitals by six ligands

The energy difference Δ_O between these two sets of d-orbitals in a typical complex ion is equal to the energy of a photon in the visible region of the spectrum.

When white light shines through a solution of a complex ion of a transition metal, photons of a particular frequency are absorbed and their energy promotes an electron from the lower energy level to the upper energy level. This is called a *d–d* transition. A colour is *removed* from the white light and the solution has the complementary colour to the light absorbed.

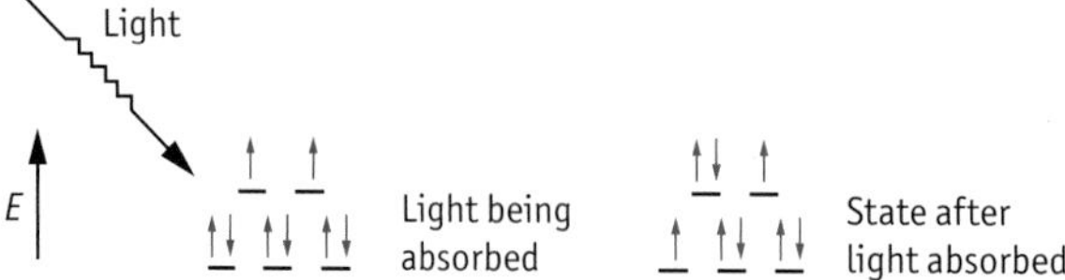

Figure 11.7
Absorption of light energy

Within a fraction of a second the ion with an electron in the upper level collides with another ion. The electron drops down and energy is released as heat. The complex ion is once again in the ground state, able to absorb more light energy.

The colour wheel shown in Figure 11.8 shows the relationship between the colour of the light absorbed and the complementary colour of the complex ion. These colours are diametrically opposite each other.

The $[Cu(H_2O)_6]^{2+}$ ion absorbs orange light, so it is blue. The $[Ni(H_2O)_6]^{2+}$ ion absorbs red light, so it is green.

Figure 11.8 *The colour wheel*

Ⓔ Ions with no *d*-electrons are not coloured simply because they do not have any *d*-electrons to promote. Therefore, Sc^{3+} and Ti^{4+} are colourless as both have the electron configuration of [Ar] $3d^0\,4s^0$.

Ⓔ Ions with ten *d*-electrons are also colourless because there is no empty space in a *d*-orbital for the electron to be promoted into. Cu^+ and Zn^{2+} both have the electron configuration [Ar] $3d^{10}\,4s^0$ and are, therefore, colourless.

Ⓔ The sequence of events for the colour of transition metal complexes is:

light energy absorbed removing colour from white light →
electron promoted to higher level

With flame colours, the opposite takes place:

heat from Bunsen → electron promoted →
electron falls back emitting coloured light

These two processes must not be confused.

ZAHOOR UL-HAQ

Solutions of some complex metal ions (left to right: $[Ti(H_2O)_6]^{3+}$, $[Co(H_2O)_6]^{2+}$, $[Ni(H_2O)_6]^{2+}$ and $[Cu(H_2O)_6]^{2+}$)

Effect of ligand on colour

Some ligands interact more strongly than others with the *d*-electrons and so cause a greater splitting. The relative strength of ligands is shown in Figure 11.9.

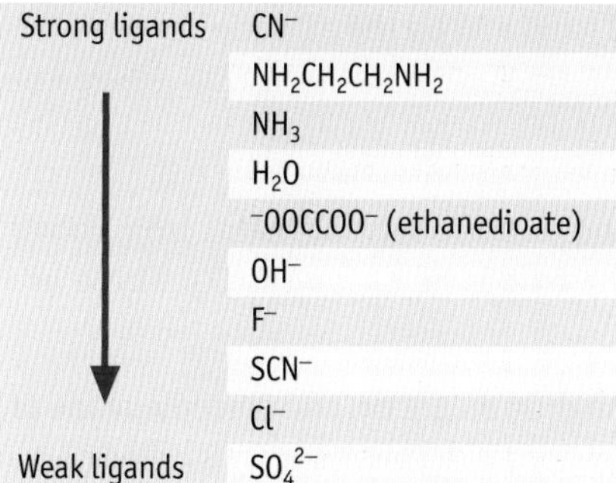

Strong ligands	CN^-
	$NH_2CH_2CH_2NH_2$
	NH_3
	H_2O
	$^-OOCCOO^-$ (ethanedioate)
	OH^-
	F^-
	SCN^-
	Cl^-
Weak ligands	SO_4^{2-}

Figure 11.9 *Relative strength of ligands*

A strong ligand splits the *d*-orbitals in the complex ion to a greater extent than a weak ligand does.

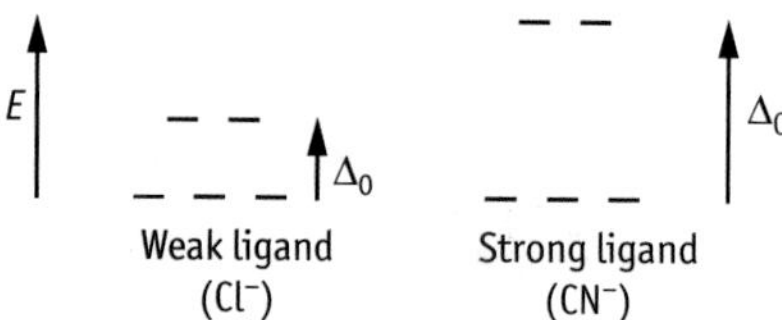

Figure 11.10 *Splitting of d-orbitals by different ligands*

If a stronger ligand replaces a weaker ligand, the colour absorbed moves towards the high-energy (violet) end of the spectrum. This can be shown by the addition of excess ammonia solution to aqueous copper sulfate. The reaction is:

$$[Cu(H_2O)_6]^{2+}(aq) + 4NH_3(aq) \rightarrow [Cu(NH_3)_4(H_2O)_2]^{2+}(aq) + 4H_2O(l)$$

Relatively weaker H_2O ligands have been replaced by stronger NH_3 ligands. The absorption moves from orange towards yellow and the colour of the complex changes from blue to violet-blue.

When concentrated HCl is added to a solution of copper(II) chloride, the equilibrium:

$$\underset{\text{blue}}{[Cu(H_2O)_6]^{2+}} + 4Cl^-(aq) \rightleftharpoons \underset{\text{yellow}}{[CuCl_4]^{2-}(aq)} + 6H_2O(l)$$

is driven to the right. The colour changes from blue via green to yellow. The green colour occurs when both $[Cu(H_2O)_6]^{2+}$ and $[CuCl_4]^{2-}$ are present. The light absorbed changes from orange (causing the ion to be blue) to violet (causing the ion to be yellow).

Oxidation state of the transition metal ion

An ion with a high charge density will attract a ligand strongly, so the splitting of the *d*-orbitals will be greater. The $[Fe(H_2O)_6]^{2+}$ ion absorbs red light which means that it is green. The $[Fe(H_2O)_6]^{3+}$ ion absorbs yellow light and is amethyst (pale purple-violet).

Summary

- The colour of transition metal complexes is caused by the ligand splitting the *d*-orbitals into two different sets of energy levels. Light is absorbed and an electron promoted. The colour of the complex is the complementary colour to that absorbed.
- The *d*-orbital splitting of the transition metal, and hence the colour of the complex, depends on the nature of the ligand and on the oxidation state of the transition metal.

Magnetic properties

Some transition metals and their compounds are strongly magnetic. Iron, cobalt and nickel can be magnetised, as can the oxides Fe_3O_4 and CrO_2. These are known as **ferromagnetic** substances.

Most transition metal complexes are very slightly magnetic, and a needle made of their crystals will lie parallel to a strong magnetic field. This property is called **paramagnetism** and is caused by the ion having one or more unpaired *d*-electrons.

Reactions of *d*-block ions with particular reference to chromium and copper

Deprotonation

By water

When a base is added to a solution of hydrated *d*-block ions, **deprotonation** takes place to a greater or lesser extent depending on the strength of the base.

Deprotonation is a reaction in which the base removes a proton (H^+ ion) from the species.

Water is a weak base and only significantly deprotonates 3+ aqua ions such as $[Cr(H_2O)_6]^{3+}$

$$[Cr(H_2O)_6]^{3+}(aq) + H_2O(l) \rightleftharpoons [Cr(H_2O)_5(OH)]^{2+}(aq) + H_3O^+(aq)$$

The formation of the H_3O^+ ion makes a solution of a chromium(III) compound acidic. Further deprotonation is not extensive.

A similar reaction takes place with hydrated iron(III) ions:

$$\underset{\text{amethyst}}{Fe(H_2O)_6]^{3+}(aq)} + H_2O(l) \rightleftharpoons \underset{\text{yellow-brown}}{[Fe(H_2O)_5(OH)]^{2+}(aq)} + H_3O^+(aq)$$

The charge density on a divalent ion such as Cu^{2+} is lower, so it attracts the ligand electrons less and the hydrogen atoms in the water ligands are less δ^+. This means that very little deprotonation takes place with hydrated copper(II) ions and water and so the solution is only slightly acidic.

By stronger bases

All hydrated *d*-block ions are deprotonated when stronger bases, such as sodium hydroxide or ammonia, are added. The deprotonation is so extensive that a neutral and hence insoluble species is produced, for example:

$$[Cr(H_2O)_6]^{3+}(aq) + 3OH^-(aq) \rightarrow [\,Cr(H_2O)_3(OH)_3](s) + 3H_2O(l)$$
$$[Cr(H_2O)_6]^{3+}(aq) + 3NH_3(aq) \rightarrow [\,Cr(H_2O)_3(OH)_3](s) + 3NH_4^+(aq)$$

The formulae of the hydroxide precipitates are sometimes written without the co-ordinated water, e.g. $Cu(OH)_2$ for copper(II) hydroxide.

Metals in the +2 oxidation state also form precipitates of the hydrated hydroxides, for example:

$$[Fe(H_2O)_6]^{2+}(aq) + 2OH^-(aq) \rightarrow [Fe(H_2O)_4(OH)_2](s) + 2H_2O(l)$$
$$[Fe(H_2O)_6]^{2+}(aq) + 2NH_3(aq) \rightarrow [Fe(H_2O)_4(OH)_2](s) + 2NH_4^+(aq)$$
$$[Cu(H_2O)_6]^{2+}(aq) + 2OH^-(aq) \rightarrow [Cu(H_2O)_4(OH)_2](s) + 2H_2O(l)$$
$$[Cu(H_2O)_6]^{2+}(aq) + 2NH_3(aq) \rightarrow [Cu(H_2O)_4(OH)_2](s) + 2NH_4^+(aq)$$

Amphoteric hydroxides

With some *d*-block metal ions, if an excess of a very strong base is added, further deprotonation can take place. These are the amphoteric hydroxides that dissolve in excess sodium hydroxide solution. For example:

$$[Cr(H_2O)_3(OH)_3](s) + 3OH^-(aq) \rightarrow [Cr(OH)_6]^{3-}(aq) + 3H_2O(l)$$
$$[Zn(H_2O)_2(OH)_2](s) + 2OH^-(aq) \rightarrow [Zn(OH)_4]^{2-}(aq) + 2H_2O(l)$$

An amphoteric hydroxide is one that can act as either an acid or a base, so chromium(III) hydroxide also reacts with acids:

$$Cr(OH)_3(s) + 3H^+(aq) \rightarrow Cr^{3+}(aq) + 3H_2O(l)$$

The formula of a metal hydroxide can be written with or without the water ligands. Thus both $Cr(OH)_3$ and $[Cr(OH)_3(H_2O)_3]$ are acceptable as the formula of chromium(III) hydroxide.

The colour of the hydrated hydroxide formed with sodium hydroxide solution and whether or not the precipitate dissolves in excess can be used to identify a transition metal and its oxidation state. The colours are listed in Table 11.6.

Ion	Colour of precipitate	Effect of adding excess NaOH
Cr^{3+} in $[Cr(H_2O)_6]^{3+}$	Green	Forms green solution
Mn^{2+} in $[Mn(H_2O)_6]^{2+}$	Off-white*	None
Fe^{2+} in $[Fe(H_2O)_6]^{2+}$	Green†	None
Fe^{3+} in $[Fe(H_2O)_6]^{3+}$	Red-brown	None
Co^{2+} in $[Co(H_2O)_6]^{2+}$	Blue‡	None
Ni^{2+} in $[Ni(H_2O)_6]^{2+}$	Green	None
Cu^{2+} in $[Cu(H_2O)_6]^{2+}$	Blue	None
Zn^{2+} in $[Zn(H_2O)_6]^{2+}$	White	Forms colourless solution

* The precipitate with Mn(II) salts darkens on exposure to air as it is slowly oxidised to manganese(IV) oxide. Darkening is much more rapid if an oxidising agent (e.g. H_2O_2) is added to the precipitate.

† The precipitate with Fe(II) salts goes brown on exposure to air as it is oxidised to iron(III) hydroxide. Darkening is much more rapid if an oxidising agent (e.g. H_2O_2) is added to the precipitate. Very pure solutions of Fe^{2+} ions give a pale green (almost white) precipitate of hydrated iron(II) hydroxide.

‡ The precipitate of $Co(OH)_2$ turns pink on standing.

***Table 11.6** Effect of adding sodium hydroxide solution to some hydrated ions*

When sodium hydroxide is added to a solution of a silver salt, deprotonation of the $[Ag(H_2O)_2]^+$ ions takes place followed by dehydration of the hydroxide to form a brown precipitate of silver oxide:

$$2[Ag(H_2O)_2]^+(aq) + 2OH^-(aq) \rightarrow Ag_2O(s) + 5H_2O(l)$$

Hydroxides of, from left: iron(II), iron(III), copper(II) and nickel(II)

Ligand exchange reactions

In most complex ions, the ligands are not irreversibly bound and can be replaced by other ligands in a **ligand exchange reaction**. In most cases, it is water ligands that are replaced. The extent to which this happens is measured by the **stability constant**.

For the ligand exchange reaction

$$M(H_2O)_6^{2+} + 4L \rightarrow ML_4^{2+} + 6H_2O$$

The equilibrium constant is given by

$$K_a = \frac{[ML_4^{2+}]}{[M(H_2O)_6]^{2+}\,[L]^4}$$

and the stability constant as

$$\text{stability constant} = \log K$$

Stability constants for some copper(II) complexes are given in Table 11.7.

$[H_2O]^6$ is omitted from the expression for K because it is the solvent and therefore its concentration is constant.

***Table 11.7** Stability constants for some copper(II) complexes*

Complex	log*K*
$[CuCl_4]^{2-}$	5.6
$[Cu(NH_3)_4]^{2+}$	13.1
$[Cu(EDTA)]^{2-}$	18.8

Addition of ammonia to *d*-block ions

The addition of excess ammonia solution to some ions results in ligand exchange. These are the ions that dissolve in excess ammonia to form an ammine. All transition metal ions are first deprotonated by basic ammonia forming precipitates of hydrated hydroxides. The hydroxides of cobalt(II), nickel(II), copper(II), zinc(II) and silver(I) readily form an ammine by ligand exchange. The precipitate of chromium(III) hydroxide reacts very slowly to form an ammonia complex.

The blue precipitate of hydrated copper(II) hydroxide redissolves in excess ammonia solution:

$$[Cu(H_2O)_4(OH)_2](s) + 4NH_3(aq) \rightarrow [Cu(NH_3)_4(H_2O)_2]^{2+} + 2OH^-(aq) + 2H_2O(l)$$

The overall reaction between hydrated copper(II) ions and ammonia solution is the exchange of four water ligands by ammonia ligands:

$$[Cu(H_2O)_6]^{2+}(aq) + 4NH_3(aq) \rightarrow [Cu(NH_3)_4(H_2O)_2]^{2+}(aq) + 4H_2O(l)$$

The results of gradually adding ammonia solution to hydrated ions of some *d*-block elements until it is in excess are shown in Table 11.8.

***Table 11.8** Effect of adding ammonia solution*

Ion	Colour of precipitate	Effect of adding excess ammonia
Mn^{2+} in $[Mn(H_2O)_6]^{2+}$	Off-white*	None
Fe^{2+} in $[Fe(H_2O)_6]^{2+}$	Green*	None
Cr^{3+} in $[Cr(H_2O)_6]^{3+}$	Green	† Dissolves slightly to form a green solution
Fe^{3+} in $[Fe(H_2O)_6]^{3+}$	Red-brown	None
Co^{2+} in $[Co(H_2O)_6]^{2+}$	Blue	Dissolves to form a brown solution
Ni^{2+} in $[Ni(H_2O)_6]^{2+}$	Green	Dissolves to form a pale blue solution
Cu^{2+} in $[Cu(H_2O)_6]^{2+}$	Blue	Dissolves to form a deep blue solution
Zn^{2+} in $[Zn(H_2O)_6]^{2+}$	White	Dissolves to form a colourless solution
Ag^+ in $[Ag(H_2O)_2]^+$	Brown‡	Dissolves to form a colourless solution

* The precipitates of $Mn(OH)_2$ and $Fe(OH)_2$ darken as they oxidise in air.

† The ammonia complex with chromium hydroxide forms very slowly. If liquid ammonia is added to solid hydrated chromium(III) chloride, the ammonia molecules take the place of water molecules around the chromium ion:

$$[Cr(H_2O)_6]Cl_3 + 6NH_3 \rightarrow [Cr(NH_3)_6]Cl_3 + 6H_2O$$

‡ The brown precipitate of silver oxide is often not seen because the ammonia complex forms so easily. Adding ammonia solution to a precipitate of AgCl or AgBr also makes this silver complex:

$$AgCl(s) + 2NH_3(aq) \rightarrow [Ag(NH_3)_2]^+(aq) + Cl^-(aq)$$

AgBr is so insoluble that concentrated ammonia is required. AgI is even more insoluble and the complex does not form, even with concentrated ammonia.

Chromium(III) complexes

Cr^{3+} complexes are formed in ligand exchange reactions with hydrated chromium(III) ions.

Hydrated chromium(III) ions react with excess sodium hydroxide to form a solution of a bright green complex anion, $[Cr(OH)_6]^{3-}$. The overall equation for this reaction is:

$$[Cr(H_2O)_6]^{3+} + 6OH^- \rightarrow [Cr(OH)_6]^{3-} + 6H_2O$$

Cr^{3+} ions also react with 1,2-diaminoethane, which may be represented by 'en':

$$[Cr(H_2O)_6]^{3+} + 3en \rightarrow [Cr(en)_3]^{3+} + 6H_2O$$

They also react with EDTA:

$$[Cr(H_2O)_6]^{3+} + EDTA^{4-} \rightarrow [Cr(EDTA)]^- + 6H_2O$$

Hydrated chromium(III) chloride, $CrCl_3.6H_2O$ exists in several isomeric forms with different numbers of chlorines within the co-ordination sphere:

Isomer I: $[Cr(H_2O)_6]Cl_3$ is grey-blue

Isomer II: $[Cr(H_2O)_5Cl]Cl_2.H_2O$ is pale green

Isomer III: $[Cr(H_2O)_4Cl_2]Cl.2H_2O$ is green

1 mol of isomer I reacts with excess silver nitrate solution to form a precipitate of 3 mol of silver chloride; 1 mol of isomer II gives rise to a precipitate of 2 mol of silver chloride; 1 mol of isomer III gives only 1 mol.

Chromium(II) complexes

Chromium is unstable in the +2 state compared with the +3 state. The +2 state can be stabilised by formation of a complex with ethanoate ions, CH_3COO^-.

A solution of potassium dichromate(VI) is placed in a flask with some zinc and 50% concentrated hydrochloric acid. A delivery tube is fitted with the bottom end below the surface of the acidified potassium dichromate(VI) solution and the other end below the surface of a solution of sodium ethanoate (Figure 11.11).

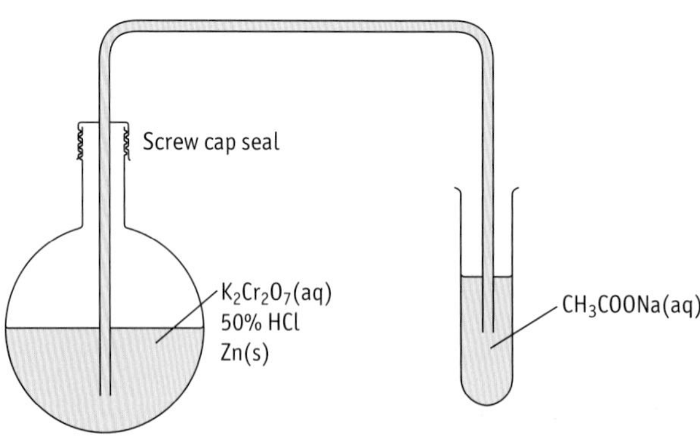

Figure 11.11 *Preparation of the chromium(II) ethanoate complex*

The seal at the top is loose to let hydrogen escape. The orange solution first turns green as Cr^{3+} ions are formed and then blue as these are reduced to Cr^{2+} ions.

At this stage, the cap of the seal is screwed shut and the pressure of hydrogen forces the solution out into the test tube.

The hydrated chromium(II) ions undergo ligand exchange and a precipitate of the red chromium(II) ethanoate complex is formed:

$$2[Cr(H_2O)_6]^{2+}(aq) + 4CH_3COO^-(aq) \rightarrow [Cr_2(CH_3COO)_4(H_2O)_2](s) + 10H_2O(l)$$

One interesting feature of this complex is that the two oxygen atoms in each ethanoate ion form dative bonds with different Cr^{2+} ions.

Copper(II) complexes

Copper forms a number of complexes.

Hydrated copper(II) ions in solution consist of an octahedral arrangement of six water molecules around the central Cu^{2+} ion. Two of the water molecules opposite each other are further away from the copper ion than the other four, which are in the same plane as the copper ion. This is why the formula is sometimes written as $[Cu(H_2O)_4]^{2+}$ rather than as $[Cu(H_2O)_6]^{2+}$. The splitting of the *d*-orbitals is such that the ion is turquoise-blue. Solid hydrated copper(II) sulfate has the formula $[Cu(H_2O)_4)]SO_4.H_2O$; the copper ion is surrounded by four water molecules. The fifth molecule of water of crystallisation is bonded to the sulfate ion.

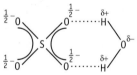

e The two π-bonds in the SO_4^{2-} ion are delocalised around the four oxygen atoms which share the 2− charge.

If excess concentrated hydrochloric acid is added to a solution of copper(II) sulfate, ligand exchange takes place and the yellow $[CuCl_4]^{2-}$ complex ion is formed.

When solid blue hydrated copper(II) sulfate is heated, it loses its water ligands. As there are now no ligands, there is no splitting of the d-orbitals and so it does not absorb visible light. Therefore, anhydrous copper(II) sulfate is white.

$$\underset{\text{blue}}{[Cu(H_2O)_6]^{2+}} + 4Cl^- \rightarrow \underset{\text{yellow}}{[CuCl_4]^{2-}} + 6H_2O$$

Solid hydrated copper(II) chloride, $CuCl_2.2H_2O$, is green because it is in fact the complex ion pair $[Cu(H_2O)_4]^{2+}.[CuCl_4]^{2-}$.

Copper(II) ions also form complexes with:

- ammonia
- amines such as ethylamine
- 1,2-diaminoethane (en)
- ethanediaminetetraacetate ($EDTA^{4-}$)

$$[Cu(H_2O)_6]^{2+} + 4NH_3 \rightarrow [Cu(NH_3)_4(H_2O)_2]^{2+} + 4H_2O$$
$$[Cu(H_2O)_6]^{2+} + 4C_2H_5NH_2 \rightarrow [Cu(C_2H_5NH_2)_4(H_2O)_2]^{2+} + 4H_2O$$
$$[Cu(H_2O)_6]^{2+} + 3en \rightarrow [Cu(en)_3]^{2+} + 6H_2O$$
$$[Cu(H_2O)_6]^{2+} + EDTA^{4-} \rightarrow [CuEDTA]^{2-} + 6H_2O$$

The relative strengths of these ligands is shown by the following experiments:

(1) Take a solution of copper(II) sulfate and add concentrated ammonia until a clear blue solution is formed. Note the colour and divide the solution into two portions.

(2) To one portion, add excess 1,2-diaminoethane solution; to the other portion, add excess of a solution of the disodium salt of EDTA. Note the colours in each case.

(3) To the solution of the 1,2-diaminoethane complex add a solution of the disodium salt of EDTA. Note any colour change.

(4) Relate the reactions to the stability constants given in Table 11.7 (p. 223) and suggest a value for the stability constant for $[Cu(en)_2]^{2+}$.

Fehling's and Benedict's solutions are complexes of copper(II) ions with tartrate ions. These complexes are reduced to a red precipitate of copper(I) oxide by aldehydes, but not by ketones.

NIGEL EVANS

Complexes of Cu^{2+} ions with water, ammonia, EDTA and 1,2-diamino-ethane

Copper(I) complexes

Hydrated copper(I) ions are unstable in water and disproportionate spontaneously:

$$2Cu^+(aq) \rightarrow Cu(s) + Cu^{2+}(aq)$$

However, they can be stabilised by complexation. Copper(I) chloride is a white solid and reacts with ammonia to form a colourless solution of a copper(I) ammine complex:

$$CuCl(s) + 2NH_3(aq) \rightarrow [Cu(NH_3)_2]^+(aq) + Cl^-(aq)$$

The CuCl solid also dissolves in concentrated hydrochloric acid, forming the colourless linear complex ion, $[CuCl_2]^-$:

$$CuCl(s) + HCl(aq) \rightarrow [CuCl_2]^-(aq) + H^+(aq)$$

Copper(I) complex ions are colourless because the electronic structure of the ion is [Ar] $3d^{10}\,4s^0$. All the d-orbitals in the d-shell are full. Therefore, even though the energy levels of the d-orbitals are split, promotion of an electron from one of the lower split d-orbitals to a higher one cannot take place.

Entropy changes

The stability constants of complexes with bi- and poly-dentate ligands are high because of the entropy changes in the system. When 1,2-diaminoethane reacts with hydrated copper(II) ions, four particles are converted into seven particles.

With EDTA there is a greater increase in the number of particles — from two to seven. An increase in the number of particles increases ΔS_{system} and is the main driving force of the ligand exchange reaction.

$$[Cu(H_2O)_6]^{2+} + 3en \rightarrow [Cu(en)_3]^{2+} + 6H_2O$$
$$[Cu(H_2O)_6]^{2+} + EDTA^{4-} \rightarrow [Cu(EDTA)]^{2-} + 6H_2O$$

Redox reactions

It is because transition metals have variable oxidation states that redox reactions are common.

The ease with which a transition metal ion is oxidised or reduced is given by its standard reduction potential, $E^{\ominus}$.

- The more negative this potential, the worse the ion on the left-hand side of the half-equation is as an oxidising agent and the better the ion on the right-hand side of the half-equation is as a reducing agent.
- The more positive the value of $E^{\ominus}$, the better the ion on the left-hand side of the half-equation is as an oxidising agent and the worse the ion on the right-hand side is as a reducing agent.

For example:

$$Cr^{3+} + e^- \rightleftharpoons Cr^{2+} \qquad E^{\ominus} = -0.41\,V$$
$$Cu^{2+} + e^- \rightleftharpoons Cu^+ \qquad E^{\ominus} = +0.15\,V$$
$$\tfrac{1}{2}Cr_2O_7^{2-} + 7H^+ + 3e^- \rightleftharpoons Cr^{3+} + 3\tfrac{1}{2}H_2O \qquad E^{\ominus} = +1.33\,V$$

Dichromate(VI) ions are the best oxidising agent and chromium(III) ions are the worst.

Chromium(III) ions are the worst reducing agent and chromium(II) ions are the best.

The feasibility of a redox reaction is indicated by the value of the standard potential of the cell, $E^{\ominus}_{cell}$.

Feasibility is worked out by first reversing the half-equation that has one of the reactants on the right-hand side and changing the sign of its $E^{\ominus}$ value. This half-equation is then added to the half-equation that has the other reactant on the left-hand side. The sum of the $E^{\ominus}$ values for the two half-equations gives $E^{\ominus}_{cell}$. If the value is positive, the reaction is thermodynamically feasible and will take place providing that the activation energy is not too high.

Worked example

Will dichromate(VI) ions oxidise Fe^{2+} ions? If so, write the ionic equation for the reaction

Answer

$$Cr_2O_7^{2-} + 14H^+ + 6e^- \rightleftharpoons 2Cr^{3+} + 7H_2O \qquad E^{\ominus} = +1.33\,V$$
$$Fe^{3+} + e^- \rightleftharpoons Fe^{2+} \qquad E^{\ominus} = +0.77\,V$$

Reverse the second half-equation, multiply it by six and add it to the first half-equation.

$$6Fe^{2+} \rightleftharpoons 6Fe^{3+} + 6e^- \qquad E^{\ominus} = -(+0.77) = -0.77\,V$$
$$\underline{Cr_2O_7^{2-} + 14H^+ + 6e^- \rightleftharpoons 2Cr^{3+} + 7H_2O} \qquad E^{\ominus} = +1.33\,V$$

$Cr_2O_7^{2-} + 14H^+ + 6Fe^{2+} \rightleftharpoons 2Cr^{3+} + 7H_2O + 6Fe^{3+}$

$E^{\ominus}_{cell} = -0.77 + (+1.33) = +0.56\,V$

$E^{\ominus}_{cell}$ is positive, so the reaction is feasible and the reactants are thermodynamically unstable relative to the products.

In some cases the activation energy of the reaction could be too high for the reaction to take place rapidly enough to be observed at room temperature. The reactants are then said to be **kinetically stable** relative to the products.

The standard reduction potentials of some transition metal ions, arranged in order of increasing oxidising power, are given in Table 11.9. The ion on the top right (Cr^{2+}) is the best reducing agent in the list and that at the bottom right (Fe^{3+}) is the worst reducing agent.

***Table 11.9** Standard reduction potentials for some transition metal ions*

Half-equation	$E^{\ominus}$/V
$Cr^{3+} + e^- \rightarrow Cr^{2+}$	-0.41
$V^{3+} + e^- \rightarrow V^{2+}$	−0.26
$CrO_4^{2-} + 4H_2O + 3e^- \rightarrow Cr(OH)_3 + 5OH^-$	−0.13
$[Co(NH_3)_6]^{3+} + e^- \rightarrow [Co(NH_3)_6]^{2+}$	+0.10
$Cu^{2+} + e^- \rightarrow Cu^+$	+0.15
$VO^{2+} + 2H^+ + e^- \rightarrow V^{3+} + H_2O$	+0.34
$[Fe(CN)_6]^{3-} + e^- \rightarrow [Fe(CN)_6]^{4-}$	+0.36
$Cu^+ + e^- \rightarrow Cu$	+0.52
$MnO_4^{2-} + 2H_2O + 2e^- \rightarrow MnO_2 + 4OH^-$	+0.59
$Fe^{3+} + e^- \rightarrow Fe^{2+}$	+0.77
$VO_2^+ + 2H^+ + e^- \rightarrow VO^{2+} + H_2O$	+1.00
$MnO_2 + 4H^+ + 2e^- \rightarrow Mn^{2+} + 2H_2O$	+1.23
$Cr_2O_7^{2-} + 14H^+ + 6e^- \rightarrow 2Cr^{3+} + 7H_2O$	+1.33
$MnO_4^- + 8H^+ + 5e^- \rightarrow Mn^{2+} + 4H_2O$	+1.51
$Co^{3+} + e^- \rightarrow Co^{2+}$	+1.81
$FeO_4^{2-} + 8H^+ + 3e^- \rightarrow Fe^{3+} + 4H_2O$	+2.20

In Table 11.9, the water ligands have been omitted to simplify the equations. Ligands other than H_2O are included in the formulae.

Notice the effect of the ligand on the value of $E^{\ominus}$:

- The reduction potential of the hydrated iron(II) ion is more positive than that of the hexacyanoferrate(III) ion. This means that the +3 oxidation state becomes stabilised relative to the +2 state by the change to the strong cyanide ligand. Therefore, hydrated iron(III) is a stronger oxidising agent than hexacyanoiron(III).
- This is even more noticeable with cobalt(III). Almost nothing will oxidise hydrated Co^{2+} ions to hydrated Co^{3+} ions, but if the Co^{2+} is complexed with ammonia, the oxidation can take place.

Notice also the effect of pH. The definition of standard reduction potential assumes that all soluble species are at a concentration of $1.0\,mol\,dm^{-3}$. Therefore, if OH^- appears on either side of a half-equation, the solution is assumed to be at pH 14; if H^+ is in the equation, the pH is assumed to be 0.

e A reduction potential measures the ease of reduction of a substance. The more positive the reduction potential, the more likely it is that the substance will be reduced and, therefore, the stronger it is as an oxidising agent.

If it is required to oxidise chromium from the +3 state to +6, it is easier if the solution is alkaline. The *oxidation* potentials are the negative of the reduction potentials:

$$Cr(OH)_3 + 5OH^- \rightarrow CrO_4^{2-} + 4H_2O + 3e^- \qquad E^\ominus = -(-0.13) = +0.13\,V$$
$$Cr^{3+} + 3\tfrac{1}{2}H_2O \rightarrow \tfrac{1}{2}Cr_2O_7^{2-} + 7H^+ + 3e^- \qquad E^\ominus = -(+1.33) = -1.33\,V$$

In alkaline conditions (pH 14), an oxidising agent with a reduction potential greater than –0.13 V will oxidise chromium hydroxide to chromate(VI) ions.

In acid conditions (pH 0), oxidising agents with reduction potentials greater than +1.33 V will oxidise Cr^{3+} ions to dichromate(VI).

e An oxidation potential measures the ease of oxidation of a species. The more positive the oxidation potential is, the more likely it is that the species will be oxidised.

Worked example 1

Use the following data and those in Table 11.9 to predict whether hydrogen peroxide will oxidise chromium(III) hydroxide to chromate(VI) ions in alkaline solution. If so, write the overall equation.

$$H_2O_2 + 2e^- \rightleftharpoons 2OH^- \quad E = +1.24V$$

Answer

The reduction potential equation of $CrO_4^{2-}/Cr(OH)_3$ must be reversed (and multiplied by 2) and then added to the H_2O_2/OH^- half-equation (multiplied by 3 to get $6e^-$ in both half-equations).

$$2Cr(OH)_3 + 10OH^- \rightarrow 2CrO_4^{2-} + 8H_2O + 6e^- \qquad E^\ominus = -(-0.13) = +0.13\,V$$
$$3H_2O_2 + 6e^- \rightleftharpoons 6OH^- \qquad E^\ominus = +1.24\,V$$
$$2Cr(OH)_3 + 3H_2O_2 + 4OH^- \rightleftharpoons 2CrO_4^{2-} + 8H_2O \qquad E^\ominus_{cell} = +0.13 + (+1.24) = +1.37\,V$$

The value of $E^\ominus_{cell}$ is positive, so the reaction is feasible.

Worked example 2

Use the following data and those in Table 11.9 to predict whether dichromate(VI) ions, $Cr_2O_7^{2-}$, in acid solution, will oxidise hydrogen peroxide to oxygen. If so, write the overall equation.

$$O_2 + 2H^+ + 2e^- \rightleftharpoons H_2O_2 \qquad E^\ominus = +0.68\,V$$

Answer

The reduction potential of O_2/H_2O_2 must be reversed (and multiplied by 3) and then added to the $Cr_2O_7^{2-}/Cr^{3+}$ half-equation:

$$3H_2O_2 \rightleftharpoons 3O_2 + 6H^+ + 6e^- \qquad E^\ominus = -(+0.68) = -0.68\,V$$
$$Cr_2O_7^{2-} + 14H^+ + 6e^- \rightarrow 2Cr^{3+} + 7H_2O \qquad E^\ominus = +1.33\,V$$
$$3H_2O_2 + Cr_2O_7^{2-} + 8H^+ \rightleftharpoons 3O_2 + 2Cr^{3+} + 7H_2O \qquad E^\ominus_{cell} = +1.33 + (-0.68) = +0.65\,V$$

The value of $E^\ominus_{cell}$ is positive, so the reaction is feasible. Hydrogen peroxide should be oxidised to oxygen by acidified dichromate(VI) ions.

Whether hydrogen peroxide oxidises Cr^{3+} to chromate(VI) or reduces chromate(VI) to Cr^{3+} depends on the pH of the solution.

e Note that hydrogen peroxide can be either a reducing agent or an oxidising agent because the oxygen is in the –1 oxidation state and can be oxidised to 0 in O_2 or reduced to –2 in H_2O.

Redox reactions of chromium compounds

+2 → +3

In the numerical list in Table 10.1 (p. 187), all the ions or molecules on the left-hand side of the equations below the Cr^{3+}/Cr^{2+} half-equation, will oxidise Cr^{2+} ions.

This explains why oxygen (air) has to be excluded from solutions of chromium(II) compounds.

$+3 \rightarrow +2$

Chromium(III) ions can be reduced by zinc in acid solution if a Bunsen valve is used to exclude the air:

$$\begin{array}{ll} Zn \rightleftharpoons Zn^{2+} + 2e^- & E^\ominus = -(-0.76) = +0.76\,V \\ Cr^{3+} + e^- \rightleftharpoons Cr^{2+} & E^\ominus = -0.41\,V \\ \hline 2Cr^{3+} + Zn \rightleftharpoons 2Cr^{2+} + Zn^{2+} & E^\ominus_{cell} = +0.76 + (-0.41) = +0.35\,V \end{array}$$

$E^\ominus_{cell}$ is positive, so it is thermodynamically feasible to reduce chromium(III) ions with zinc.

$+3 \rightarrow +6$

This is achieved by heating a solution of chromium(III) sulfate containing excess alkali (NaOH) with excess hydrogen peroxide solution (see worked example 1 above)

If a solution of sodium dichromate(VI) is required, the excess hydrogen peroxide has to be decomposed by heating before acidification. This prevents the formation of a blue peroxycompound which, on acidification, would undergo a self-redox reaction to form oxygen and chromium(III) ions.

$+6 \rightarrow +3$

In acid solution dichromate(VI) ions are a powerful oxidising agent. Among other reactions, they will oxidise Fe^{2+} to Fe^{3+}, Sn^{2+} to Sn^{4+}, Br^- to Br_2, and I^- to I_2.

$$\begin{array}{ll} Cr_2O_7{}^{2-} + 14H^+ + 6e^- \rightleftharpoons 2Cr^{3+} + 7H_2O & E^\ominus = +1.33\,V \\ 3Sn^{2+} \rightleftharpoons 3Sn^{4+} + 6e^- & E^\ominus = -(+0.15) = -0.15\,V \\ \hline Cr_2O_7{}^{2-} + 14H^+ + 3Sn^{2+} \rightleftharpoons 2Cr^{3+} + 3Sn^{4+} + 7H_2 & E^\ominus_{cell} = +1.33 + (-0.15\,V = +1.18\,V \end{array}$$

$E^\ominus_{cell}$ is positive so the oxidation of tin(II) ions by acidified dichromate(VI) ions is thermodynamically feasible.

Disproportionation

Copper(I) ions are not stable in aqueous solution. The reduction potentials in Table 11.9 show that copper(I) ions will disproportionate into copper metal and copper(II) ions as $E^\ominus_{cell}$ for the reaction is positive:

$$\begin{array}{ll} Cu^+ + e^- \rightleftharpoons Cu & E^\ominus = +0.52\,V \\ Cu^+ \rightleftharpoons Cu^{2+} + e^- & E^\ominus = -(+0.15)\,V = -0.15\,V \\ \hline 2Cu^+ \rightleftharpoons Cu + Cu^{2+} & E^\ominus_{cell} = +0.52 + (-0.15) = +0.37\,V \end{array}$$

Insoluble copper(I) compounds and copper(I) complex ions exist.

- Copper(I) oxide, Cu_2O, is the red precipitate formed when an aldehyde reduces Fehling's or Benedict's solution.
- Copper(I) iodide is precipitated when iodide ions reduce copper(II) ions:

$$2Cu^{2+} + 4I^- \rightarrow 2CuI + I_2$$

The addition of ammonia solution to this precipitate produces a colourless solution of the copper(I) ammonia complex:

$$CuI(s) + 2NH_3(aq) \rightarrow [Cu(NH_3)_2]^+(aq) + I^-(aq)$$

A blue colour begins to form on the surface of this colourless solution as the copper(I) complex is slowly oxidised by the oxygen in the air to form the copper(II) complex. This blue colour gradually spreads through the whole solution.

Chromium(III) ions do not disproportionate.

$$6Cr^{3+} + 6e^- \rightleftharpoons 6Cr^{2+} \qquad E^\ominus = -0.41\,V$$

$$2Cr^{3+} + 7H_2O \rightleftharpoons Cr_2O_7 + 14H^+ + 6e^- \qquad E^\ominus = -(+1.33) = -1.33\,V$$

$$8Cr^{3+} + 7H_2O \rightleftharpoons 6Cr^{2+} + Cr_2O_7^{2-} + 14H^+ \qquad E^\ominus_{cell} = -0.41 + (-1.33) = -1.74\,V$$

$E^\ominus_{cell}$ is very negative and so the reaction will not happen.

Uses of some *d*-block metals and their compounds

Many transition metals or their alloys are used for structural or decorative purposes.

An alloy is a mixture of a metal with at least one other element, usually another metal, but sometimes carbon.

Transition metals are often used as catalysts in industrial processes.

Titanium

Titanium forms a protective oxide layer and does not corrode, even at high temperatures. It also has a significantly high melting point and a good strength-to-weight ratio. It is used as a structural material where weight is an important factor (such as in aeroplanes and racing bicycles) or where high temperatures have to be withstood, such as in jet engines.

It is very difficult to weld, so articles made from titanium are expensive.

Titanium(IV) oxide is used as the main component of domestic paints. It is white but can be mixed with dyes to form coloured paints.

Titanium(IV) chloride is used as a catalyst in the Ziegler–Natta process for polymerisation of alkenes.

JAMES KING-HOLMES/SPL

Titanium fan blades for the engines used to power jet aircraft

Chromium

Chromium also forms a protective oxide layer, but does not have as good a strength-to-weight ratio as titanium. It is itself not used as a structural metal, but is found, alloyed with iron and nickel, in stainless steel.

Iron

Iron is the second most abundant metal in the Earth's crust. It is also easily extracted from its ore by reducing iron oxide with carbon monoxide made in the blast furnace from pre-heated air and coke. It is by far the cheapest metal and has many structural uses: from frameworks in industrial buildings to car bodies. It is normally alloyed with a small percentage of carbon, which increases its strength. Cast iron has a higher percentage of carbon (about 2%) and is extremely strong, but brittle. Iron is used as the catalyst in the Haber process for manufacturing ammonia.

ONTANET/R. G. PARKES

The iron bridge at Ironbridge

Copper

Copper is a very good electrical conductor; only silver is better.

Copper can be alloyed with zinc to form brass and with tin to form bronze. Ores containing tin and copper are easily reduced to the metal, so early civilisations were able to manufacture bronze for weapons and artefacts. Higher technological skills are needed to reduce iron ore to iron, so the Iron Age followed the Bronze Age.

SUSANNAH HEAD

A bronze chariot from the tombs at Xian in China

Copper(II) compounds are used as fungicides and copper metal is used as a catalyst in the manufacture of hydrogen from methane.

Copper(I) chloride is used in some photochromic sunglasses. The glass contains small amounts of colourless copper(I) chloride and silver chloride. When UV light shines on the glasses, the copper(I) chloride reduces the silver chloride to silver, which darkens the glass. So not only are harmful UV rays removed, light intensity is also reduced. When no longer exposed to UV light, the reaction reverses and the glass becomes transparent.

$$CuCl + AgCl \rightleftharpoons CuCl_2 + Ag$$

Copper(I) chloride cannot be used in plastic sunglasses made from polycarbonate. These are coated with a complex organic substance based on adamatylidene, which darkens reversibly in bright sunlight.

Questions

1 Give the electron configurations of vanadium and the V^{3+} ion.

V $1s^2$.........

V^{3+} $1s^2$.........

2 Explain why molybdenum has the electron configuration [Kr] $4d^5\ 5s^1$.

3 Explain why the first ionisation energies of the *p*-block elements increase considerably from left to right whereas those of the *d*-block elements hardly alter.

4 Explain why the melting point of titanium is higher than that of calcium.

5 Draw a dot-and-cross diagram, showing the outer electrons only, for:

a CrO_4^{2-} **b** VO_2^+

6 Explain why titanium can form both 2+ and 3+ ions whereas calcium and zinc do not form 3+ ions.

7 Write equations for the reactions, if any, of dilute hydrochloric acid and of aqueous sodium hydroxide with:

a iron(III) oxide, Fe_2O_3

b iron(VI) oxide, FeO_3

8 Draw a diagram of the $[Cr(H_2O)_6]^{3+}$ ion to show its shape. Mark on your diagram the bond angles. Name the types of bond in the ion and say where in the structure these bonds occur.

9 Give the formula of a compound of chromium in the following oxidation states:

a +2

b +3

c +6

10 Give the name and formula of a copper compound in:

a the +1 oxidation state

b the +2 oxidation state

11 Give the formula of:

a a chromium(III) complex ion

b a copper(II) complex ion

12 What colour of light is absorbed by:

a $[Cr(H_2O)_4Cl_2]^+$, which is green?

b $[Fe(H_2O)_5NCS]^{2+}$, which is red?

13 Explain why hydrated Ti^{4+} ions are colourless, but hydrated Ti^{3+} ions are coloured.

14 Explain why, when excess ammonia is added to copper(II) sulfate solution, the colour changes from turquoise (blue-green) to violet-blue.

15 What factors affect the colours of transition metal ions?

16 Why is anhydrous copper(II) sulfate white and why are copper(I) complexes also white?

17 State what you would see, and write the equations for, the reactions of sodium hydroxide and of ammonia solution with:

a a solution of nickel(II) ions

b a solution of copper(II) ions

18 When sodium hydroxide solution is added to a solution of iron(II) ions, a green precipitate is produced. When hydrogen peroxide solution is then added, the precipitate turns red-brown. Identify the green and red-brown precipitates and explain what is happening in the two reactions.

19 Hydrated chromium(III) ions can be deprotonated. Write equations to show the deprotonation that takes place when:

a the hydrated ions are dissolved in water

b sodium hydroxide is added to the hydrated ions, slowly and then in excess

20 Use the data in Table 11.9 to predict:

a whether an acidic solution of Fe^{3+} ions will disproportionate into Fe^{2+} and FeO_4^- ions.

b whether hexacyanoferrate(III) ions, $[Fe(CN)_6]^{3-}$, will be reduced by cobalt(II) ammine, $[Co(NH_3)_6]^{2+}$.

21 A $25.0\,cm^3$ sample of a $0.100\,mol\,dm^{-3}$ solution of V^{3+} ions was placed in a conical flask and excess dilute sulfuric acid added. Potassium manganate(VII) of concentration $0.0500\,mol\,dm^{-3}$ was added from a burette until a faint pink colour was seen. The titre was $20.0\,cm^3$.

a Calculate the amount (moles) of potassium manganate(VII) ions in the titre.

b Calculate the number of moles of electrons that the manganate(VII) has received from the vanadium(III) ions.

c Calculate the amount (moles) of vanadium(III) ions in the sample.

d Calculate the oxidation state of the vanadium after reaction with potassium manganate(VII).

Use the E values in Table 11.9 on p. 228 and Table 10.1 on p. 187 for data for the following questions.

22 a Predict whether hydrogen sulfide will reduce VO_2^+ ions to VO^{2+} or V^{3+} ions.

b Write the overall equation for the reaction that takes place.

23 a Predict the oxidation state of vanadium when oxygen gas is blown through an acidified solution of V^{2+} ions.

b Write the equation for the overall reaction that takes place.

24 a Predict whether iodine will oxidise V^{3+} ions to VO^{2+} or to VO_2^+ or not react at all.

b Write the equation for any reaction that takes place.

25 Predict whether chloride and bromide ions will be oxidised by manganese(IV) oxide, MnO_2, in acid solution.

26 The reduction potential for Cr^{2+} ions being reduced to chromium metal is $-0.91\,V$ and that for Cr^{3+} ions being reduced to Cr^{2+} ions is $-0.41\,V$:

$$Cr^{2+} + e^- \rightleftharpoons Cr \qquad E^\ominus = -0.91\,V$$

$$Cr^{3+} + e^- \rightleftharpoons Cr^{2+} \qquad E^\ominus = -0.41\,V$$

Predict whether Cr^{2+} ions will disproportionate in aqueous solution to Cr metal and Cr^{3+} ions. If so, write the equation for the reaction that takes place.

27 Describe how you would prepare, from a solution of chromium(III) sulfate:

a a solution of a chromium(III) complex

b a solution containing chromate(VI) ions

c a solution containing chromium(II)ions

28 What do you understand by the following terms? Give an example of each.

a ligand exchange

b heterogeneous catalyst

Arenes and their derivatives

Chapter 12

Introduction

Organic chemistry can be divided into three categories:

- **Aliphatic chemistry**: this is the study of simple compounds that have straight or branched carbon chains or rings of carbon atoms. Alkenes, ethanol, propanone, ethylamine, cyclohexane and cyclohexene are examples of aliphatic compounds.

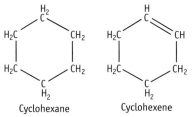

Cyclohexane   Cyclohexene

- **Arenes**: these are also called **aromatic compounds.** They all contain a benzene ring. This is a ring of six carbon atoms in which each forms two σ-bonds with its neighbouring carbon atoms and has one *p*-electron in a π-bond. The fourth valence electron is in a σ-bond with an atom that is joined to the ring.
- **Natural products**: these compounds are complex organic substances found in nature. They include colouring matter, poisons, flavourings, proteins, hormones and compounds with specific odours that may be used to attract pollinating insects or are pheromones.

The benzene ring

Benzene has the formula C_6H_6. There was doubt about its structure until the German chemist Kekulé suggested that the carbon atoms were arranged in a ring with alternate single and double bonds.

There are four major problems with the Kekulé structure:

Problem 1 In aliphatic compounds a C=C is shorter than a C–C bond, but all the bonds between the carbon atoms in benzene are the same length.

X-ray diffraction shows the position of the centre of atoms. Analysis of the diffraction pattern of benzene shows clearly that all the bond lengths between the carbon atoms are the same (Table 12.1).

SPL

August Kekulé 1829–96

Bond	Bond length/nm
Carbon–carbon single bond in cyclohexene	0.15
Carbon–carbon double bond in cyclohexene	0.13
Carbon–carbon bond in benzene	0.14

Table 12.1
Bond lengths

This is also shown by the electron density map of benzene. The map also shows that there is a significant electron density in rings above and below the plane of the six carbon atoms (Figure 12.1).

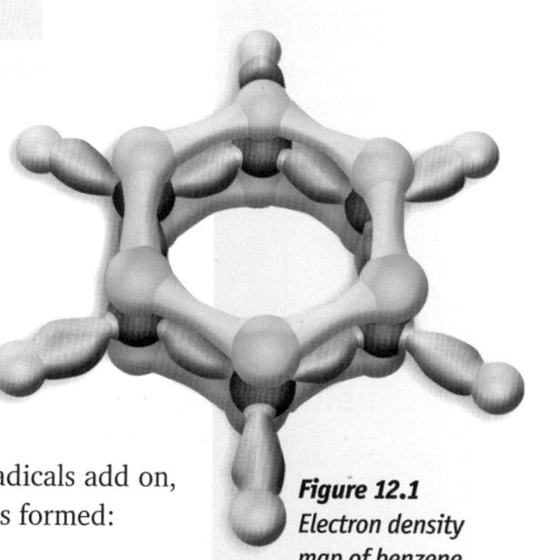

Figure 12.1
Electron density map of benzene

Problem 2 The enthalpies of hydrogenation are not as would be expected if benzene had three separate double bonds.

When benzene vapour and hydrogen are passed over a heated nickel catalyst, 3 mol of hydrogen per mole of benzene add on and cyclohexane is formed. Hydrogen is adsorbed as atoms at the active sites on the surface of the catalyst. These hydrogen radicals add on, one at a time, breaking the π-bonds until saturated cycloalkane is formed:

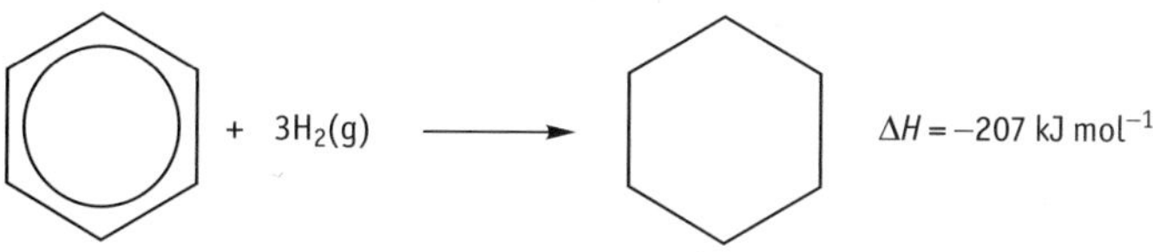

The reaction is similar to the addition of hydrogen to cyclohexene, which has one localised π-bond:

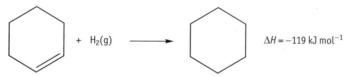

$$+ \ H_2(g) \longrightarrow \qquad \Delta H = -119 \text{ kJ mol}^{-1}$$

If benzene had three *localised* π-bonds, the enthalpy change for the addition of 3mol of hydrogen would be $3 \times -119 = -357\,\text{kJ mol}^{-1}$. The difference between this value and the actual enthalpy change for the addition of 3 mol of hydrogen to benzene is $150\,\text{kJ mol}^{-1}$. This is the value by which the benzene molecule is stabilised because of the delocalisation of the π-electrons (resonance stabilisation energy).

The theoretical molecule with three localised double bonds is called 'cyclohexatriene'. The energy levels of this molecule and the actual molecules of benzene and cyclohexane are shown in Figure 12.2.

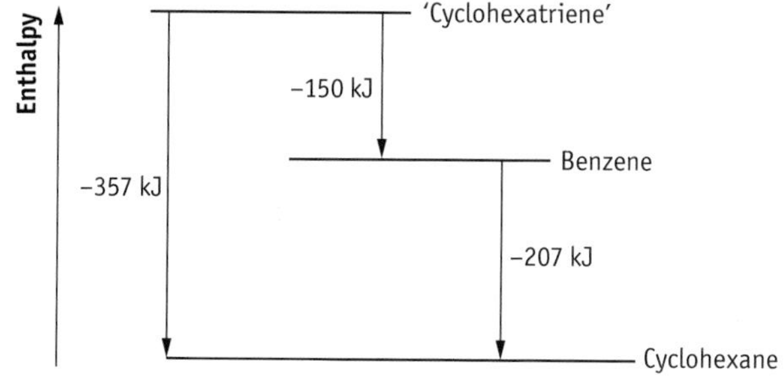

Figure 12.2
Energy level diagram showing delocalisation energy of 150 kJ

Benzene is at an energy level 150 kJ mol^{-1} lower than that of 'cyclohexatriene'. Therefore, the resonance stabilisation energy is 150 kJ mol^{-1}.

Problem 3 Comparison of the infrared spectra of aromatic compounds with those of aliphatic compounds containing a C=C group shows slight differences. The C–H stretching vibration in benzene is 3036 cm^{-1} and that for C=C stretching is 1479 cm^{-1}; the equivalents in an aliphatic compound such as cyclohexene are 3023 cm^{-1} and 1438 cm^{-1} (Figures 12.3 and 12.4).

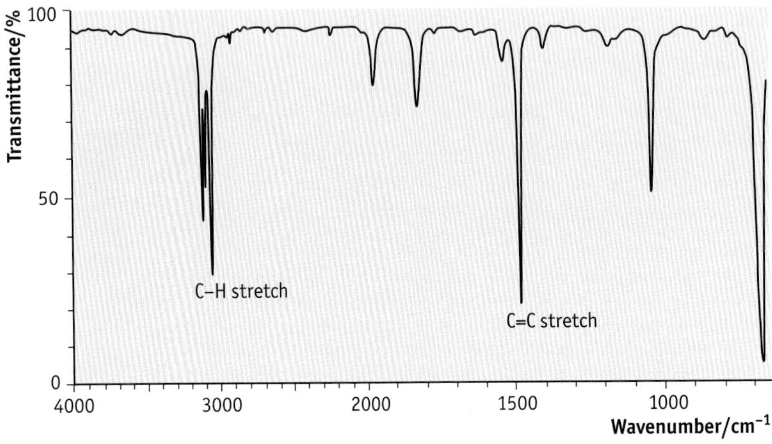

Figure 12.3 *Infrared spectrum of benzene*

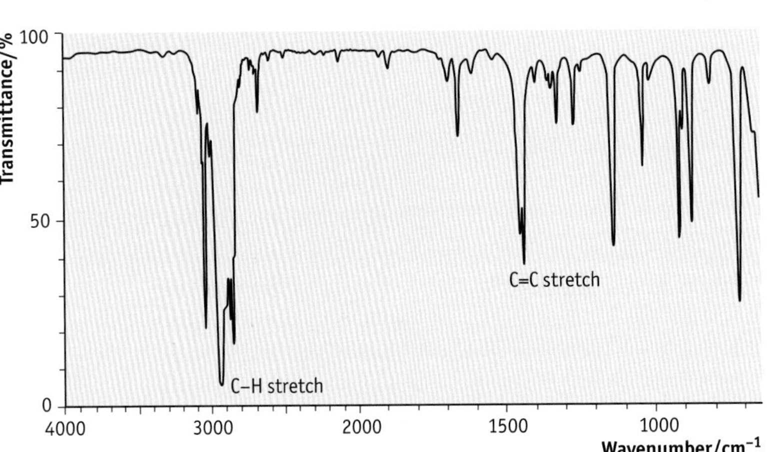

Figure 12.4 *Infrared spectrum of cyclohexene*

Problem 4 Benzene does not show the typical electrophilic addition reactions of unsaturated compounds such as ethene and cyclohexene. For example, it reacts with bromine by substitution rather than by addition.

One solution to these problems was to apply the concept of a **resonance structure**. This was first used to account for the identical bond lengths in ozone, O=O→O. In theory, the double bond should be shorter than the dative bond. However, both bonds are the same length. The reason is that the molecule of ozone is a unique structure that appears to resonate between O=O→O and O←O=O. The idea of resonance is that the actual molecule is a definite structure that is a hybrid of two theoretical structures. An ozone molecule is neither

O←O=O nor O=O→O and it does not oscillate between the two. Resonance can be shown as a double-headed arrow between the two theoretical structures, both of which must obey the normal rules of bonding.

Ozone can be written as:

O=O→O ↔ O←O=O

Benzene can be written as:

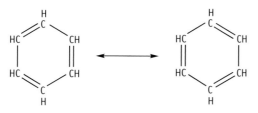

It is usual to write the formula of benzene without showing either the carbon atoms or the hydrogen atoms. Thus, a hexagon with alternate double and single bonds can be used to represent benzene, although it would be better to show that it is a resonance structure, as in Figure 12.5.

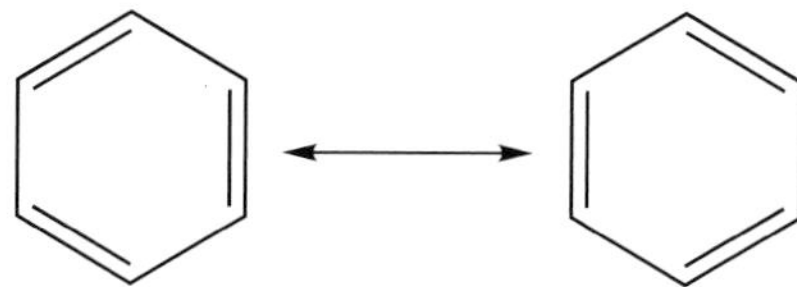

Figure 12.5 *The Kekulé formulae of benzene*

A double bond is a σ-bond (an overlap of atomic orbitals between two atoms) and a π-bond (an overlap of *p*-orbitals above and below the σ-bond).

The best representation of a benzene molecule is six carbon atoms in a hexagonal plane bonded to each other and each bonded to a hydrogen atom by σ-bonds. The fourth valence electron on each carbon atom is in a p_z-orbital at right angles to the plane of the six carbon atoms. The p_z-orbital of one carbon atom overlaps equally with the p_z-orbitals of *both* adjacent carbon atoms, forming a continuous π-bond above and below the ring of carbon atoms. These six p_z-electrons are **delocalised** over the ring. This gives the molecule greater stability compared with a theoretical molecule in which the π-electrons are localised between individual atoms.

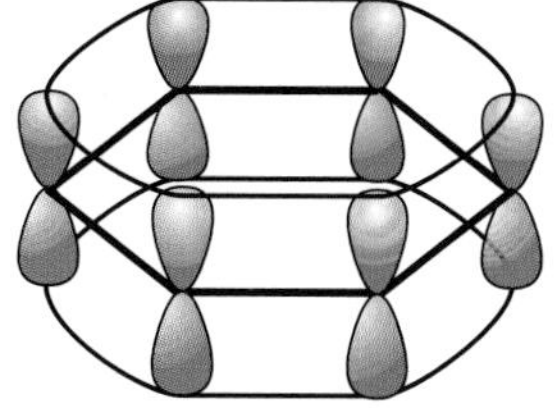

Figure 12.6 *Overlap of P_z-orbitals in benzene*

This gain in stability is called the **resonance stabilisation energy**. In benzene, its value is 150 kJ mol^{-1} less than that of the theoretical molecule that has three single and three localised double bonds (p. 237).

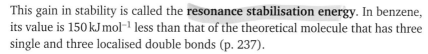

The way in which a delocalised π-system is written in a formula is to draw a continuous curve round the atoms that are part of that system — for benzene, a circle is drawn within the hexagon representing the six carbon atoms (Figure 12.7).

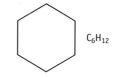

Figure 12.7 *The skeletal formula for benzene*

Figure 12.8 *The skeletal formula for cyclohexane*

The formula of cyclohexane, which has no π-electrons, is drawn as in Figure 12.8.

Remember that there is a hydrogen atom attached to each carbon atom in benzene. When benzene reacts by substitution, a hydrogen atom is replaced by another group. Therefore, one of the products is a simple molecule such as HBr, HCl or H_2O.

Nomenclature of arenes

Single substituents

If a single atom or group replaces one hydrogen atom in the ring, the name of the compound has the stem –benzene with a prefix of the group or atom entering, thus:

- C_6H_5Cl is called chlorobenzene.
- $C_6H_5CH_3$ is called methylbenzene.
- $C_6H_5NO_2$ is called nitrobenzene.
- $C_6H_5NH_2$ is called aminobenzene.

Methylbenzene used to be called toluene because it was first isolated from the resin of the South American tree *Tolu balsam*.

An additional naming system is based on the **phenyl group**, $-C_6H_5$ group, thus:

- C_6H_5OH is called phenol.
- $C_6H_5NH_2$ is called either phenylamine or aminobenzene.
- $CH_3CH(C_6H_5)CH_3$ is called 2-phenylpropane.
- $C_6H_5COCH_3$ has a carbon chain of two atoms (ethan-) and is a ketone (-one) with a phenyl group substituted into it, so it is called phenylethanone.

Aminobenzene is called aniline.

Two or more substituents

If there are two or more groups in a benzene ring, positional isomerism occurs. For example, there are three isomers of $C_6H_4Cl_2$. The relative positions of the two chlorine atoms are defined by numbers. One substituent is always named as being in the 1-position and the second substituent by the lower of the possible numbers.

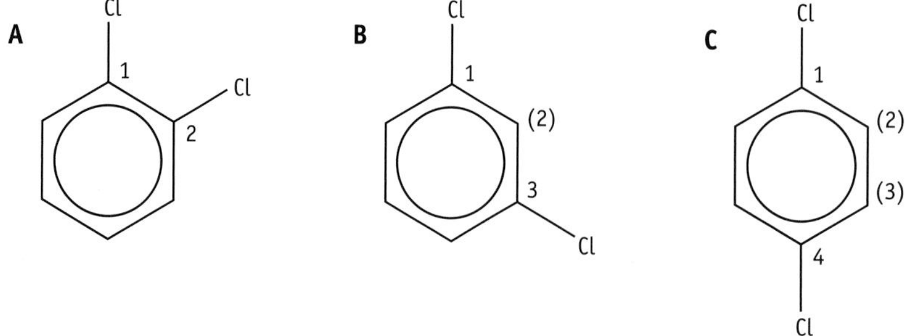

- Structure A is 1,2-dichlorobenzene (not 1,6-dichlorobenzene).
- Structure B is 1,3-dichlorobenzene (not 1,5-dichlorobenzene).
- Structure C is 1,4-dichlorobenzene.

The old name for A was *ortho*-dichlorobenzene, the prefix *ortho*- indicating that the two substituents are on adjacent carbon atoms; B was *meta*-dichlorobenzene (the substituents being one carbon apart); C was *para*-dichlorobenzene (the substituents being opposite each other).

The compound 2,4,6-tribromophenol has the formula:

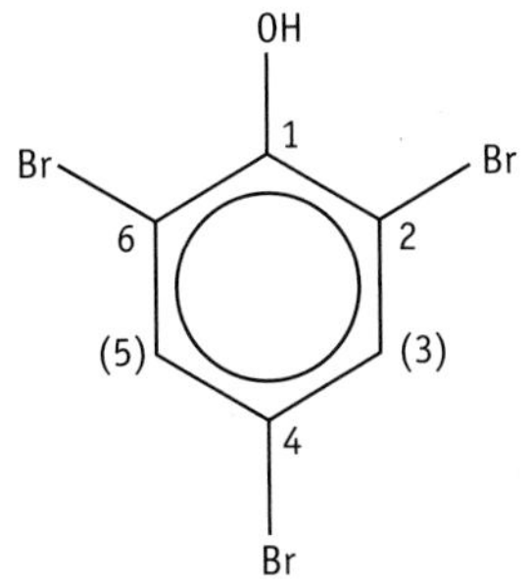

TNT is 2,4,6-trinitrotoluene and has the formula:

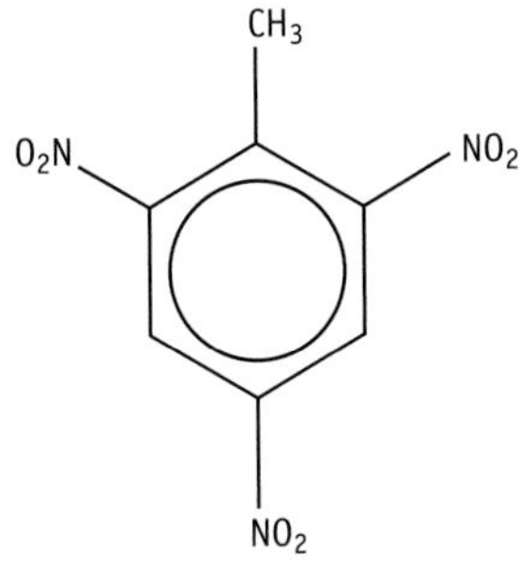

An exception to the rules for naming aromatic compounds is C_6H_5COOH, which is called benzoic acid, rather than the cumbersome phenylmethanoic acid.

Many aromatic substances are still known by their old names, which were derived from the source of the chemical. For example, the compound with the formula

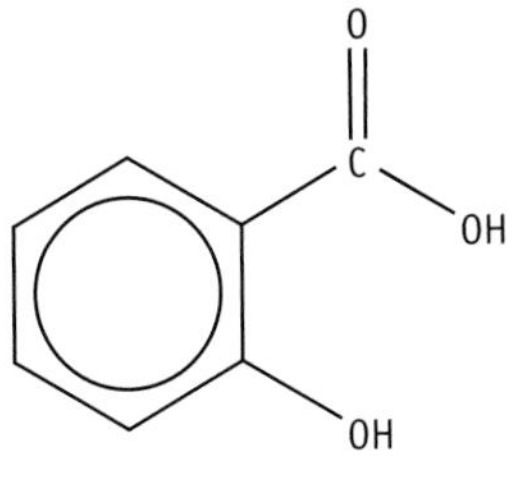

is commonly referred to as salicylic acid, rather than 2-hydroxybenzoic acid.

Salicylic acid is derived from *salix*, the Latin name for a willow tree.

Benzene, C_6H_6

Physical properties

Benzene and many compounds that contain a substituted benzene ring have a characteristic smell — hence the name aromatic. The modern name for compounds that contain a benzene ring is **arene.**

Arene is derived from *aromatic* and *-ene*, representing a carbon–carbon double bond.

Benzene is a non-polar liquid at room temperature. The main forces between molecules are instantaneous induced dipole (dispersion) forces and as the molecule has 42 electrons, these forces are strong enough for benzene to be a liquid. It cannot form hydrogen bonds with water and so it is immiscible.

The physical properties of benzene are:

- melting temperature, 5.5°C
- boiling temperature, 80.1°C
- density, $0.878\,g\,cm^{-3}$

Benzene and some other compounds that contain a benzene ring are carcinogenic.

Chemical reactions

Benzene, like all organic compounds, burns in air. It also undergoes addition and substitution reactions.

Combustion

In excess air, benzene burns to form carbon dioxide and water:

$$C_6H_6(l) + 7\tfrac{1}{2}O_2(g) \rightarrow 6CO_2(g) + 3H_2O(l) \qquad \Delta H_c^\ominus = -3273\,kJ\,mol^{-1}$$

It has an octane rating of 101 and is used, in small quantities, as a petrol additive. Its main problem is that it is carcinogenic.

In a limited amount of air, it burns with a smoky flame since carbon is formed, rather than carbon dioxide:

$$C_6H_6(l) + 1\tfrac{1}{2}O_2(g) \rightarrow 6C(s) + 3H_2O(g)$$

The smoky flame can be used as a test to suggest the presence of a benzene ring in an organic compound.

Free-radical addition

Reaction with chlorine

When chlorine is bubbled into boiling benzene in the presence of ultraviolet light, addition takes place and 1,2,3,4,5,6-hexachlorocyclohexane is produced:

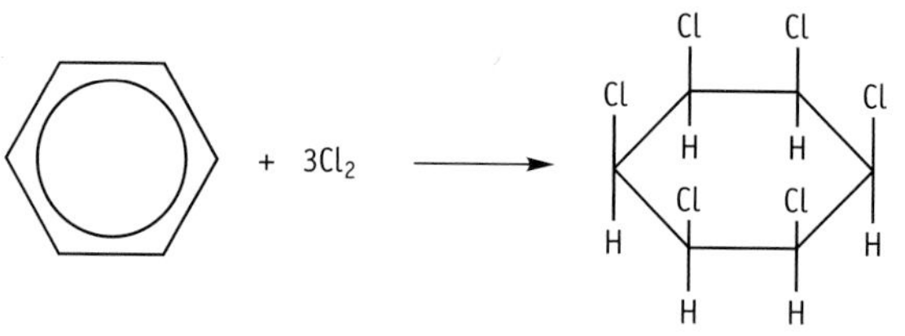

+ $3Cl_2$ →

A mixture of the geometric isomers of $C_6H_6Cl_6$ is obtained with different *cis* and *trans* arrangements on adjacent carbon atoms, only one of which is shown in this formula.

The mechanism is that ultraviolet light splits a chlorine molecule into two radicals:

$$Cl_2 + \text{light energy} \rightarrow 2Cl\bullet$$

The chlorine radicals add on, one at a time, to the benzene ring until all the π-bonds have been broken.

Bromine reacts in a similar way.

Reaction with hydrogen

When heated with a nickel catalyst, hydrogen adds on and cyclohexane is formed:

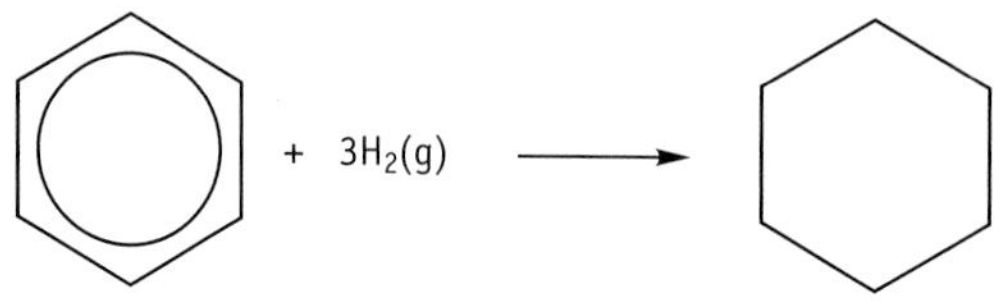

+ $3H_2(g)$ →

Electrophilic substitution reactions

Electrophiles attack the high-electron density in the delocalised π-ring of the benzene molecule.

- The first step is the addition of the electrophile to form an intermediate that is positively charged and in which the full delocalisation has been partially broken.
- The second step is the elimination of an H^+ ion. The fully delocalised ring and stabilisation energy are thus regained.

The reaction is:

addition + elimination = substitution

The delocalised π-system is stable compared with the localised π-bond in alkenes, so the activation energy required to break it is fairly high. Therefore, the electrophilic substitution reactions of benzene always require a catalyst.

The mechanism is that an electrophile, E^+, adds on to a carbon atom in the benzene ring:

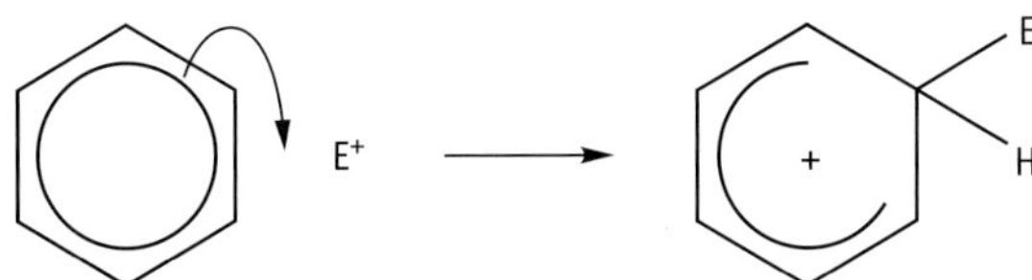

This step is identical to the addition of an electrophile to an alkene.

However, a fully delocalised ring of π-electrons stabilises the benzene ring. Therefore, rather than the second step being the subsequent addition of an anion (as with alkenes), it is loss of an H^+:

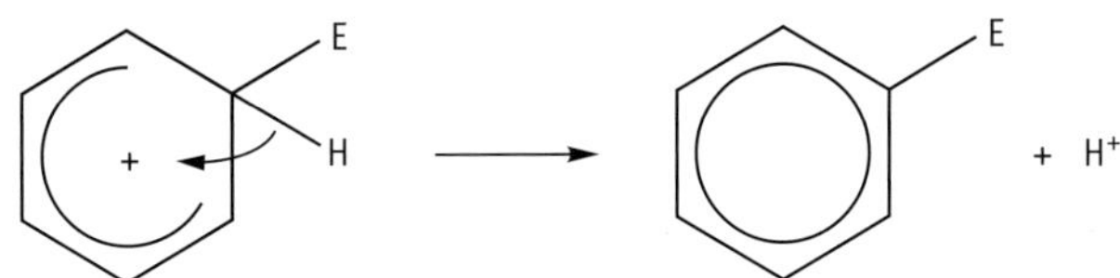

Nitration: the reaction with nitric acid

When concentrated nitric acid, benzene and a catalyst of concentrated sulfuric acid are warmed together at a temperature of 60°C in a flask fitted with a reflux condenser, nitrobenzene and water are formed:

+ HNO_3 ⟶ NO_2 + H_2O

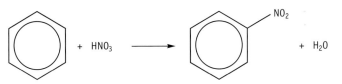

Below 55°C the reaction is too slow; above 65°C a second $-NO_2$ group is substituted into the ring.

The reaction proceeds in three steps.

Step 1: concentrated sulfuric acid protonates a nitric acid molecule, forming $H_2NO_3^+$. This loses water, forming NO_2^+, which is a powerful electrophile.

$$H_2SO_4 + HNO_3 \rightarrow H_2NO_3^+ + HSO_4^-$$
$$H_2NO_3^+ \rightarrow NO_2^+ + H_2O$$

The H_2O is then protonated by another H_2SO_4 molecule.

Step 2: the electrophile NO_2^+ ion draws a pair of electrons from the π-system and forms a covalent bond:

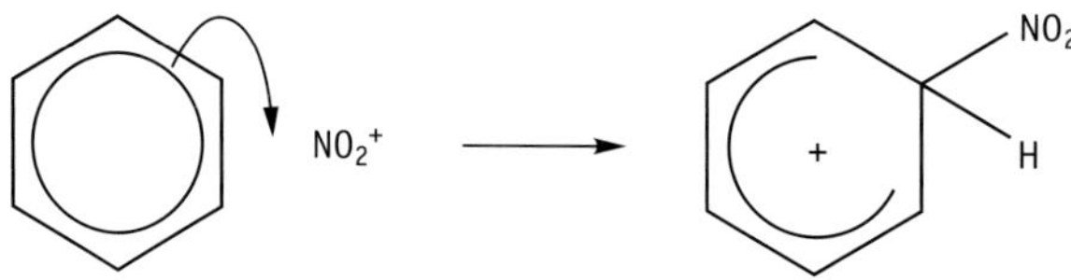

Step 3: the intermediate loses an H^+ ion to a HSO_4^- ion. The stability of the benzene ring is regained and the H_2SO_4 catalyst is regenerated:

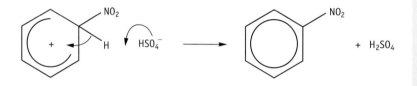

NO2 · + · H · HSO4⁻ ⟶ NO2 · + H2SO4

Do not forget to put the positive charge in the ring in the formula of the intermediate. Also, make sure that the delocalised circle goes round five carbon atoms but not the one to which the NO_2 group is attached.

If the temperature is too high, a second $-NO_2$ group is substituted in the 3-position:

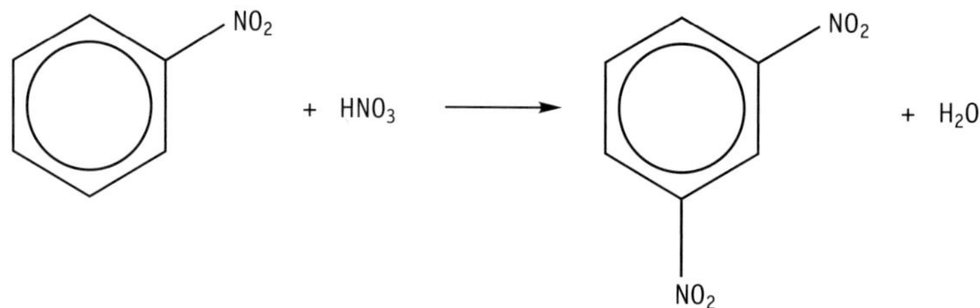

The products are 1,3-dinitrobenzene and water.

Reaction with halogens

Under certain conditions, chlorine and bromine react rapidly with benzene. A catalyst of anhydrous iron(III) halide (or aluminium halide) must be used and all the reagents must be dry. For the bromination of benzene, the catalyst is made *in situ* by adding iron filings to a mixture of benzene and liquid bromine. Clouds of hydrogen bromide fumes are given off in this exothermic reaction:

In situ means in the same apparatus.

+ Br2 ⟶ Br + HBr

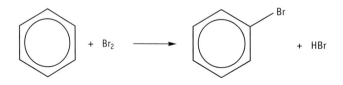

Step 1: if iron filings are added (rather than iron(III) bromide), they react to form the catalyst:

$$2Fe + 3Br_2 \rightarrow 2FeBr_3$$

This then reacts with more bromine to form the electrophile, Br^+:

$$FeBr_3 + Br_2 \rightarrow Br^+ + FeBr_4^-$$

Step 2: the Br^+ electrophile attacks the benzene ring:

Step 3: The $FeBr_4^-$ removes H^+ from the intermediate. HBr is formed and the catalyst, $FeBr_3$, is regenerated:

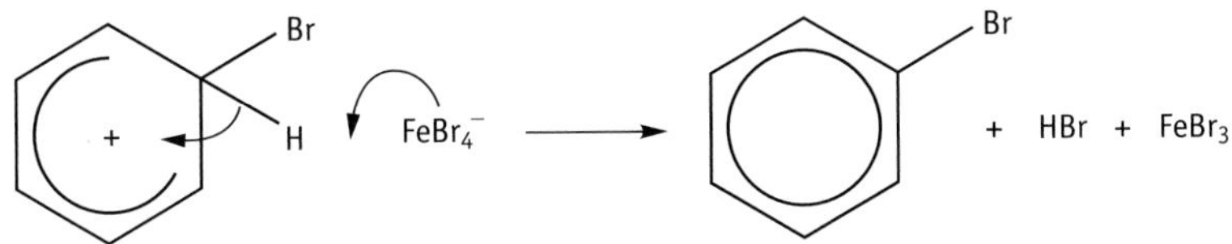

ⓔ Alkenes, such as ethene, react with bromine water. There is no reaction between bromine water and benzene.

Friedel–Crafts reactions

The French chemist Charles Friedel and the American James Crafts discovered the reaction of benzene with organic halogen compounds.

- The reaction can be represented by the equation:

$$C_6H_6 + RCl \rightarrow C_6H_5R + HCl$$

- The catalyst has to be a covalent anhydrous metal chloride, for example aluminium chloride, $AlCl_3$, or iron(III) chloride, $FeCl_3$. All water must be excluded from the reaction.
- The reaction can be:
  - alkylation, with a halogenoalkane
  - acylation, with an acid chloride

In **alkylation reactions**, benzene reacts with a halogenoalkane in the presence of a catalyst of anhydrous aluminium chloride, under dry conditions, to form a hydrocarbon and gaseous hydrogen halide. For example, benzene and chloroethane react to form ethylbenzene and hydrogen chloride:

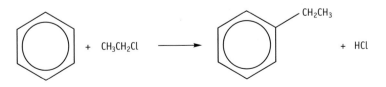

It is difficult to stop the reaction at this stage, as further alkylation to 1,2- and 1,4-diethylbenzene takes place.

Step 1: the catalyst reacts with the halogenoalkane to form $CH_3CH_2^+$, which is the electrophile:

$$CH_3CH_2Cl + AlCl_3 \rightarrow CH_3CH_2^+AlCl_4^-$$

Step 2: the electrophile accepts a pair of π-electrons from the ring, forming a covalent bond:

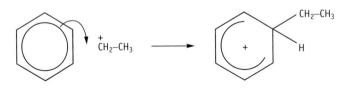

Step 3: the $AlCl_4^-$ removes an H^+ from the intermediate. The stability of the benzene ring is regained and the catalyst is regenerated:

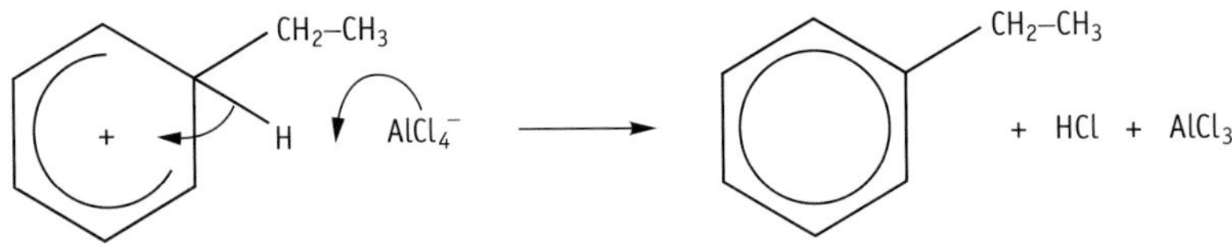

ⓔ When drawing this mechanism, make sure that the ethyl group is bonded to the benzene ring by the CH_2 carbon and not by the carbon of the CH_3.

The reaction profile for the two-step substitution reactions of benzene is shown in Figure 12.9.

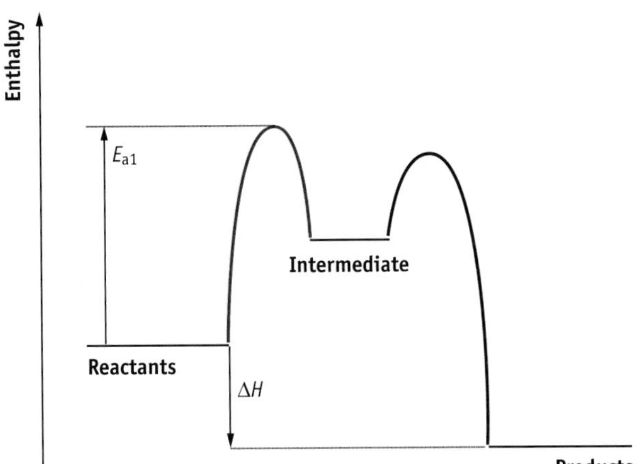

Figure 12.9 *Reaction profile for the substitution reactions of benzene*

The intermediate is at a higher energy level than the reactants due to the loss of stability of the delocalised π-system.

Step 1 has the higher activation energy and so is rate-determining.

Friedel–Crafts **acylation reactions** are carried out using an acid chloride such as ethanoyl chloride, CH_3COCl. The catalyst is anhydrous aluminium chloride, which produces the electrophile $CH_3–C^+=O$.

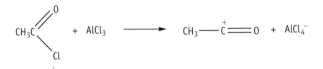

$$CH_3COCl + AlCl_3 \longrightarrow CH_3–\overset{+}{C}\equiv O + AlCl_4^-$$

The remainder of the mechanism is similar to that of Friedel–Crafts alkylation reactions.

The overall reaction is:

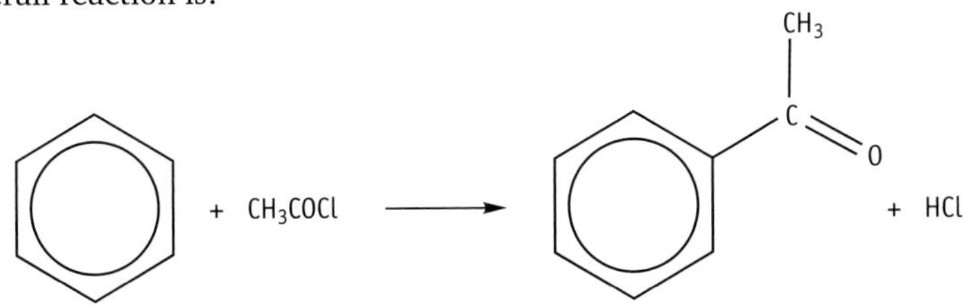

$$C_6H_6 + CH_3COCl \longrightarrow C_6H_5COCH_3 + HCl$$

The products are the ketone (phenylethanone) and hydrogen chloride.

Further substitution does not take place, because the $COCH_3$ group deactivates the benzene ring.

All substitution reactions of benzene must be carried out in dry conditions with a catalyst that produces a powerful electrophile.

Sulfonation

Benzene will also react with sulfuric acid:

The **sulfonation** of benzene requires gentle heating with sulfuric acid that contains some dissolved sulfur trioxide (fuming sulfuric acid or oleum). The attack is by the electrophile SO_3. The products are benzenesulfonic acid and water:

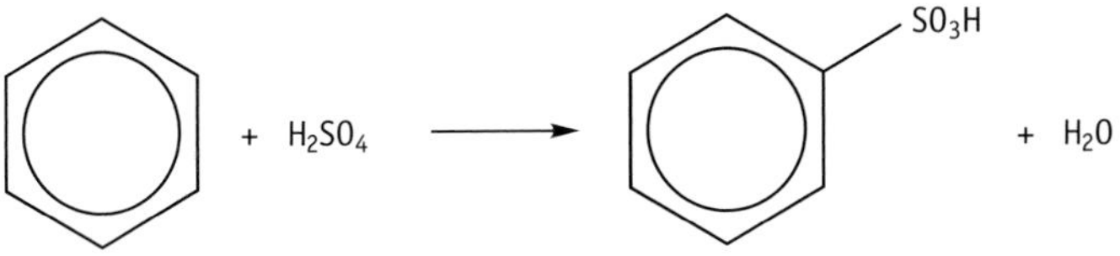

$$C_6H_6 + H_2SO_4 \longrightarrow C_6H_5SO_3H + H_2O$$

Benzenesulfonic acid is a strong acid, so its salts are neutral.

Reaction with alkenes

Benzene adds on to alkenes under Friedel–Crafts conditions. The reaction can also be thought of as substitution into a benzene ring. The product with alkenes (other than ethene) is a branched-chain alkane substituted into the benzene ring, for example:

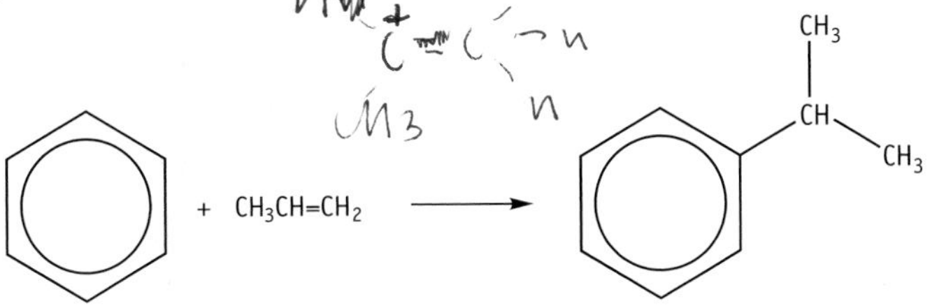

$$C_6H_6 + CH_3CH{=}CH_2 \longrightarrow C_6H_5CH(CH_3)CH_3$$

These last two reactions are important in the manufacture of detergents. A long-chain alkene (C_{10} or greater) is reacted with benzene and the product is sulfonated. This product is then neutralised by sodium hydroxide. It is an excellent detergent because its calcium salt is soluble, so no scum is formed in hard water.

+ $CH_3(CH_2)_8CH{=}CH_2$ → (with benzene) $C_6H_5CH(CH_3)(CH_2)_8CH_3$ + H_2SO_4 → $HO_3S\text{-}C_6H_4\text{-}CH(CH_3)(CH_2)_8CH_3$ + NaOH → $Na^+\ {}^-O_3S\text{-}C_6H_4\text{-}CH(CH_3)(CH_2)_8CH_3$

A detergent

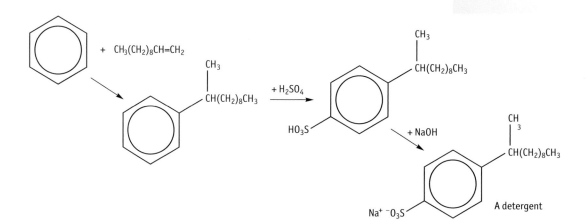

Aromatic compounds with a carbon side chain

If a group is attached via a carbon atom to the benzene ring, the compound has all the properties of an aliphatic compound with that group, as well as the reactions due to the benzene ring.

Alkane side chain

Methylbenzene, $C_6H_5CH_3$, reacts similarly to both alkanes and benzene.

Methylbenzene used to be called toluene

Reaction with chlorine

When chlorine gas is bubbled into methylbenzene in the presence of UV light, the hydrogen atoms of the $-CH_3$ group are replaced one at a time:

$$C_6H_5CH_3 + Cl_2 \rightarrow C_6H_5CH_2Cl + HCl$$
$$C_6H_5CH_2Cl + Cl_2 \rightarrow C_6H_5CHCl_2 + HCl \quad \text{etc.}$$

Reaction with concentrated nitric acid

If methylbenzene and concentrated nitric acid are warmed to 50°C in the presence of concentrated sulfuric acid, a mixture of 2- and 4-nitromethylbenzene and water is produced:

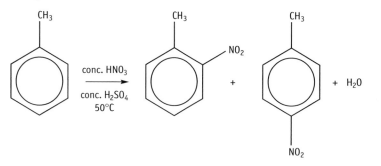

$$C_6H_5CH_3 \xrightarrow[\text{conc. } H_2SO_4,\ 50^\circ C]{\text{conc. } HNO_3} 2\text{-}NO_2C_6H_4CH_3 + 4\text{-}NO_2C_6H_4CH_3 + H_2O$$

If the nitration or other reactions of arenes are to be carried out in the laboratory, methylbenzene or methoxybenzene, $C_6H_5OCH_3$, should be used, rather than the carcinogenic benzene.

If the temperature rises above 50°C, 2,4-dinitromethylbenzene will be formed and if methylbenzene is heated under reflux with the nitrating mixture, 2,4,6-trinitrotoluene (TNT) is produced.

Ketone side chain

Phenylethanone, $C_6H_5COCH_3$, has similar reactions to propanone, including the iodoform reaction.

Reaction with 2,4-dinitrophenylhydrazine

Phenylethanone reacts with 2,4-dinitrophenylhydrazine to produce a 2,4-dinitrophenylhydrazone:

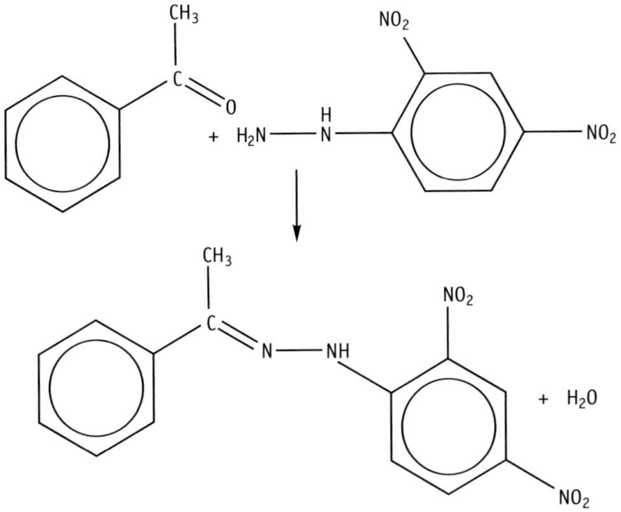

$$C_6H_5C(=O)CH_3 + H_2N{-}NH{-}C_6H_3(NO_2)_2 \rightarrow C_6H_5C(CH_3){=}N{-}NH{-}C_6H_3(NO_2)_2 + H_2O$$

Reaction with HCN

In the presence of a trace of base, hydrogen cyanide adds on to the C=O group of phenylethanone:

$$C_6H_5COCH_3 + HCN \rightarrow C_6H_5C(OH)(CN)CH_3$$

Reaction with $LiAlH_4$

Phenylethanone is reduced by lithium tetrahydridoaluminate(III) in dry ether, followed by the addition of dilute acid. The product is a secondary alcohol, 1-phenylethanol:

$$C_6H_5COCH_3 + 2[H] \rightarrow C_6H_5CH(OH)CH_3$$

Reaction with iodine and sodium hydroxide

This is the iodoform reaction.

$$C_6H_5COCH_3 + 3I_2 + 4NaOH \rightarrow CHI_3 + 3NaI + C_6H_5COONa + 3H_2O$$

Phenol, C_6H_5OH

Phenol is an aromatic alcohol. It contains an –OH group bonded directly to a carbon atom in a benzene ring:

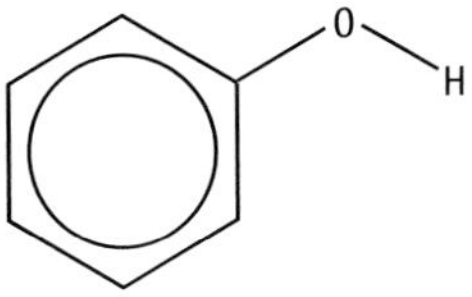

$$C_6H_5{-}O{-}H$$

The lone pair of electrons in the p_z-orbital on the oxygen atom interacts with the delocalised π-electrons in the ring (Figure 12.10). This increases the electron density inside the ring, making it easier for phenol to be attacked by an electrophile. It also decreases the δ^- charge on the oxygen atom, making it less reactive as an alcohol. Pulling the p_z electrons from the oxygen atom into the ring causes the hydrogen atom to be more δ^+ than it is in aliphatic alcohols. This means that it is much more easily lost from phenol than it is from aliphatic alcohols, so phenol has stronger acid properties than ethanol.

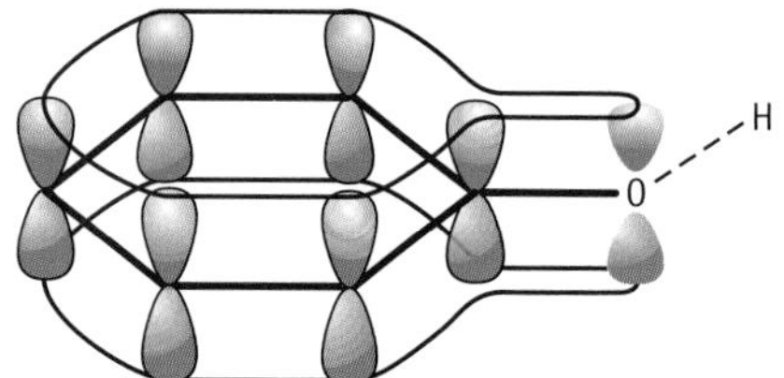

Figure 12.10 *Interaction of the oxygen electrons with the aromatic ring*

Physical properties

Phenol is a white solid that absorbs water from damp air.

Melting temperature

Phenol is a polar molecule with a δ^- oxygen atom and a δ^+ hydrogen atom. It can, therefore, form hydrogen bonds with other phenol molecules (intermolecular hydrogen bonding):

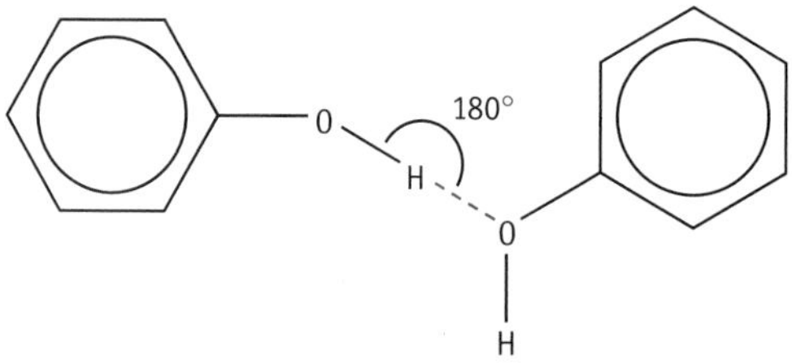

This is why phenol melts at a higher temperature than benzene and methylbenzene, which have only weaker dispersion forces between their molecules.

Pure phenol melts at 41°C and boils at 182°C.

Solubility

When phenol is added to water at room temperature, two liquid layers are formed. One is a solution of phenol in water and the other is a solution of water in phenol. If the temperature is raised to above 66°C, a single layer is formed because phenol and water are totally miscible above this temperature.

Phenol dissolves in water because hydrogen bonds are formed between phenol molecules and water molecules:

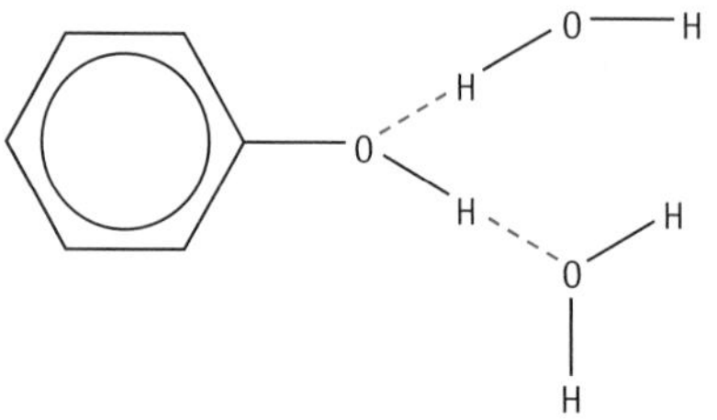

It is not very soluble in water (unlike ethanol, which is totally miscible) because of the large hydrophobic benzene ring.

Phenol is soluble in a variety of organic solvents because of the existence of dispersion (van der Waals) forces between its molecules and those of the solvent.

Chemical reactions

Phenol undergoes some of the reactions of aromatic compounds and some of the reactions of alcohols.

Electrophilic aromatic substitution reactions

The electron density of the delocalised π-system is greater in phenol than in benzene. This makes phenol more reactive towards electrophiles. The substituents go into the 2- and/or 4-positions.

The substitution reactions of benzene require anhydrous conditions and a catalyst. Phenol reacts in aqueous conditions and no catalyst is needed. This is best explained by looking at the Kekulé resonance structures of phenol:

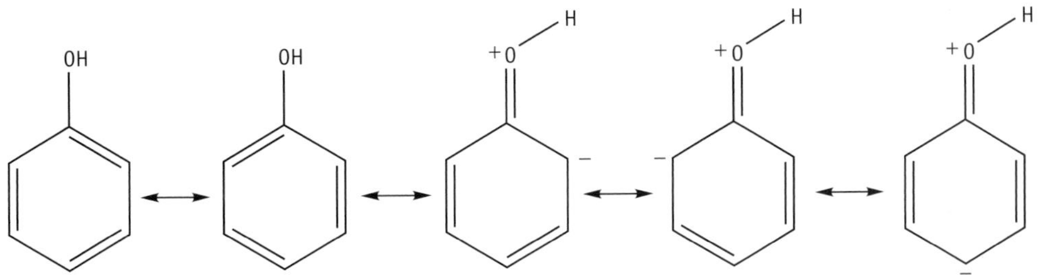

Not only are there the two benzene-type resonance structures, there are three others that have a negative charge on the 2-, 4- or 6-positions. Although the contribution of these structures is less than that of the two benzene-type structures, it is enough to make the incoming electrophile more likely to attack these electron-rich positions instead of the 3- and 5-positions.

Reaction with bromine

The conditions for the substitution of bromine into benzene are liquid bromine and a catalyst of iron(III) bromide (usually made *in situ* by adding iron filings to the bromine and benzene mixture).

The conditions for the bromination of phenol are quite different. When bromine *water* is added to a mixture of phenol and water, the red-brown bromine colour disappears and a white antiseptic-smelling precipitate of 2,4,6-tribromophenol in a solution of hydrogen bromide is immediately formed. The molecular equation for this reaction is:

$$C_6H_5OH(aq) + 3Br_2(aq) \rightarrow C_6H_2Br_3(OH)(s) + 3HBr(aq)$$

The equation, showing skeletal formulae, is:

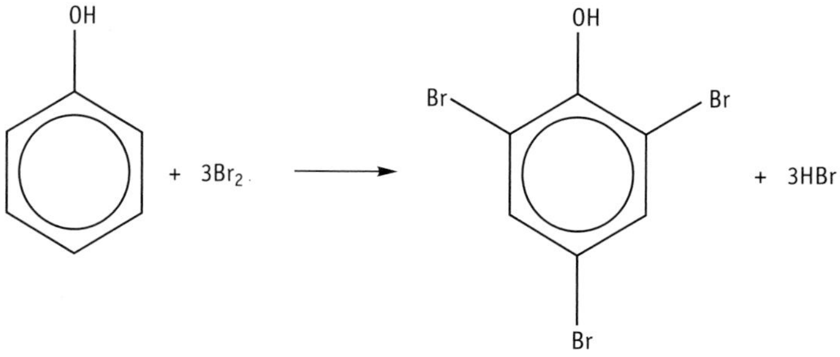

OH + $3Br_2$ ⟶ (Br, Br, OH, Br substituted ring) + 3HBr

> Do not forget that these are substitution reactions, so there must be two products on the right-hand side of the equation. The second product is a simple molecule, such as HBr or H_2O.

The electrophile is the δ^+ bromine atom in an HOBr molecule, made by the reaction of bromine and water:

$$Br_2 + H_2O \rightleftharpoons HOBr + HBr$$

2,4,6-tribromophenol, like many halogenated phenols, is an antiseptic. Dettol is 2,4-dichloro-3,5-dimethylphenol and TCP is 2,4,6-trichlorophenol. Phenol itself is also an antiseptic and was used in the nineteenth century by the surgeon Lister to reduce deaths from infection after surgery.

Nitration

Phenol reacts with dilute nitric acid to form a mixture of 2-nitrophenol and 4-nitrophenol plus water.

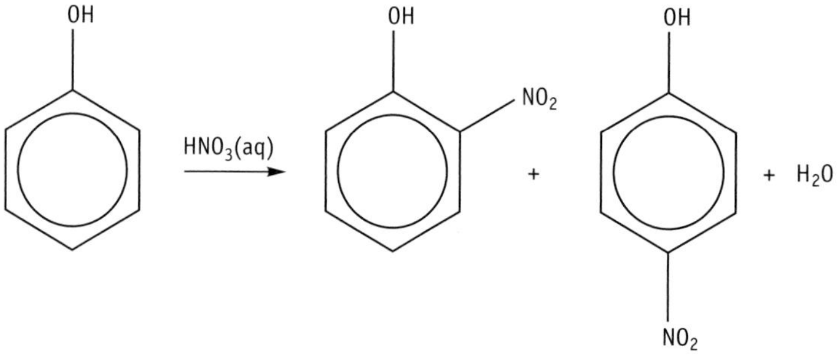

OH ⟶ $HNO_3(aq)$ ⟶ (OH, NO_2 ring) + (OH, NO_2 ring) + H_2O

> Remember that nitration of benzene requires concentrated nitric acid with a catalyst of concentrated sulfuric acid.

Reaction with diazonium ions

In alkaline solution, phenol is substituted by diazonium ions (p. 265).

Friedel–Craft reactions

Phenol reacts with acid chlorides, but *not* in a Friedel–Crafts reaction.

Instead the oxygen atom in phenol acts as a nucleophile and attacks the δ^+ carbon atom in the acid chloride, forming an ester (p. 254).

Phenol as an acid

The p_z-lone pair on the oxygen atom is partially drawn into the ring and this makes the hydrogen atom much more δ^+ than in alcohols. Thus, phenol reacts with water to form an acidic solution and with alkalis to form a salt. However, phenol is much weaker than a carboxylic acid and so does not give carbon dioxide when reacted with sodium carbonate or sodium hydrogencarbonate.

Reaction with water

$$C_6H_5OH + H_2O \rightleftharpoons H_3O^+ + C_6H_5O^- \qquad K_a = 1.3 \times 10^{-10}\,mol\,dm^{-3}$$

The pH of a 0.10 mol dm^{-3} solution of phenol is 5.44.

The phenate ion formed is stabilised by resonance:

O^- O^- O O O

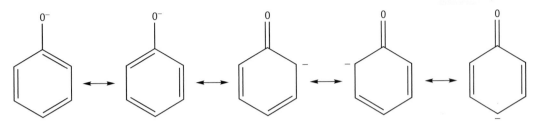

In all reactions involving the –OH group, the simplest way to write the formula of phenol is C_6H_5OH. The benzene ring needs to be drawn in full only for aromatic (electrophilic substitution) reactions.

Reaction with sodium hydroxide

Aqueous hydroxide ions are a strong enough base to deprotonate phenol to form the salt sodium phenate and water:

$$C_6H_5OH + NaOH \rightarrow C_6H_5ONa + H_2O$$

The ionic equation is:

$$C_6H_5OH(s) + OH^-(aq) \rightarrow C_6H_5O^-(aq) + H_2O(l)$$

Solid phenol 'dissolves' in aqueous sodium hydroxide.

The word 'dissolves' is often used in this context. A more correct description would be that solid phenol reacts to form a solution.

Reaction with sodium

Phenol is a strong enough acid to react with metallic sodium to form hydrogen:

$$C_6H_5OH + Na \rightarrow C_6H_5ONa + \tfrac{1}{2}H_2$$

Water, alcohols and carboxylic acids all give hydrogen gas with sodium metal.

Reaction with sodium hydrogencarbonate and sodium carbonate

Phenol is a weaker acid than carbonic acid, which has $K_a = 3.4 \times 10^{-7}\,mol\,dm^{-3}$. Therefore, it will not liberate carbon dioxide from either sodium hydrogencarbonate or sodium carbonate. This is one way to distinguish phenol from a carboxylic acid.

***Table 12.2** Acidity of alcohols, phenol and carboxylic acids*

	Add blue litmus	Add sodium hydroxide	Add sodium carbonate	Add sodium
Alcohols	Stays blue	No reaction	No reaction	H_2 evolved
Phenol	Goes red	Salt formed	No reaction	H_2 evolved
Carboxylic acids	Goes red	Salt formed	CO_2 evolved	H_2 evolved

Phenol as an alcohol

Phenol does not react as an alcohol as readily as aliphatic alcohols, for example ethanol and propan-2-ol.

Reaction with acid chlorides

Phenol reacts slowly with aliphatic acid chlorides, for example ethanoyl chloride, to form an ester and fumes of hydrogen chloride.

$$C_6H_5OH + CH_3COCl \rightarrow CH_3COOC_6H_5 + HCl$$

The structural formula of the ester phenylethanoate is:

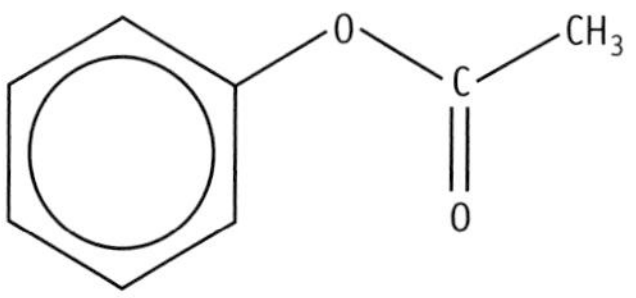

The oxygen atom in phenol acts as a nucleophile and attacks the δ^+ carbon atom in the acid chloride.

Aromatic acid chlorides, such as benzoyl chloride, C_6H_5COCl, are less reactive. Sodium hydroxide has to be added to create $C_6H_5O^-$ ions, which are a stronger nucleophile than phenol molecules. The ester, phenyl benzoate, is formed along with chloride ions:

$$C_6H_5O^- + C_6H_5COCl \rightarrow C_6H_5COOC_6H_5 + Cl^-$$

Remember that aliphatic alcohols, such as ethanol, react rapidly with both aliphatic and aromatic acid chlorides.

Reaction with carboxylic acids

Phenol does not form esters with carboxylic acids, unlike aliphatic alcohols, which react reversibly.

Benzoic acid, C_6H_5COOH

The structural formula of benzoic acid is:

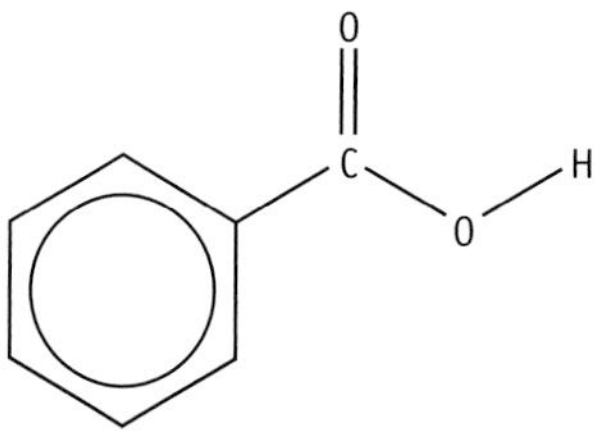

- It is a solid with a melting temperature of 122°C.
- It is almost insoluble in cold water.
- The dissolving of benzoic acid is endothermic, so the position of the equilibrium:

  $$C_6H_5COOH(s) + aq \rightleftharpoons C_6H_5COOH(aq)$$

  is driven to the right by an increase in temperature. Benzoic acid is soluble in hot water and so can be purified by recrystallisation using hot water (p. 295).
- Benzoic acid has the typical reactions of a carboxylic acid. It is a slightly stronger acid than ethanoic acid:
  - K_a of benzoic acid = 6.31×10^{-5} mol dm^{-3}; pH of 0.10 mol dm^{-3} solution = 2.60
  - K_a of ethanoic acid = 1.74×10^{-5} mol dm^{-3}; pH of 0.10 mol dm^{-3} solution = 2.88

- On warming with a mixture of concentrated nitric and sulfuric acids, 3-nitrobenzoic acid and water are formed:

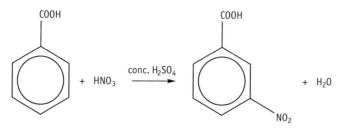

$$C_6H_5COOH + HNO_3 \xrightarrow{\text{conc. } H_2SO_4} C_6H_4(COOH)NO_2 + H_2O$$

Ⓔ The rules that determine the position of an aromatic substitution are as follows:

- If the atom that is bonded by a σ-bond to the benzene ring also has a π-bond, the ring is less easily substituted and the substituent goes into the 3- (or 5-) position. Nitration of benzoic acid forms 3-nitrobenzoic acid; nitration of nitrobenzene forms 1,3-dinitrobenzene.
- If the atom which is bonded to the benzene ring does not have any π-bonding, the ring is activated to substitution and the incoming substituent goes into the 2-, 4- and/or 6- positions. Methylbenzene and phenylethanone both give a mixture of 2- and 4- nitro compounds on nitration. Phenol gives 2,4,6-tribromophenol with bromine water.

2-methylnitrobenzene, $CH_3C_6H_4NO_2$

2-methylnitrobenzene is an oily liquid that smells of almonds. It boils at 220°C.

Preparation and purification

2-methylnitrobenzene can be prepared from methylbenzene and nitric acid:

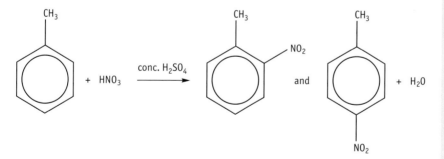

$$C_6H_5CH_3 + HNO_3 \xrightarrow{\text{conc. } H_2SO_4} \text{2-}CH_3C_6H_4NO_2 \text{ and } \text{4-}CH_3C_6H_4NO_2 + H_2O$$

The procedure is as follows:

- The nitrating mixture, which consists of equal amounts of concentrated nitric and sulfuric acids, is mixed with benzene and the mixture is heated in a water bath maintained at 50°C. The temperature must not rise above this or a significant amount of 2,4-dinitromethylbenzene will be formed. If the temperature is allowed to fall below 50°C, the reaction becomes very slow.
- After refluxing, the mixture is cooled and water is added to dilute the acids. The mixture is then poured into a separating funnel and the bottom layer, which contains the methylnitrobenzenes, is run off. The top layer is discarded.

- The impure mixture of methylnitrobenzenes is washed with sodium carbonate solution, to remove residual acid, and then with water, to remove residual sodium carbonate. This is also carried out in the separating funnel.
- The oily liquid is dried by adding lumps of anhydrous calcium chloride and leaving for several hours.
- It is then decanted into a flask and distilled. The fraction that boils between 218°C and 222°C is collected. This is pure 2-methylnitrobenzene. The final distillation separates 2-nitromethylbenzene (boiling temperature 220°C) from 4-nitromethylbenzene (boiling temperature 238°C)

Questions

1 Some enthalpy data are given below.

ΔH_a of carbon = +715 kJ mol^{-1}
ΔH_a of hydrogen = +218 kJ mol^{-1}
ΔH_f of benzene(g) = +83 kJ mol^{-1}

Average bond enthalpy	ΔH/kJ mol^{-1}
C–C	+348
C=C	+612
C–H	+412

Use the data to calculate:

a the enthalpy of formation of the theoretical gaseous molecule, cyclohexatriene

b the resonance stabilisation energy of benzene

c Hence, draw a labelled energy-level diagram showing both the formation of cyclohexatriene and benzene from solid carbon and hydrogen gas and the resonance stabilisation energy.

2 Draw the resonance structures of the ethanoate, CH_3COO^-, ion.

3 Draw and name all the aromatic isomers of $C_6H_4ClNO_2$.

4 Draw and name all the aromatic isomers of $C_6H_3Br_2OH$.

5 State the conditions for the conversion of nitrobenzene, $C_6H_5NO_2$, to:

a $C_6H_5Br_5NO_2$

b $C_6H_4BrNO_2$

6 Write equations for the production of the electrophiles for the following reactions of aromatic compounds:

a nitration

b bromination

c alkylation

7 Write the mechanism, including the production of the electrophile, of the reaction of benzene with ethanoyl chloride.

8 Write equations and state the conditions for the reactions of methylbenzene with:

a ethanoyl chloride

b bromine

c nitric acid

9 Explain why phenol:

a is a stronger acid than ethanol

b has a higher boiling temperature than ethanol

c is less soluble in water than ethanol

10 Write the structural formulae of the organic products, if any, of attempting to react phenol with:

a aqueous potassium hydroxide

b chlorine water

c benzoyl chloride, C_6H_5COCl, in alkaline solution

11 Calculate the pH of a 0.20 mol dm^{-3} solution of phenol, $K_a = 1.3 \times 10^{-10}$ mol dm^{-3}.

12 Calculate:

a the mass of bromine that would react with 1.23 g of phenol

b the percentage yield, if 4.25 g of 2,4,6-tribromophenol were produced

13 Give an example of phenol reacting as a nucleophile.

14 Draw a diagram to show how two molecules of benzoic acid are hydrogen bonded. Mark in the bond angles around the hydrogen bond.

15 Explain the importance of the conditions in the nitration of benzene.

16 Write the mechanism, including the formation of the electrophile, of the nitration of methylbenzene.

17 An organic compound **X** burns with a smoky flame. It gives an orange precipitate with Brady's reagent but does not form a silver mirror with Tollens' reagent. It gives a yellow precipitate when warmed gently with a solution of iodine and sodium hydroxide. Draw the structural formula of a compound that could be **X**.

18 Search the internet and write a brief account of the discovery in the nineteenth century of phenol (carbolic acid) and its use as an antiseptic by the surgeon Joseph Lister.

Organic nitrogen compounds

Chapter 13

Amines

Amines contain the group:

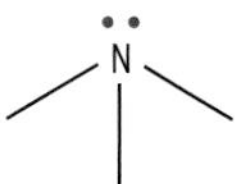

There are three types of amine:

- A **primary amine** has only one carbon atom bonded to the nitrogen atom and, therefore, has an $-NH_2$ group. Methylamine, CH_3NH_2, is an example.
- A **secondary amine** has two carbon atoms directly joined to the nitrogen atom. Dimethylamine, $(CH_3)_2NH$, is an example. All secondary amines have an $>NH$ group.
- A **tertiary amine** has three carbon atoms and no hydrogen atoms attached directly to the nitrogen atom. Trimethylamine, $(CH_3)_3N$, is an example.

The nitrogen atom in all amines has three σ-bonds and a lone pair of electrons. The four pairs of electrons are arranged in a tetrahedron around the nitrogen. Therefore, the three bonding pairs are arranged pyramidally, with the H–N–H bond angle less than the tetrahedral angle. This is because the lone pair/bond pair repulsion is greater than the bond pair/bond pair repulsion.

Amines can be named by:

- adding amine to the stem of the alkyl group. Thus, $C_2H_5NH_2$ is ethylamine and $CH_3CH_2CH_2NH_2$ is 1-propylamine; or
- adding the prefix amino- to the alkane from which the amine is derived. $C_2H_5NH_2$ is called aminoethane and $CH_3CH_2CH_2NH_2$ is 1-aminopropane.

Physical properties

Methylamine is a gas at room temperature and pressure. Ethylamine boils around room temperature and the next members of the homologous series of primary amines are liquids (Table 13.1). Amines have a characteristic fish-like smell.

Name	Formula	Boiling temperature/°C
Methylamine (aminomethane)	CH_3NH_2	−6
Ethylamine (aminoethane)	$CH_3CH_2NH_2$	+17
1-propylamine (1-aminopropane)	$CH_3CH_2CH_2NH_2$	+49
2-propylamine (2-aminopropane)	$CH_3CH(NH_2)CH_3$	+33
Dimethylamine	$(CH_3)_2NH$	+7
Trimethylamine	$(CH_3)_3N$	+4
Phenylamine (aminobenzene, aniline)	$C_6H_5NH_2$	+184

***Table 13.1** Boiling temperatures of amines*

Hydrogen bonding occurs between amine molecules:

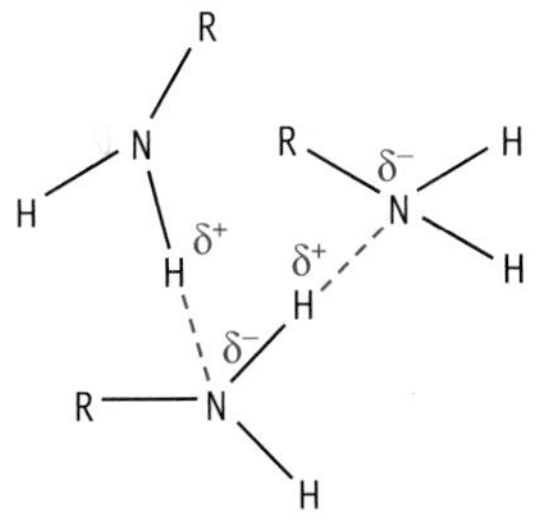

This explains why the boiling temperatures of amines are higher than those of the parent alkane. Methylamine (18 electrons) boils at −6°C whereas the non-hydrogen bonded ethane (also 18 electrons) boils at −89°C.

Amines are water-soluble because they form hydrogen bonds with water molecules. The nitrogen atom is δ^- and the hydrogen atom attached to it is δ^+:

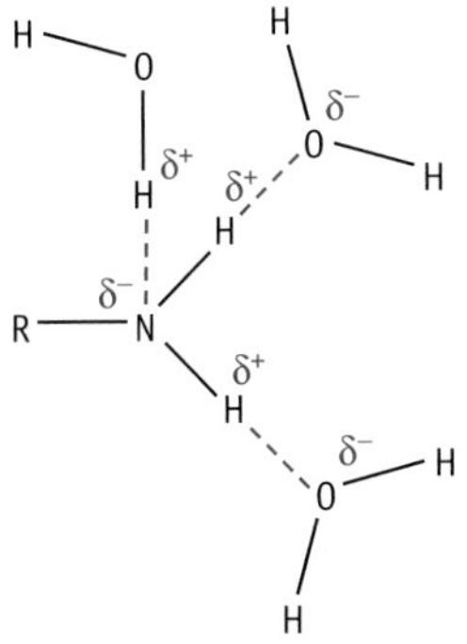

Remember that the strength of the inter-molecular dispersion (van der Waals) forces depends mainly on the total number of electrons in the molecules.

Note that the angle around a hydrogen-bonded hydrogen atom is 180°.

Preparation

Reaction of ammonia with a halogenoalkane

If a halogenoalkane is mixed with excess concentrated ammonia in aqueous ethanolic solution and left for several minutes, a primary amine is obtained, for example:

$$C_2H_5Cl + 2NH_3 \rightarrow C_2H_5NH_2 + NH_4Cl$$

An excess of the halogenoalkane will yield the secondary amine.

Another condition for this reaction, often quoted in textbooks and accepted by examiners, is heating the mixture in a sealed tube.

When the base ammonia is a reactant, do not write an equation that has an acid such as HCl as a product. With excess ammonia, ammonium chloride is produced. If the ammonia is not in excess, the amine salt is formed (see the reaction of amines with acids on p. 260).

Reduction of a nitrile or amide

A nitrile (e.g. ethanenitrile) or an amide (e.g. ethanamide) can be reduced by warming with a solution of lithium tetrahydridoaluminate(III) in dry ether (ethoxyethane), followed by hydrolysis of the adduct with dilute acid:

Lithium tetrahydrido-aluminate(III) is also called lithium aluminium hydride

$$\underset{\text{ethanenitrile}}{CH_3CN} + 4[H] \rightarrow \underset{\text{ethylamine}}{CH_3CH_2NH_2}$$

$$\underset{\text{ethanamide}}{CH_3CONH_2} + 4[H] \rightarrow \underset{\text{ethylamine}}{CH_3CH_2NH_2} + H_2O$$

This is an example of the reduction of a polar π bond by the nucleophilic H^- ion in lithium tetrahydridoaluminate(III). This reagent does not reduce the non-polar C=C in alkenes.

Chemical reactions

The nitrogen atom has a lone pair of electrons and this causes it to be a base and a nucleophile.

As a base

The base reactions of amines are similar to those of ammonia. Ammonia and amine molecules have a lone pair of electrons on the nitrogen atom that can be used to form a dative covalent bond with an H^+ ion.

Reaction with water

Amines react reversibly with water, for example:

$$C_2H_5NH_2 + H_2O \rightleftharpoons C_2H_5NH_3^+ + OH^-$$

As OH^- ions are produced, the solution is alkaline, with a pH of about 11.

This reaction is similar to:

$$NH_3 + H_2O \rightleftharpoons NH_4^+ + OH^-$$

The $-C_2H_5$ group pushes electrons slightly towards the nitrogen atom making it more δ^- than it is in ammonia. Therefore, ethylamine is a stronger base than ammonia:

- K_b of ethylamine = $5.6 \times 10^{-4}\,mol\,dm^{-3}$
- K_b of ammonia = $1.8 \times 10^{-5}\,mol\,dm^{-3}$

Diethylamine, $(C_2H_5)_2NH$, is a slightly stronger base than ethylamine because it contains two electron-pushing groups.

Reaction with acids

When an amine reacts with a strong acid, such as dilute hydrochloric acid, a salt (similar to an ammonium salt) is formed. For example, ethylamine and dilute hydrochloric acid produce ethylammonium chloride:

$$C_2H_5NH_2 + HCl \rightarrow C_2H_5NH_3^+Cl^-$$

e Ammonia reacts with acids to form ammonium salts such as ammonium chloride: $NH_3 + HCl \rightarrow NH_4^+Cl^-$

This reaction can be reversed by adding a strong base, such as sodium hydroxide:

$$C_2H_5NH_3^+ + OH^- \rightarrow C_2H_5NH_2 + H_2O$$

The amine can then be distilled off from the mixture of the amine salt and sodium hydroxide.

Reaction with acid chlorides

The lone pair of electrons on the nitrogen atom in the amine acts as a nucleophile and attacks the δ^+ carbon atom in the acid chloride in an addition–elimination reaction. For example, when ethanoyl chloride is added to ethylamine, a secondary amide, *N*-ethylethanamide, and misty fumes of hydrogen chloride are produced:

$$C_2H_5NH_2 + CH_3COCl \rightarrow CH_3CONHC_2H_5 + HCl$$

Paracetamol can be made by reacting 4-hydroxyphenylamine with ethanoyl chloride:

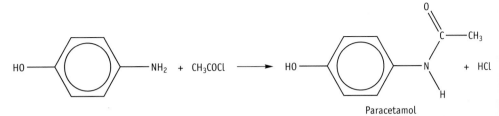

$$HO-C_6H_4-NH_2 + CH_3COCl \longrightarrow HO-C_6H_4-N(H)-C(=O)-CH_3 + HCl$$

Paracetamol

The product is *N*-4-hydroxyphenylethanamide or paracetamol.

Paracetamol is also known as acetaminophen or as Tylenol.

Amines with two $-NH_2$ groups in the molecule form polyamides with acid chlorides with two –COCl groups (see p. 268).

Ammonia reacts with acid chlorides to give amides (p. 159).

Reaction with halogenoalkanes

A primary amine reacts with a halogenoalkane to produce a mixture of the salts of a secondary and a tertiary amine:

$$C_2H_5NH_2 + C_2H_5Cl \rightarrow (C_2H_5)_2NH_2^+Cl^-$$
$$(C_2H_5)_2NH_2^+Cl^- + C_2H_5Cl \rightarrow (C_2H_5)_3NH^+Cl^- + HCl$$

The reagents are heated in a solution in ethanol in a sealed tube.

If the free amine is required, a strong base is added to the solution and any unreacted primary amine and the secondary and tertiary amines formed are separated by fractional distillation.

$$(C_2H_5)_2NH_2^+Cl^- + OH^- \rightarrow (C_2H_5)_2NH + H_2O + Cl^-$$
$$(C_2H_5)_3NH^+Cl^- + OH^- \rightarrow (C_2H_5)_3N + H_2O + Cl^-$$

Reaction with *d*-block metal ions

Amines can act as ligands with transition metal ions. The lone pair of electrons on the amine group can form a bond with the empty orbital of some *d*-block metal ions. The reaction is similar to the formation of ammines with excess ammonia.

A precipitate of the metal hydroxide is produced first. With Co^{2+}, Ni^{2+}, Cu^+, Cu^{2+}, Zn^{2+} and Ag^+ ions the precipitate dissolves because a complex ion is formed. The colour of the complex is similar to that formed with ammonia solution (see p. 223).

The first reaction, forming the precipitate, is an example of a deprotonation reaction. For example, the reaction between hexaaquacopper(II) ions and ethylamine produces a pale blue precipitate:

$$[Cu(H_2O)_6]^{2+}(aq) + 2C_2H_5NH_2(aq) \rightarrow [Cu(H_2O)_4(OH)_2](s) + 2CH_3NH_3^+(aq)$$

With excess methylamine solution, a soluble complex ion is formed:

$$[Cu(H_2O)_4(OH)_2](s) + 4C_2H_5NH_2(aq) \rightarrow [Cu(NH_2C_2H_5)_4(H_2O)_2]^{2+}(aq) + 2OH^-(aq) + 2H_2O(l)$$

The overall reaction is:

$$[Cu(H_2O)_6]^{2+}(aq) + 4C_2H_5NH_2(aq) \rightarrow [Cu(NH_2C_2H_5)_4(H_2O)_2]^{2+}(aq) + 4H_2O(l)$$

This complex is deep blue.

The equivalent equations with ammonia are:

$$[Cu(H_2O)_6]^{2+}(aq) + 2NH_3(aq) \rightarrow [Cu(H_2O)_4(OH)_2](s) + 2NH_4^+(aq)$$

$$[Cu(H_2O)_4(OH)_2](s) + 4NH_3(aq) \rightarrow [Cu(NH_3)_4(H_2O)_2]^{2+}(aq) + 2OH^-(aq) + 2H_2O(l)$$

Phenylamine, $C_6H_5NH_2$

The structural formula of phenylamine is:

Phenylamine used to be called aniline.

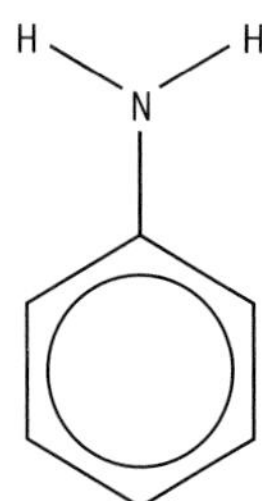

The lone pair of electrons on the nitrogen atom is in the same plane as the delocalised π-electrons of the ring and so, to some extent, becomes part of the delocalised system. This pulling of the electrons into the ring and away from the nitrogen atom has two effects — the ring becomes more susceptible to electrophilic attack and the nitrogen atom becomes less δ^- and so less effective as a base.

Physical properties

- Phenylamine is a liquid at room temperature. It boils at 184°C and is slightly denser than water.
- The nitrogen atom is δ^- and the hydrogen atoms attached to the nitrogen are δ^+. Therefore, phenylamine can form intermolecular hydrogen bonds as well as strong dispersion (van der Waals) forces between molecules. It is slightly soluble in water because it forms hydrogen bonds with water molecules. However, the benzene ring inhibits solubility.
- It dissolves readily in many organic solvents such as ether (ethoxyethane) and ethanol.
- It is a toxic substance that can be absorbed through the skin.

Laboratory preparation

The procedure for the laboratory preparation of phenylamine is as follows:

- Nitrobenzene and tin are mixed in a round-bottomed flask fitted with a reflux condenser. Concentrated hydrochloric acid is carefully added and, after the rapid evolution of hydrogen has ceased, the mixture is heated to 100°C in a bath of boiling water for 30 minutes.
- The reaction mixture is cooled to room temperature, the condenser removed and an *excess* of sodium hydroxide is carefully added with cooling.
- The mixture is steam distilled to remove the phenylamine from the sludge of tin hydroxides. This is carried out by blowing steam into the mixture and condensing the phenylamine and steam mixture that comes off.

- The distillate is placed in a separating funnel and some solid sodium chloride is added to reduce the solubility of phenylamine. The phenylamine layer is run off into a flask and some anhydrous potassium carbonate added to remove any traces of water.
- The phenylamine is decanted off from the solid potassium carbonate and distilled, using an air condenser. The fraction that boils between 180°C and 185°C is collected.

The equation for the process can be represented by:

$$C_6H_5NO_2 + 6[H] \xrightarrow{\text{Sn and conc. HCl}} C_6H_5NH_2 + 2H_2O$$

Chemical reactions

Phenylamine reacts as both an aromatic compound and as a primary amine. It also reacts with nitrous acid.

Electrophilic substitution reactions

In neutral solutions the benzene ring is activated by the lone pair of electrons on the nitrogen atom. In acidic conditions, the $-NH_2$ group is protonated to form an NH_3^+ group, which withdraws electrons from the ring and deactivates it.

Bromination

Phenylamine reacts with bromine water to form a white precipitate of 2,4,6-tribromophenylamine and hydrogen bromide. The molecular equation is:

$$C_6H_5NH_2 + 3Br_2 \rightarrow C_6H_2Br_3NH_2 + 3HBr$$

This is similar to the reaction between phenol and bromine water to form a white precipitate of 2,4,6-tribromophenol.

Nitration

Concentrated nitric and sulfuric acids protonate the $-NH_2$ group in phenylamine, which deactivates the benzene ring. Electrophilic substitution does not occur. Phenylamine is oxidised to a black solid of indeterminate structure.

Friedel–Crafts reaction

The $-NH_2$ group reacts preferentially with acid chlorides and halogenoalkanes, so phenylamine does not undergo Friedel–Crafts reactions.

Reactions of the NH_2 group

The pulling of the lone pair of electrons from the nitrogen atom by the benzene ring makes the nitrogen atom less δ^- and, therefore, less effective as a base and as a nucleophile.

Phenylamine as a base

Phenylamine is a weak base, so it reacts reversibly with water:

$$C_6H_5NH_2 + H_2O \rightleftharpoons C_6H_5NH_3^+ + OH^-$$

e If you draw the *full* structural formula or the displayed formula of the phenylammonium ion (showing the three hydrogen atoms separately attached to the nitrogen) make sure that you put the positive charge on the nitrogen atom and not on a hydrogen atom. For the ordinary structural formula, writing the group as NH_3^+ is acceptable.

It reacts with strong acids to form phenylammonium salts. For example, the molecular equation for the reaction of phenylamine with hydrochloric acid to form phenylammonium chloride is:

$$C_6H_5NH_2 + HCl \rightarrow C_6H_5NH_3Cl$$

The ionic equation is:

$$C_6H_5NH_2(l) + H^+(aq) \rightarrow C_6H_5NH_3^+(aq)$$

Liquid phenylamine dissolves in aqueous acids because it forms an ionic salt.

Phenylamine is a weaker base than ammonia, because of the electron-withdrawing effect of the benzene ring.

These reactions are similar to those of ammonia:

$NH_3 + H_2O \rightleftharpoons NH_4^+ + OH^-$

$NH_3 + HCl \rightarrow NH_4Cl$

or

$NH_3(aq) + H^+(aq) \rightarrow NH_4^+(aq)$

Phenylamine as a nucleophile

Phenylamine, like an aliphatic primary amine, reacts with an acid chloride to form a secondary amide:

$$C_6H_5NH_2 + CH_3COCl \rightarrow C_6H_5NHCOCH_3 + HCl$$

Reaction with nitrous acid, HNO_2

Nitrous acid is unstable and has to be made *in situ* by mixing a solution of sodium nitrite, $NaNO_2$, with excess dilute hydrochloric acid. It reacts with phenylamine to form benzenediazonium chloride in an exothermic reaction. The reaction mixture must be cooled so as to maintain a temperature between 0°C and 10°C.

Nitrous acid is also known as nitric(III) acid.

$$C_6H_5NH_2 + HNO_2 + HCl \rightarrow C_6H_5N_2Cl + 2H_2O$$

The temperature must be controlled carefully. If it drops below 0°C, the reaction is too slow; if it rises above 10°C the product decomposes.

Benzenediazonium compounds

Benzenediazonium compounds contain the $C_6H_5N_2^+$ ion. The structure of this ion is:

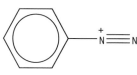

$$C_6H_5-\overset{+}{N}\equiv N$$

These compounds are very unstable. If the temperature is allowed to rise above 10°C, or if the solution becomes too concentrated, they decompose giving off nitrogen gas.

e If you draw the full structural formula of a diazonium ion (with the nitrogen atoms shown separately) make sure that you put the positive charge on the nitrogen atom that is attached to the benzene ring.

Laboratory preparation

The procedure for the preparation of benzenediazonium compounds is as follows:

- Phenylamine and concentrated hydrochloric acid are mixed and the mixture is cooled to 0°C in an ice bath.
- A solution of sodium nitrite (cooled to 0°C) is slowly added, making sure that the temperature neither rises above 10°C nor falls below 0°C.
- The solution is kept at 5°C and used as necessary.

e If the temperature rises above 10°C, the benzenediazonium ions decompose. If the temperature falls below 0°C, the reaction does not take place.

Coupling reactions of diazonium ions

One nitrogen atom in the diazonium ion is positively charged and, as nitrogen is a very electronegative element, it is a powerful electrophile. It will substitute into a benzene ring if the ring is activated by an –OH or $-NH_2$ group. The resulting compound can form hydrogen bonds with groups in cotton and wool and so binds strongly to the fabrics. This makes diazonium compounds useful as dyes.

The organic products of these coupling reactions have two benzene rings joined by an –N=N– group.

Reaction with phenol

When phenol is mixed with sodium hydroxide, the phenate ion $C_6H_5O^-$ is produced. A diazonium ion attacks this species by an electrophilic substitution reaction. A yellow precipitate of 4-hydroxyazobenzene is obtained:

The term azo refers to compounds in which two nitrogen atoms are covalently bonded together. It is derived from the French for nitrogen, which is *azote*.

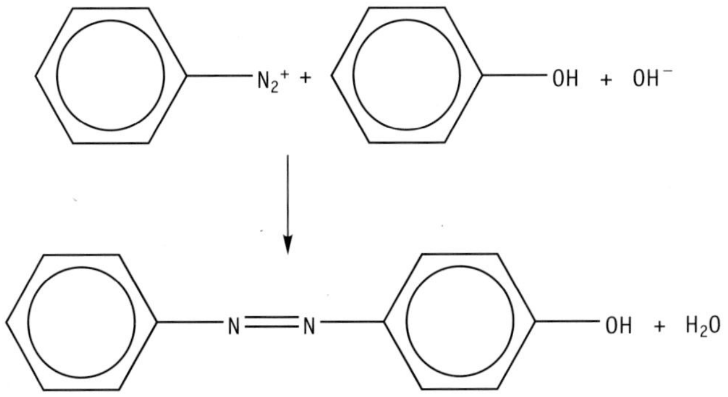

N_2^+ + OH + OH^-

N=N OH + H_2O

Reaction with 2-naphthol

Naphthols are compounds with two fused benzene rings and an –OH group on one of the rings. 2-naphthol couples with diazonium ions to form a red precipitate that can be used as a Turkey-red dye:

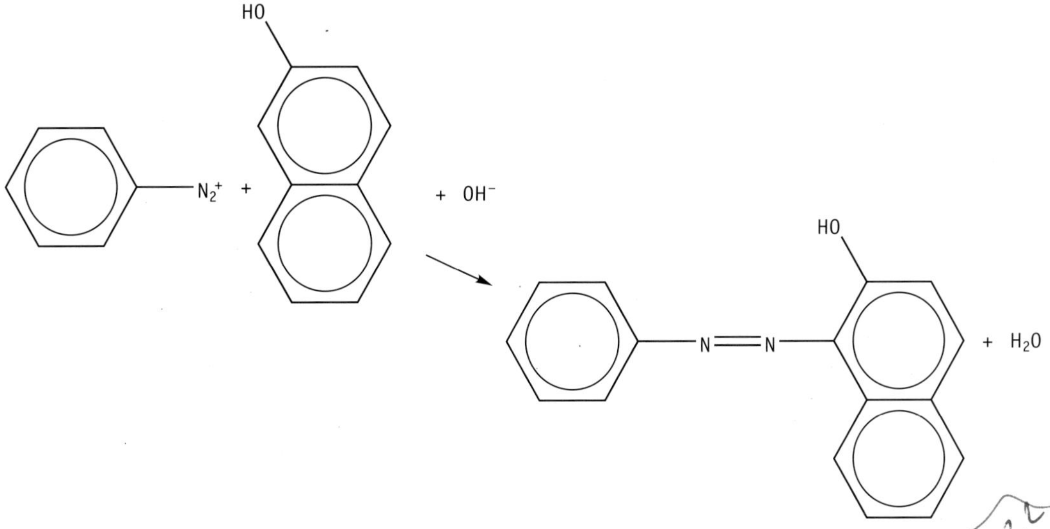

HO

N_2^+ + + OH^-

HO

N=N + H_2O

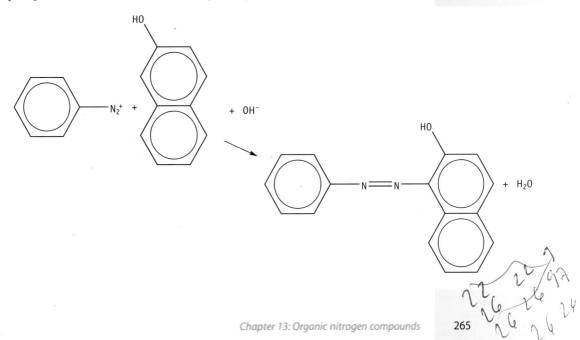

Reaction with phenylamine

If a solution of diazonium ions is added to phenylamine, a yellow precipitate of 4-aminoazobenzene is formed:

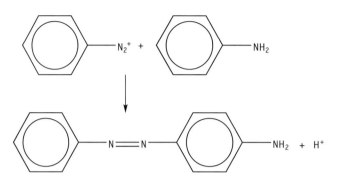

$$C_6H_5N_2^+ + C_6H_5NH_2 \rightarrow C_6H_5N{=}NC_6H_4NH_2 + H^+$$

Amides

Amides contain the group:

$$-C(=O)NH_2$$

Physical properties

An amide molecule contains two δ^+ hydrogen atoms, a δ^- oxygen atom and a δ^- nitrogen atom. Therefore, the molecules can form several *inter*molecular hydrogen bonds. This means that amides have higher melting temperatures than acid chlorides derived from the same acid.

Name	Formula	Melting temperature/°C	Boiling temperature/°C
Ethanamide	CH_3CONH_2	82	222
Propanamide	$CH_3CH_2CONH_2$	79	222
Benzamide	$C_6H_5CONH_2$	132	290

Table 13.2 *Some amides*

Apart from those with a large hydrophobic group, for example benzamide, amides are water-soluble. This is because the molecules can form several hydrogen bonds with water molecules.

Preparation

Amides can be prepared by the reaction between ammonia and an acid chloride:

$$CH_3COCl + NH_3 \rightarrow CH_3CONH_2 + HCl$$

Chemical reactions

Hydrolysis

When boiled under reflux with either aqueous acid or aqueous alkali, amides are hydrolysed.

For example, ethanamide is hydrolysed in acid solution to ethanoic acid and ammonium ions:

$$CH_3CONH_2 + H^+ + H_2O \rightarrow CH_3COOH + NH_4^+$$

With alkali, the salt of the carboxylic acid is formed and ammonia gas is evolved. For example, when hydrolysed by sodium hydroxide, benzamide gives sodium benzoate:

$$C_6H_5CONH_2 + OH^- \rightarrow C_6H_5COO^- + NH_3$$

If the carboxylic acid is required, the product of alkaline hydrolysis must be acidified with a solution of a strong acid:

$$C_6H_5COO^- + H^+ \rightarrow C_6H_5COOH$$

Benzoic acid is insoluble in cold water and so on cooling it precipitates out.

Polyamides

The reaction between an acid chloride and an amine is described above. If a compound that has two COCl groups reacts with a substance with two NH_2 groups a polyamide is formed.

The first commercial polyamide was nylon-6,6. This can be made in the laboratory by the reaction of hexane-1,6-dioyl dichloride, $ClOC(CH_2)_4COCl$ and 1,6-diaminohexane, $NH_2(CH_2)_6NH_2$. The former is soluble in 1,1,2-trichloroethane and the latter in water. If the two solutions are carefully added together so that the aqueous layer floats on top of the organic layer, a thread of nylon can be drawn from the interface of the two liquids.

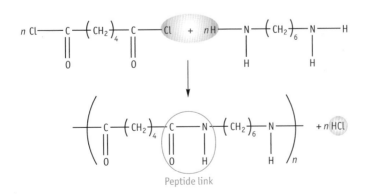

A thread of nylon being pulled from the interface of the two solutions

The equation for the reaction is:

$$n\,\mathrm{Cl{-}\underset{\underset{O}{\|}}{C}{-}(CH_2)_4{-}\underset{\underset{O}{\|}}{C}{-}Cl} + n\,\mathrm{H{-}\underset{\underset{H}{|}}{N}{-}(CH_2)_6{-}\underset{\underset{H}{|}}{N}{-}H}$$

$$\downarrow$$

$$\mathrm{\left(\underset{\underset{O}{\|}}{C}{-}(CH_2)_4{-}\underset{\underset{O}{\|}}{C}{-}\underset{\underset{H}{|}}{N}{-}(CH_2)_6{-}\underset{\underset{H}{|}}{N} \right)_n} + n\,\mathrm{HCl}$$

Peptide link

Nylon is a condensation polymer — a molecule of hydrogen chloride is eliminated as each acid chloride group reacts with an amino group.

The strength of nylon is increased by cold-drawing. This involves putting the threads under tension. The polymer chains become more aligned. This results in hydrogen bonding between the δ^- oxygen in the C=O groups and the δ^+ hydrogen atoms in the NH groups.

Uses of nylon include

- Nylon fibres are used in stockings and carpets.
- Nylon is used to make a number of machine parts such as bearings and rollers. This is because it is a very tough material that has a melting point above 250°C.
- It has a high electrical resistance and so is used to make switches.

Other polyamides

The general equation for the formation of polyamides is:

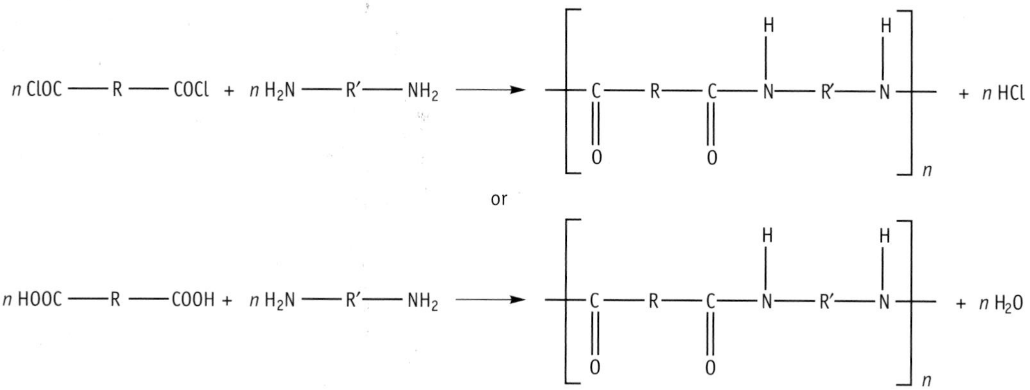

$$n\,ClOC—R—COCl + n\,H_2N—R'—NH_2 \longrightarrow \left[—\underset{\underset{O}{\|}}{C}—R—\underset{\underset{O}{\|}}{C}—\overset{\overset{H}{|}}{N}—R'—\overset{\overset{H}{|}}{N}— \right]_n + n\,HCl$$

or

$$n\,HOOC—R—COOH + n\,H_2N—R'—NH_2 \longrightarrow \left[—\underset{\underset{O}{\|}}{C}—R—\underset{\underset{O}{\|}}{C}—\overset{\overset{H}{|}}{N}—R'—\overset{\overset{H}{|}}{N}— \right]_n + n\,H_2O$$

Nomex® is a long straight-chain polymer with flame-resistant properties. The equation for its formation is:

n Cl + nH
O H H
Cl—C
O
+ n HCl
O H H
C
O
n

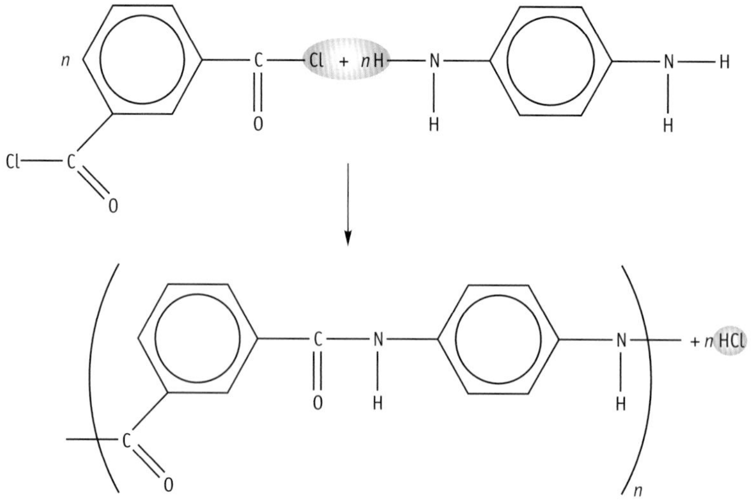

Kevlar® is another tough polyamide. It is used in bulletproof vests. The equation for its formation is:

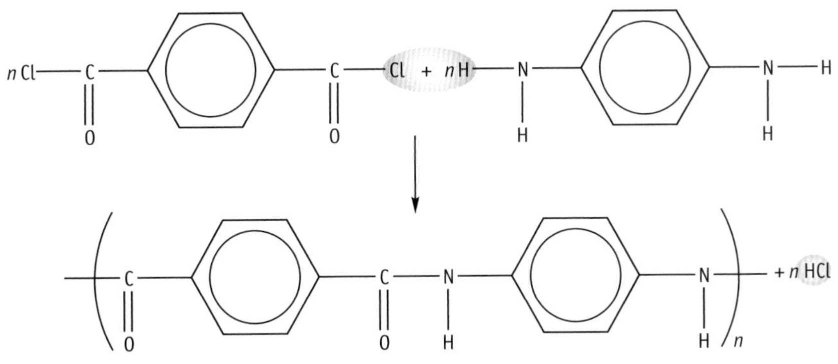

All polyamides contain the peptide link. This is the same link that joins α-amino acid residues in proteins.

Natural polyamides

Silk and wool are made up of polyamides. Silk contains a high proportion of peptide links between serine molecules. Serine is an amino acid of formula $NH_2CH(CH_2OH)COOH$. The –COOH group at one end forms a peptide link with the $-NH_2$ group of another serine molecule and its –COOH group forms a peptide link with a $-NH_2$ group of a third serine molecule and so on.

The structural formula showing three repeat units is:

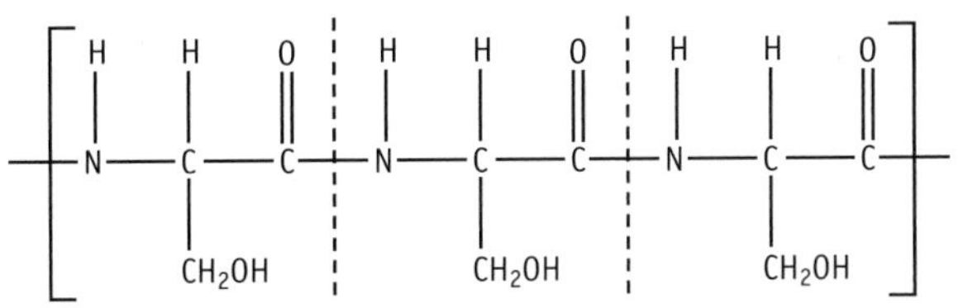

Repeat unit Repeat unit Repeat unit

Wool has a high proportion of aminoethanoic acid (glycine) NH_2CH_2COOH, residues

Physical properties of polyamides

Nylon, silk and wool have a β-pleated sheet structure. The chains are arranged in such a manner that the fibre is like a stack of corrugated cardboard with hydrogen bonds between the δ^- oxygen of the >C=O group in one repeat unit and the δ^+ hydrogen atom in an >NH group in another repeat unit. This makes the fibres strong.

Condensation polymerisation

The formation of nylon, Nomex and Kevlar are all examples of condensation polymerisation.

In most cases, two different monomers, each having two identical functional groups, react. However, sometimes a polymer is formed in a condensation reaction from a single monomer that has two different functional groups.

An example is the polymerisation of 6-aminohexanoic acid, $NH_2(CH_2)_5COOH$.

$$n\,NH_2(CH_2)_5COOH \longrightarrow \left[-\overset{\displaystyle H}{\overset{|}{N}} - (CH_2)_5 - \overset{\displaystyle O}{\overset{\|}{C}} - \right]_n + n\,H_2O$$

Another example of condensation polymerisation is the formation of polyesters such as Terylene (p. 156).

Addition polymerisation

Addition polymerisation has been covered at AS. The monomer is a compound with a >C=C< group. The π-bond in a monomer breaks and a new σ-bond forms with another monomer. As a π-bond is weaker than a σ-bond, and as all bond making is exothermic, such polymerisations are exothermic.

A simple example is the production of poly(propene) from propene:

$$n\,H_2C{=}CH{-}CH_3 \longrightarrow \left[-CH_2 - \overset{\displaystyle CH_3}{\overset{|}{CH}} - \right]_n$$

e In all addition polymers the repeat unit consists of a chain of two carbon atoms.

Some special addition polymers have been invented that show unusual properties in the presence of water.

Poly(propenamide) and poly(propenoic acid)

$$n\,H_2C{=}CH{-}C(=O)NH_2 \longrightarrow \left[-CH_2 - \overset{\displaystyle C(=O)NH_2}{\overset{|}{CH}} - \right]_n$$

Propenamide — Repeat unit of poly(propenamide)

$$n\,H_2C{=}CH{-}COOH \longrightarrow \left[-CH_2 - \overset{\displaystyle C(=O)OH}{\overset{|}{CH}} - \right]_n$$

Propenoic acid — Repeat unit of poly(propenoic acid)

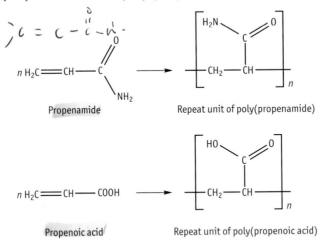

These two polymers, particularly when cross-linked, absorb water through hydrogen bonding. They are used in disposable nappies and as water-holding gels in the soil of pot plants and hanging baskets. As they absorb water, the polymers swell to many times their original volumes.

Some toys make use of this property. For example, there is a toy dinosaur egg in which the shell does not have cross-linking and is, therefore, water-soluble. When the egg is placed in water the shell dissolves. The 'yolk' is cross-linked and as it absorbs water it expands. Therefore, the egg appears to hatch and grow into a tiny toy dinosaur.

Poly(ethenol)

This polymer is also called (polyvinylalcohol). Its repeat unit is:

$$\left[\!-CH_2-\underset{}{\overset{OH}{\overset{|}{C}H}}-\right]$$

The prefix vinyl is sometimes used as the name for the $CH_2{=}CH$ group.

Poly(ethenol) has the unusual property of being water-soluble. It is used as the coating in liquitabs, which contain liquid washing detergent, and in soluble bags to hold soiled hospital laundry.

The polymer is water-soluble because of the many hydrogen bonds that can form between the –OH groups that occur on alternate carbon atoms and water molecules.

This polymer is also unusual in that its monomer does not exist. Attempts to make ethenol, $CH_2{=}CHOH$, result in the production of ethanal, CH_3CHO. The polymer is manufactured by reacting the polyester, polyvinylacetate (PVA) with methanol in a transesterification reaction.

$$n\,H_3C-C(=O)-O-CH{=}CH_2 \longrightarrow \left[-\underset{O-C(=O)-CH_3}{CH}-CH_2-\right]_n \xrightarrow{CH_3OH} \left[-\overset{OH}{CH}-CH_2-\right]_n + n\,CH_3COOCH_3$$

Vinyl acetate — PVA — Poly(ethenol)

Natural products containing the amine group

The bases in DNA and RNA contain a nitrogen atom with a lone pair of electrons. Each base is attached to a carbohydrate (deoxyribose in DNA) in the carbohydrate–phosphate chain and forms hydrogen bonds with a particular base in the opposite DNA strand. Adenine (A) always forms hydrogen bonds with thymine (T); guanine (G) always forms hydrogen bonds with cytosine (C).

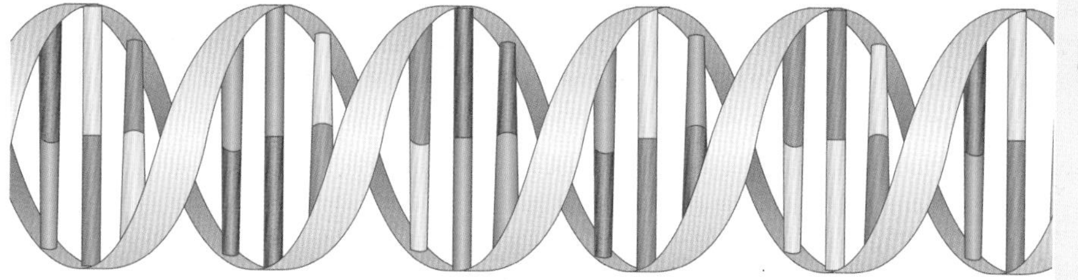

The DNA double helix

Therefore, if one DNA strand in the double helix has the base order ATGGAC, the opposite strand will have the order TACCTG. Hydrogen bonds between strands are shown in red in Figures 13.1 and 13.2. The bond that attaches the base to the carbohydrate is shown in green.

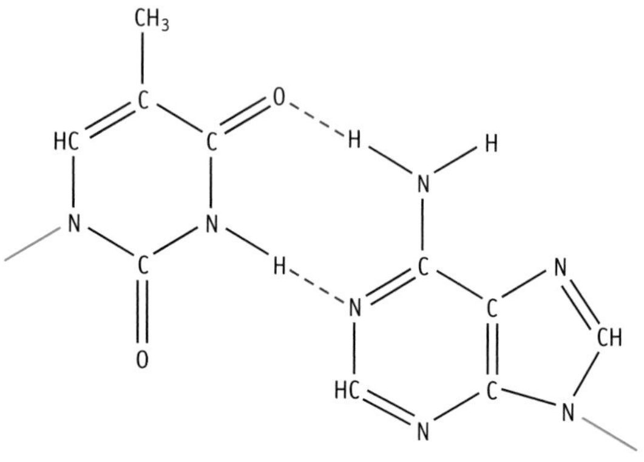

Thymine Adenine

***Figure 13.1** The base pair thymine and adenine in DNA*

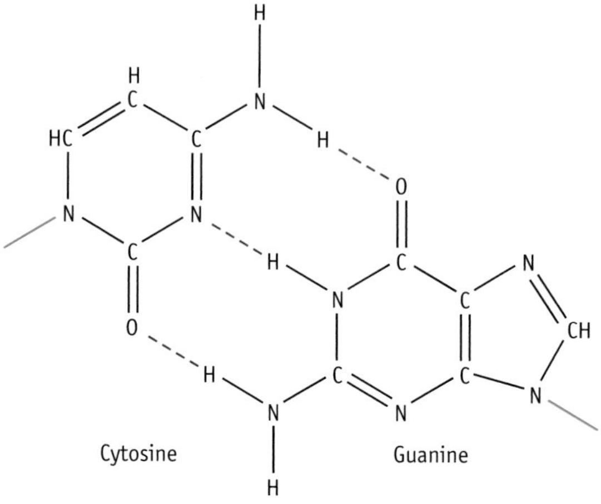

Cytosine Guanine

***Figure 13.2** The base pair cytosine and guanine in DNA*

In RNA, uracil replaces thymine and is found hydrogen-bonded to adenine. Thymine has a CH3 group on one of the doubly bonded carbon atoms whereas uracil has a hydrogen atom.

The hormone adrenaline is a secondary amine:

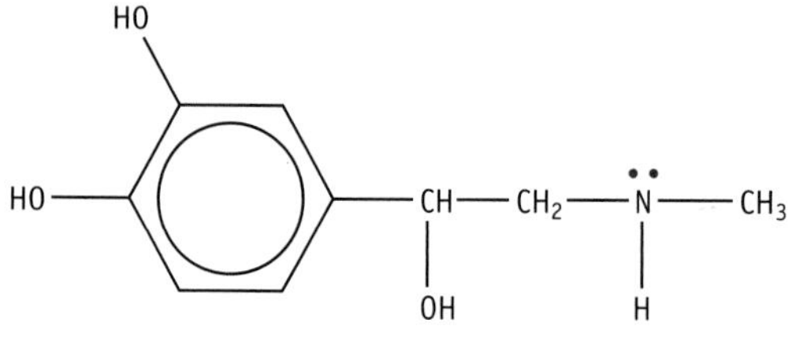

Other organic bases that contain an amine group include:

- the alkaloid poison, strychnine
- the painkillers, morphine and codeine
- the stimulants, caffeine and nicotine

Amino acids

Amino acid molecules contain at least one $-NH_2$ group and one –COOH group. Those amino acids that are the building blocks of proteins have the $-NH_2$ group attached to the carbon atom next to the –COOH group. These are called α-amino acids. The simplest is aminoethanoic acid (glycine), NH_2CH_2COOH.

Amino acids have the general formula:

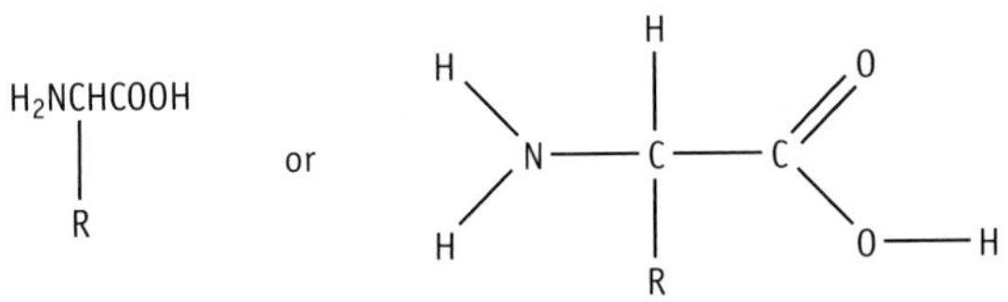

where R is either a hydrogen atom or an organic residue.

There are 20 amino acids that make up the proteins in the human body. Of these, 12 can be synthesised from other amino acids, but eight cannot. The latter are called **essential amino acids** and must be present in the diet. Two examples are valine and lysine.

Amino acids can be divided into three types:

- **neutral amino acids** — a molecule of a neutral amino acid contains only one basic $-NH_2$ group and one acidic –COOH group
- **acidic amino acids** — a molecule of an acidic amino acid contains two –COOH groups and one $-NH_2$ group
- **basic amino acids** — a molecule of a basic amino acid contains two $-NH_2$ groups and one –COOH group

Table 13.3 *Some amino acids*

Type	Common name	Systematic name	Formula
Neutral	Glycine	Aminoethanoic acid	$CH_2(NH_2)COOH$
	Alanine	2-aminopropanoic acid	$CH_3CH(NH_2)COOH$
	Valine	2-amino-3-methylbutanoic acid	$(CH_3)_2CHCH(NH_2)COOH$
	Leucine	2-amino-4-methylpentanoic acid	$(CH_3)_2CHCH_2CH(NH_2)COOH$
	Cysteine	2-amino-3-sulfanylpropanoic acid	$HSCH_2CH(NH_2)COOH$
Acidic	Aspartic acid	2-aminobutane-1,4-dioic acid	$HOOCCH_2CH(NH_2)COOH$
	Glutamic acid	2-aminopentane-1,5-dioic acid	$HOOC(CH_2)_2CH(NH_2)COOH$
Basic	Lysine	2,6-diaminohexanoic acid	$NH_2CH_2CH_2CH_2CH_2CH(NH_2)COOH$

The salt of glutamic acid in which one hydrogen atom has been replaced by sodium is monosodium glutamate, the flavour enhancer added to some foods.

Physical properties

Amino acids are solids at room temperature. This is because the molecules form **zwitterions**. The –COOH group in one molecule protonates the $-NH_2$ group in another molecule, forming a species that has a positive charge at one end and a negative charge at the other. The positive charge on one zwitterion is strongly attracted to the negative charge of an *adjacent* zwitterion. The –COOH group is weakly acidic and the $-NH_2$ group is weakly basic, so the formation of the zwitterion is reversible. Therefore, both the zwitterion and the neutral molecule are present. This can be represented for aminoethanoic acid (glycine) by the equation:

$$NH_2CH_2COOH \rightleftharpoons {}^{+}NH_3CH_2COO^{-}$$

where ${}^{+}NH_3CH_2COO^{-}$ is the zwitterion.

e It is common practice to write the formula of an amino acid as an uncharged molecule, rather than as the zwitterion. However, it should be remembered that the position of this equilibrium is well to the right. Equations for reactions of amino acids should have the zwitterion on the left. However, an equation with the molecular species on the left will score full marks at A-level.

The full structural formula of the zwitterion of glycine is:

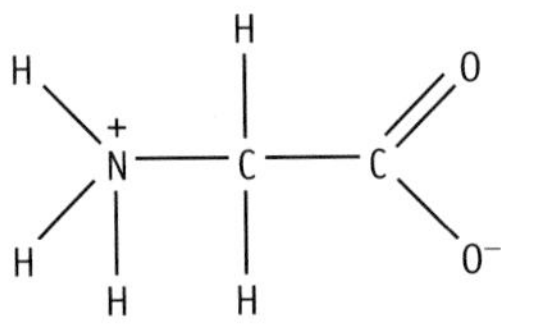

The name 'zwitterion' is derived from the German for the word 'mongrel', which is the offspring of two different breeds of dog. In the chemical sense, it is the mongrel of a cation and an anion.

Chemical reactions

The two most important reactions are those of amino acids with acids and with bases.

Reaction with acids

The COO^- group in the zwitterion acts as a base and accepts an H^+ ion from the acid:

$$^+NH_3CH_2COO^- + H^+ \rightarrow {}^+NH_3CH_2COOH$$

This reaction can also be written with the uncharged molecule as the starting species:

$$NH_2CH_2COOH + H^+ \rightarrow {}^+NH_3CH_2COOH$$

The basic $-NH_2$ group reacts with the H^+ ions in a strong acid.

e A base is a species that accepts a proton. An acid is a species that protonates a base.

Reaction with bases

The NH_3^+ group in the zwitterion acts as an acid and protonates the base:

$$^+NH_3CH_2COO^- + OH^- \rightarrow NH_2CH_2COO^- + H_2O$$

The $-NH_3^+$ group, which is the conjugate acid of the weakly basic $-NH_2$ group, gives a proton to the OH^- ion.

The alternative equation shows the acidic $-COOH$ group in the uncharged molecule protonating hydroxide ions:

$$NH_2CH_2COOH + OH^- \rightarrow NH_2CH_2COO^- + H_2O$$

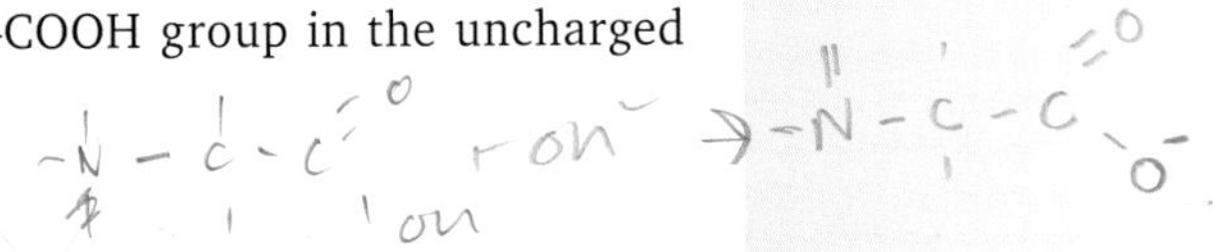

Other reactions as an acid

Amino acids have $-COO^-$ groups, rather than $-COOH$, because they exist mainly as zwitterions. This means that they do not undergo reactions of the $-OH$ group, unless the reaction is carried out in the presence of an acid. In this case the $-COO^-$ group becomes protonated, forming a $-COOH$ group. Thus amino acids can be esterified in the presence of concentrated sulfuric acid.

Other reactions as an amine

Amino acids have $-NH_3^+$ groups, rather than $-NH_2$ groups, because of the presence of zwitterions. This means that an amino acid molecule does not have a lone pair of electrons on the nitrogen atom and so will not act as a nucleophile, unless the pH of the reaction mixture is significantly above 7.

Stereoisomerism

All the natural amino acids, except glycine, are optically active. The carbon atom next to the $-COOH$ is a chiral centre because it has four different groups attached to it. Alanine (2-aminopropanoic acid) has two optical isomers (Figure 13.3). The isomers can be written as zwitterions or as uncharged molecules. The latter is more common, especially in biochemistry.

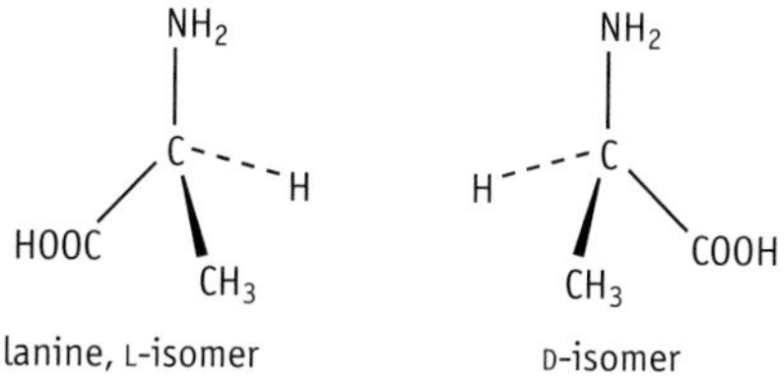

***Figure 13.3** The two optical isomers of 2-aminopropanoic acid*

Alanine, which occurs in proteins, is only one of the two optical isomers shown in Figure 13.3. Alanine rotates the plane of polarisation of plane-polarised light clockwise.

e Almost all the amino acids in proteins have the same absolute configuration in space. This is the L-configuration, as shown in the diagram of alanine (Figure 13.3).

Glycine (aminoethanoic acid, NH_2CH_2COOH) is optically inactive because it does not have a chiral centre. The carbon atom next to the –COOH group has two hydrogen atoms bonded to it, so it cannot be chiral. Therefore, a solution of glycine has no effect on the plane of polarisation of plane-polarised light.

Separation and identification of amino acids

These processes can be carried out in school laboratories using thin-layer chromatography, TLC. The principles of TLC are the same as for other chromatographic techniques (see p. 176). The stationary phase is either silica gel or aluminium oxide immobilised on a flat inert sheet that is made usually from glass or plastic.

- The unknown amino acid or mixture of amino acids is dissolved in a suitable solvent and a spot of the test solution is placed about 2 cm from the bottom of the plate.
- Spots of dissolved known amino acids are placed separately on the same plate at the same level.
- The plate is then dipped in a suitable eluent (the mobile phase) with the spots above the level of the liquid eluent. The plate is placed in a sealed container. The eluent is drawn up the plate by capillary action.
- The plate is left until the eluent has risen to the top of the plate.
- The plate is removed and sprayed with a solution of ninhydrin and then heated. Ninhydrin reacts with amino acids (and amines) producing a blue-purple colour.
- The height that the unknown reached is compared with the heights reached by the known amino acids. Spots at the same height are caused by the same amino acid and so the amino acids in the unknown can be identified. Alternately, an amino acid can be identified by measuring its R_f value and comparing it with values in a book of data.

$$R_f \text{ value} = \frac{\text{distance moved by amino acid}}{\text{distance moved by solvent}}$$

The separation works because different amino acids have different R_f values.

This is because the interaction with the stationary phase compared with solubility in the eluent differs for different amino acids.

Proteins

Proteins consist of a number of amino acids joined together by peptide links.

If two amino acids join together in a condensation reaction, the product is called a **dipeptide**. The artificial sweetener aspartame is a dipeptide formed from phenylalanine and aspartic acid:

$$H_2N{-}CH(CH_2C_6H_5){-}COOH \; + \; H_2N{-}CH(CH_2COOH){-}COOH$$

$$\downarrow$$

$$H_2N{-}CH(CH_2C_6H_5){-}\underbrace{C({=}O){-}N(H)}{-}CH(CH_2COOH){-}COOH \; + \; H_2O$$

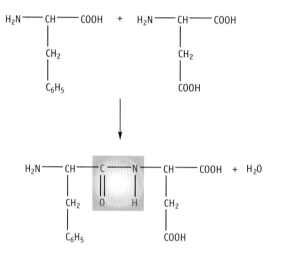

The peptide link is shown in red.

e The structural formula of an amino acid can also be written as $NH_2CHRCOOH$, where R is a hydrogen atom (in glycine) or an organic group.

An amino acid in a peptide chain is called a **residue**. Thus, aspartame has the residues of phenylalanine and aspartic acid.

Natural proteins are **polypeptides** formed by condensation polymerisation of many amino acids. Bovine insulin has 51 amino acids in one specific sequence. The number of different ways of arranging these 51 amino acids is about the same as there are atoms in the whole of the Milky Way galaxy, but only one way produces insulin.

It is remarkable that so few mutations occur to produce rogue polypeptides. However, sickle cell anaemia is a hereditary disease that results when two glutamic acid

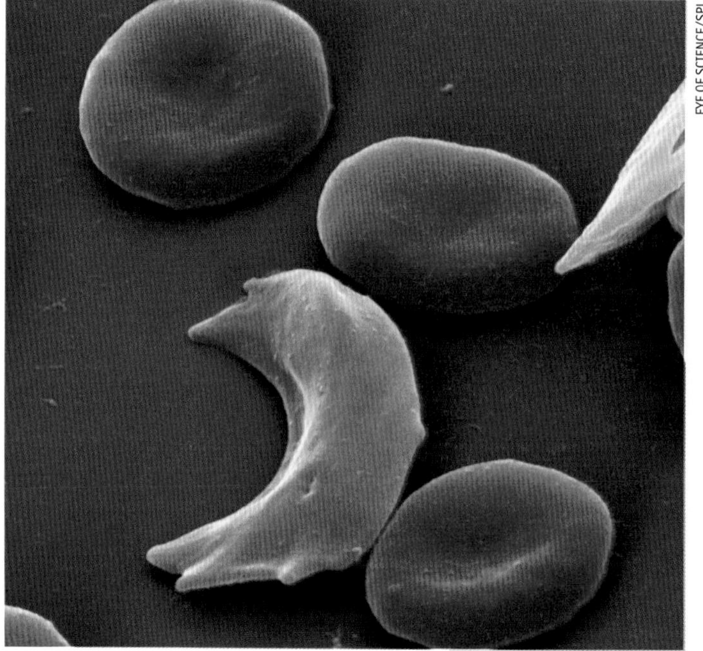

False-colour scanning electron micrograph of normal red blood cells (rounded) and sickle cells (crescent-shaped)

EYE OF SCIENCE/SPL

molecules (one in each β-polypeptide sequence in haemoglobin) are replaced by valine molecules. The resulting haemoglobin molecule is a much poorer oxygen carrier, which tends to shorten the life of the sufferer. However, this mutation does provide protection against malaria, and the gene is passed on from generation to generation, in regions where malaria is endemic.

Structure of proteins

The **primary structure** is the order of the amino acids in the chain. This is specific to the protein. However, the peptide chains are not randomly arranged like a pile of spaghetti. They are organised into a **secondary structure**, of which there are two types:

- **α-helix** — the peptide is coiled in a spiral and the structure held together by *intramolecular* hydrogen bonds between one group in one part of the chain and another group four residues away along the chain. There are also ionic bonds between the residues of basic and acidic amino acids. Myoglobin is an example of a protein with an α-helical structure.
- **β-pleated sheet** — the polypeptide chains are almost fully extended and are held by hydrogen bonds to other polypeptide chains. The result is rather like a stack of corrugated cardboard. Silk has this structure. A thread of silk is strong because of the many intermolecular hydrogen bonds between its polypeptide chains.

e Nylon is a polyamide in which all the pairs of residues are identical. Its secondary structure is a β-pleated sheet, which is why nylon is so strong.

Nitriles

Nitriles contain the –C≡N group. They used to be called cyanides. However, unlike ionic cyanides such as KCN, they are not poisonous.

The names of nitriles depend on the longest carbon chain, including the carbon of the –CN group. Therefore, CH_3CN is ethanenitrile and CH_3CH_2CN is propanenitrile.

Some nitriles contain another group as well as –CN. The product of the reaction of HCN with propanone — 2-hydroxy-2-methylpropanenitrile — is an example of this type of compound:

$$(H_3C)_2C{=}O + HCN \longrightarrow (H_3C)_2C(OH)C{\equiv}N$$

This is a nucleophilic addition reaction.

Preparation

Simple nitriles

Simple nitriles are prepared by the reaction between a halogenoalkane and potassium cyanide. For example, the product of the reaction between bromoethane and potassium cyanide is propanenitrile:

$$C_2H_5Br + KCN \rightarrow C_2H_5CN + KBr$$

This is a nucleophilic substitution reaction. The halogenoalkane and potassium cyanide are heated under reflux in a solution of ethanol and water.

Hydroxynitriles

Hydroxynitriles are made by the nucleophilic addition of HCN to an aldehyde or ketone, for example:

$$CH_3CHO + HCN \rightarrow CH_3CH(OH)CN$$

The pH of the solution must be maintained at about 8 so that there is a significant amount of CN^- for the first step of the addition, and enough HCN for the second step (p. 135). The conditions are either to add hydrogen cyanide, HCN, plus some base or to use a mixture of hydrogen cyanide and potassium cyanide.

Questions

1 Name the reagents and conditions for the two-step preparation of ethylamine from ethene.

2 $C_4H_{11}N$ has several structural isomers. Write the structural formula of the isomer of $C_4H_{11}N$ that is:

- **a** a primary amine with a branched chain
- **b** a symmetrical secondary amine with a straight chain
- **c** a secondary amine with a branched chain
- **d** a tertiary amine

3 Give the H–N–H bond angle in methylamine. Justify your answer.

4 Give the N–C–C bond angle in ethanamide. Justify your answer.

5 Explain why ethylamine is soluble in water but chloroethane and ethane are both insoluble.

6 Explain why ethanoic acid, CH_3COOH, has a higher boiling point than 1-propylamine, $CH_3CH_2CH_2NH_2$, and why both these compounds have a higher boiling point than butane, $CH_3CH_2CH_2CH_3$.

7 Ethylamine, $C_2H_5NH_2$, can be prepared from ethanoic acid by a three-step synthesis. Outline this process.

8 An organic compound contains 39.3% carbon, 11.5% hydrogen, 26.2% oxygen and 23.0% nitrogen by mass.

- **a** Calculate its empirical formula. (You may need to refer to p. 283.)
- **b** The molecular formula of this compound is the same as its empirical formula. It reacts with phosphorus pentachloride, giving off misty fumes. Suggest a structural formula for this compound.

9 Write equations for the reaction of 2-amino-ethanol, $NH_2CH_2CH_2OH$ with:

- **a** dilute hydrochloric acid
- **b** ethanoyl chloride
- **c** iodomethane

10 Explain why methylamine is a stronger base than ammonia which itself is a stronger base than phenylamine.

11 Explain why aminoethanoic acid is a solid at room temperature and is soluble in water.

12 Ethylamine has a distinctive smell. When some dilute hydrochloric acid is added, the smell disappears. If aqueous sodium hydroxide is now added in excess, the smell returns. Write equations for the reactions involved.

13 State the reagents for the conversion of nitrobenzene to phenylamine.

14 Phenylamine and chlorobenzene are equally polar molecules. Phenylamine boils at 184°C; chlorobenzene boils at 132°C. Phenylamine is slightly soluble in water and forms a clear solution when dilute hydrochloric acid is added. Chlorobenzene is immiscible with both water and dilute hydrochloric acid.

Explain:

a the difference in boiling temperatures of the two substances

b why phenylamine is more soluble in hydrochloric acid than in water

c why chlorobenzene is insoluble both in water and in hydrochloric acid

15 Write equations for the reaction of phenylamine with:

a hydrochloric acid

b ethanoyl chloride

16 When phenylamine is treated with an acidified solution of sodium nitrite at a temperature between 0°C and 10°C, benzenediazonium ions are formed. Explain why the temperature must not

a drop below 0°C

b rise above 10°C

17 Write:

a the structural formula of the $C_6H_5N_2^+$ ion

b the equation for its reaction with phenol in alkaline solution

18 Explain why ethanamide is soluble in water.

19 Write the equation for the formation of the polymer made from 1,5-pentanedioic acid and diaminoethane.

20 Explain why a polyamide such as Terylene has a higher melting temperature than an addition polymer such as poly(propene).

21 Explain why poly(ethenol) is water soluble.

22 The formula of all naturally occurring amino acids can be written as $NH_2CHRCOOH$.

a Rewrite this formula as the zwitterion.

b Write an equation for the reaction of the zwitterion with aqueous acid.

c Write an equation for the reaction of the zwitterion with aqueous alkali.

23 Draw the two optical isomers of aminobutane-1,4-dioic acid. Identify which of the two is aspartic acid.

24 Write the structure of a dipeptide formed from aminoethanoic acid and 2-aminopropanoic acid.

25 RNA contains the base uracil:

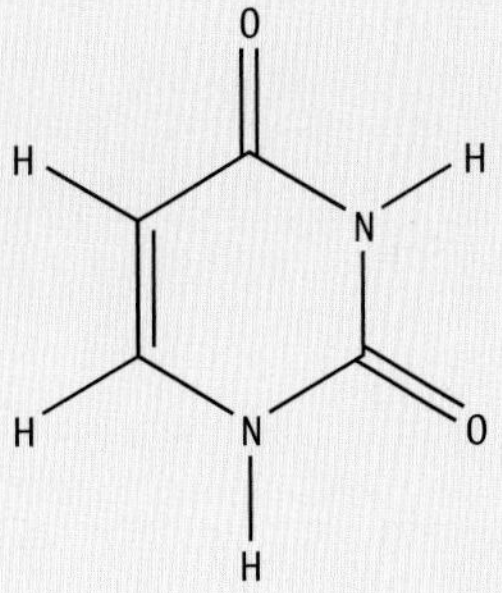

Draw a diagram of a uracil molecule in one strand of mRNA when hydrogen-bonded to an adenine molecule in a strand of DNA (see pp. 271–272).

The information needed for questions 26–29 is not found in this textbook. You are advised to do your own research to find answers to these questions. Possible sources are the internet, biology textbooks and other students who are studying biochemistry.

26 Explain the link between a DNA molecule, an mRNA molecule and the polypeptide formed. What is the sequence of basesneeded to put a leucine residue in a polypeptide?

27 Use the internet to find out about Chargaff's rule.

28 Write brief notes on the discoveries about DNA structure made by Wilkins and Franklin in London and Crick and Watson in Cambridge.

29 Write short notes on DNA fingerprinting

Organic analysis and synthesis

Chapter 14

Introduction

Methods of analysis must be sufficiently sensitive to be able to detect minute quantities of impurities. This is particularly important in forensic chemistry, when monitoring the purity of a pharmaceutical drug and when checking on the composition of effluent from an industrial plant. Analytical techniques must also be accurate. This is fundamental to forensic chemistry and is vital when following the progress of an experimental multi-step synthesis of a potential new drug.

To produce the drug ibuprofen:

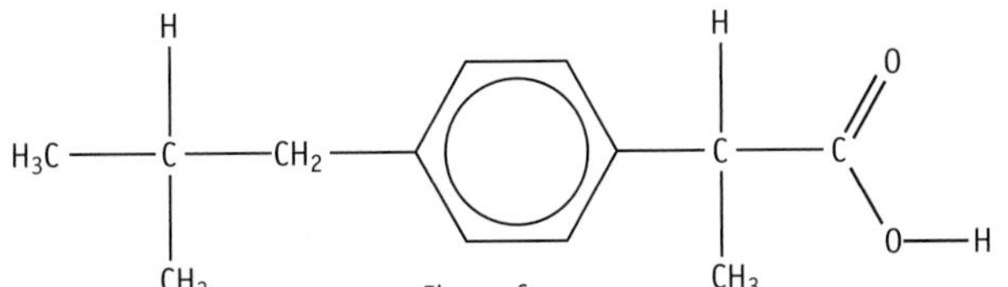

$$H_3C-CH(CH_3)-CH_2-C_6H_4-CH(CH_3)-COOH$$

Ibuprofen

a $(CH_3)_2CHCH_2$ group has to be introduced into a benzene ring. A possible multi-step synthesis involves a Friedel–Crafts reaction between benzene and 1-chloro-2-methylpropane, $(CH_3)_2CHCH_2Cl$ in the presence of anhydrous aluminium chloride as catalyst. The first step is the production of the electrophile:

$$(CH_3)_2CHCH_2Cl + AlCl_3 \longrightarrow (CH_3)_2CH\overset{+}{C}H_2 + AlCl_4^-$$

However, this is a primary carbocation and may rearrange to form the more stable secondary carbocation, which is then the electrophile:

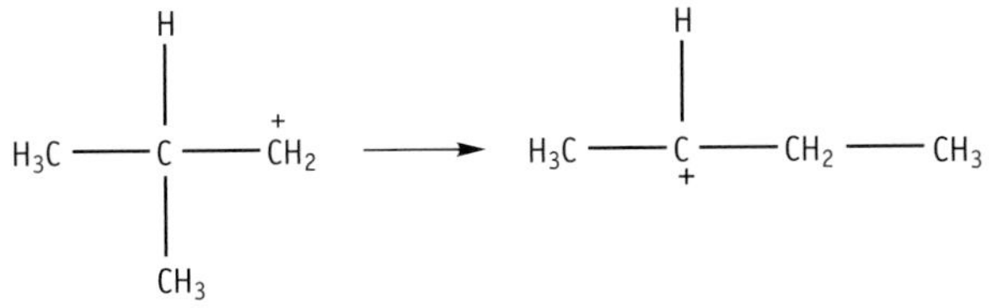

$$H_3C-CH(CH_3)-\overset{+}{C}H_2 \longrightarrow H_3C-\underset{+}{C}H-CH_2-CH_3$$

Thus the reaction of benzene with 1-chloro-2-methylpropane gives rise to one, or both, of the two products labelled I and II overleaf.

Accurate analysis of the reaction product mixture is essential. If compound I is the major product, this will not provide a route for the production of ibuprofen and this process will have to be abandoned. In fact, compound I is the major product and so the reaction with a halogenoalkane is not used in ibuprofen manufacture.

The actual synthesis uses 2-methylpropanoyl chloride, $(CH_3)_2CHCOCl$, in a Friedel–Crafts reaction. The C=O group is later reduced to a CH_2 group.

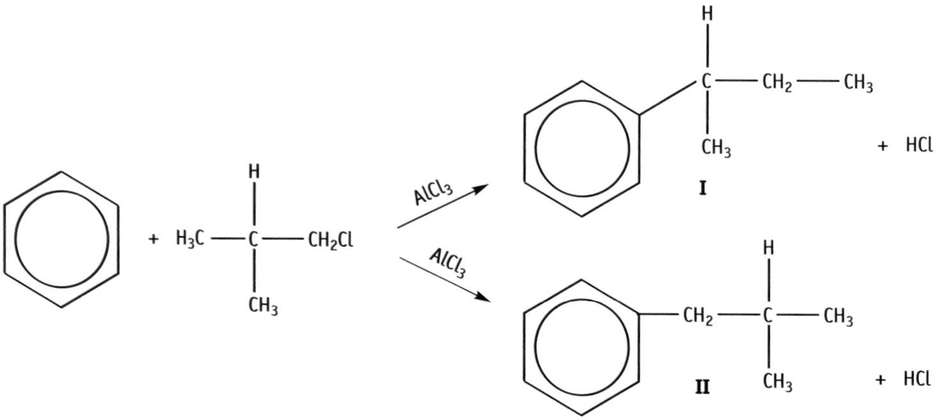

$$C_6H_6 + H_3C\text{—}CH(CH_3)\text{—}CH_2Cl \xrightarrow{AlCl_3} C_6H_5\text{—}C(H)(CH_3)\text{—}CH_2\text{—}CH_3 \ (\text{I}) + HCl$$

$$C_6H_6 + H_3C\text{—}CH(CH_3)\text{—}CH_2Cl \xrightarrow{AlCl_3} C_6H_5\text{—}CH_2\text{—}C(H)(CH_3)\text{—}CH_3 \ (\text{II}) + HCl$$

Whether compound I or II is the product depends on whether rearrangement of the electrophile takes place.

Organic analysis

To find the molecular formula

Percentage composition

This is normally carried out by combustion analysis. A known mass of the substance is burned in excess dry oxygen and the masses of water and carbon dioxide produced are measured. The mass of water is found by adsorbing the water vapour on either silica gel or absorbing it into anhydrous calcium chloride; the mass of carbon dioxide is found by reacting it with calcium oxide.

The route is:

- mass of carbon dioxide → mass of carbon → % carbon
- mass of water → mass of hydrogen → % hydrogen
- % oxygen = 100 – (% of carbon + % hydrogen)

Worked example

2.90 g of an organic compound X, containing hydrogen, carbon and oxygen only, was burnt in excess oxygen. 6.60 g of carbon dioxide and 2.70 g of water were produced. Calculate the percentage composition of compound X.

Answer

Method 1: using moles

$$\text{moles of } CO_2 = \frac{\text{mass}}{M_r} = \frac{6.60}{44.0} = 0.150 = \text{moles carbon}$$

$$\text{mass carbon} = \text{moles} \times A_r = 12.0 \times 0.150 = 1.80\,\text{g}$$

$$\%\ \text{carbon} = 100 \times \frac{1.80}{2.90} = 62.1\%$$

$$\text{moles of } H_2O = \frac{\text{mass}}{M_r} = \frac{2.70}{18.0} = 0.150$$

moles of H = 2 × 0.150 = 0.300

mass of hydrogen = moles × A_r = 0.300 × 1.0 = 0.300 g

$$\% \text{ hydrogen} = 0.300 \times \frac{100}{2.90} = 10.3\%$$

% oxygen = 100 − (62.1 + 10.3) = 27.6%

Method 2: by mass ratio

44.0 g of CO_2 contains 12.0 g of carbon

$$6.60\text{ g of } CO_2 \text{ contains } 12 \times \frac{6.60}{44.0} = 1.80\text{ g of carbon}$$

$$\% \text{ carbon} = 100 \times \frac{1.80}{2.90} = 62.1\%$$

18.0 g of H_2O contains 2.0 g of hydrogen

$$2.70\text{ g of } H_2O \text{ contain } 2.0 \times \frac{2.70}{18.0} = 0.300\text{ g of hydrogen}$$

$$\% \text{ hydrogen} = 0.300 \times \frac{100}{2.90} = 10.3\%$$

% oxygen = 100 − (62.1 + 10.3) = 27.6%

The percentage composition by mass of compound X is: carbon 62.1%, hydrogen 10.3%, oxygen 27.6%.

Empirical formula

The empirical formula of a compound is obtained by dividing the percentage composition of each element by its atomic mass and then dividing through by the smallest.

Worked example

Find the empirical formula of compound X, which has the following percentage composition by mass: carbon 62.1%, hydrogen 10.3%, oxygen 27.6%.

Answer

	%	Divide by A_r	Divide by smallest
Carbon	62.1	62.1/12.0 = 5.175	3.0
Hydrogen	10.3	10.3/1.0 = 10.3	5.97 ≈ 6.0
Oxygen	27.6	27.6/16.0.o = 1.725	1.0

The empirical formula of compound X is C_3H_6O.

Remember:

- Do not round up the %/A_r values.
- Round the final column to 1 decimal place.
- If the numbers in the final column are not whole numbers, try multiplying by 2 or 3 to obtain integers.

Molecular formula

The empirical formula can be converted to the molecular formula if the molar mass is known.

$$\frac{\text{molar mass}}{\text{mass of empirical formula}} = x \text{ (an integer)}$$

then

molecular formula = x times the empirical formula

Worked example

An alkene has the empirical formula CH_2 and molar mass 56.0 g mol^{-1}. What is its molecular formula?

Answer

$$\frac{\text{molar mass}}{\text{mass of empirical formula}} = \frac{56.0}{14.0} = 4$$

molecular formula = $4 \times CH_2 = C_4H_8$

Finding the molar mass

- Molar mass can be found from the *m/e* value of the molecular ion (the largest) in the mass spectrum.
- If the unknown can be titrated, the number of moles of a given mass can be estimated. Then, molar mass = mass/moles
- If the unknown is a gas, molar mass = volume/molar volume, where the molar volume at 0°C and 1 atm pressure is 22.4 dm^3 (24 dm^3 under normal laboratory conditions).

Specific identification of the unknown

This is carried out in two stages. First, the functional groups are identified and then the exact formula is derived.

Tests for functional groups

In organic analysis, care must be taken to follow the logic of the tests. For example, a positive test with Brady's reagent (2,4-dinitrophenylhydrazine) indicates that the unknown is a carbonyl compound but does not distinguish between an aldehyde and a ketone. If the unknown does not react with acidified potassium dichromate(VI) but gives a positive result with Brady's reagent then it must be a ketone. Aldehydes are oxidised by acidified potassium dichromate(VI), so a positive result in both these tests shows that the unknown is an aldehyde.

C=C group

The test for the C=C group is to add bromine water. Compounds that contain a C=C group quickly decolorise the brown bromine water and do not give a precipitate.

e Phenols also decolorise bromine water, but they give a white precipitate of a polybrominated phenol.

Alkenes rapidly turn neutral potassium manganate(VII) from a purple solution to a brown precipitate at room temperature. However, aldehydes also do this.

Halogenoalkanes

Halogenoalkanes contain a halogen atom, which, on warming with aqueous sodium hydroxide, is removed by hydrolysis. *Excess* dilute nitric acid is added to this solution to neutralise the sodium hydroxide. On addition of a solution of silver nitrate a precipitate is obtained. The equations for the reactions of a chloroalkane are:

$$RCl + OH^- \rightarrow ROH + Cl^-$$

$$Cl^-(aq) + Ag^+(aq) \rightarrow AgCl(s)$$

- A white precipitate soluble in dilute ammonia solution proves the presence of chlorine in the organic compound:

$$AgCl(s) + 2NH_3(aq) \rightarrow [Ag(NH_3)_2]^+(aq) + Cl^-(aq)$$

- A cream precipitate, insoluble in dilute ammonia but soluble in concentrated ammonia shows the presence of bromine in the organic compound:

$$AgBr(s) + 2NH_3(aq) \rightarrow [Ag(NH_3)_2]^+(aq) + Br^-(aq)$$

- A pale yellow precipitate, insoluble in both dilute and concentrated ammonia, proves the presence of iodine in the original organic compound:

$$AgI(s) + 2NH_3(aq) \rightarrow \text{no reaction}$$

The distinction between primary, secondary and tertiary halogenoalkanes cannot be made by a simple chemical test. Analysis by mass spectrometry or spin coupling in NMR spectroscopy can be conclusive (pp. 174 and 172).

Alcohols

- Apart from phenols, compounds with an –OH group give steamy fumes of hydrogen chloride on the addition of solid phosphorus pentachloride:

$$ROH + PCl_5 \rightarrow HCl + RCl + POCl_3$$
$$RCOOH + PCl_5 \rightarrow HCl + RCOCl + POCl_3$$

Therefore, all alcohols and all carboxylic acids give a positive result. However, alcohols (unlike carboxylic acids) do not give bubbles of carbon dioxide on the addition of either sodium hydrogencarbonate or sodium carbonate.

- On warming with ethanoic acid in the presence of a few drops of concentrated sulfuric acid, all alcohols form esters:

$$ROH + CH_3COOH \rightleftharpoons CH_3COOR + H_2O$$

If the product is poured into a beaker of dilute sodium hydrogencarbonate solution, the characteristic odour of an ester (like nail varnish, glue or fruit) will be detected.

- Primary and secondary alcohols turn orange acidified potassium dichromate(VI) solution to a green solution of Cr(III). Tertiary alcohols are not oxidised by acidified dichromate(VI) ions, so the solution remains orange.
  - This test can be modified to distinguish between primary and secondary alcohols, by distilling off the oxidised organic product as it is formed. If this is mixed with Tollens' reagent, only the oxidised product of the primary alcohol (an aldehyde) reduces the silver ion complex to give a silver mirror.
  - The oxidised product of the secondary alcohol (a ketone) has no reaction with Tollens' reagent, but (like all carbonyl compounds) gives a yellow-orange precipitate with Brady's reagent (2,4-dinitrophenylhydrazine).
- Alcohols, as well as carboxylic acids, have a broad band in the IR spectrum at approximately 3000 cm^{-1}. Carboxylic acids, but not alcohols, also have a peak at approximately 1700 cm^{-1} (p. 166).

Aldehydes and ketones

Both aldehydes and ketones give a yellow or orange precipitate when a few drops

e Do not state that the appearance of steamy fumes of HCl with PCl_5 is a test for alcohols, because carboxylic acids (and water) also give the same result.

e When a colour change indicates a positive result to a test, always give the colour of the solution *before* the test as well as after.

of 2,4-dinitrophenylhydrazine (Brady's reagent) are added. The equation for the reaction with ethanal is:

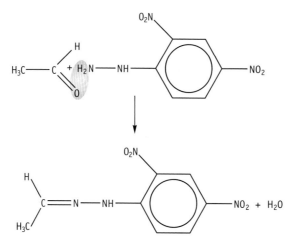

$$H_3C{-}C(H){=}O + H_2N{-}NH{-}C_6H_3(NO_2)_2 \longrightarrow (H)(H_3C)C{=}N{-}NH{-}C_6H_3(NO_2)_2 + H_2O$$

The infrared spectra of aldehydes and ketones have peaks at about 1700 cm^{-1}, as do the IR spectra of all other compounds with a C=O group. This includes carboxylic acids, esters, amides and acid chlorides.

There are several simple chemical tests to differentiate between aldehydes and ketones:

- The carbonyl compound is warmed with Fehling's or Benedict's solution. Aldehydes give a red precipitate of copper(I) oxide. Ketones do not react. Therefore, the copper(II) complex in the test solution is not reduced and the solution remains blue.

ANDREW LAMBERT PHOTOGRAPHY/SPL

Distinguishing between an aldehyde and a ketone: left, Fehling's solution; centre, Fehling's solution that has been reduced by an aldehyde; right, ketones do not react with Fehling's solution

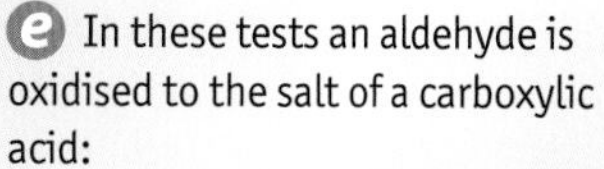

In these tests an aldehyde is oxidised to the salt of a carboxylic acid:

$$RCHO + [O] + OH^- \rightarrow RCOO^- + H_2O$$

- The unstable Tollens' reagent is made by adding a few drops of sodium hydroxide to silver nitrate solution and dissolving the precipitate formed in dilute ammonia solution. If the carbonyl compound is gently warmed with Tollens' reagent, aldehydes give a silver mirror but ketones do not.

JERRY MASON/SPL

Silver mirror test for the presence of an aldehyde. Ketones do not react and the solution remains colourless.

- On warming with potassium dichromate(VI), dissolved in dilute sulfuric acid, aldehydes turn the orange solution green. Ketones do not react, so the solution stays orange.

Aldehydes can be distinguished from ketones by examination of NMR spectra. The NMR spectrum of an aldehyde has a peak at $\delta = 9.0\text{–}10.0$, caused by the hydrogen atom in the aldehyde group, CHO. This is absent in ketones.

e Primary and secondary alcohols also give a positive result with acidified dichromate(VI) ions. Therefore, their absence must first be shown by a lack of steamy fumes with phosphorus pentachloride.

The iodoform reaction

The formula of iodoform is CHI_3. It is a pale yellow solid that is insoluble in water. It has an antiseptic smell. Iodoform is produced when an organic compound containing the $CH_3C{=}O$ or $CH_3CH(OH)$ group is gently warmed with iodine mixed with sodium hydroxide solution.

- The only aldehyde that performs the iodoform reaction is ethanal.
- All methyl ketones give a yellow precipitate. The equation for the reaction of butanone is:

$$CH_3COC_2H_5 + 3I_2 + 4NaOH \rightarrow CHI_3 + C_2H_5COONa + 3NaI + 3H_2O$$

The products are a precipitate of iodoform and a solution of sodium propanoate and sodium iodide.

e The iodoform reaction is one of the few reactions in which a carbon chain is shortened by one carbon atom.

- Alcohols undergo this reaction if they can be oxidised to give rise to a $CH_3C{=}O$ group. Ethanol is the only primary alcohol that gives a precipitate of iodoform. The alkaline solution of iodine oxidises it to ethanal, which then reacts to give the precipitate:

$$CH_3CH_2OH + [O] \rightarrow CH_3CHO + H_2O$$
$$CH_3CHO + 3I_2 + 4NaOH \rightarrow CHI_3 + HCOONa + 3NaI + 3H_2O$$

The sequence is:
*ethan*ol → ethanal → iodoform + sodium *methan*oate

Methyl secondary alcohols are oxidised to methyl ketones and, therefore, give a positive iodoform test. For example, propan-2-ol gives a yellow precipitate with an alkaline solution of iodine:

$$CH_3CH(OH)CH_3 + [O] \rightarrow CH_3COCH_3 + H_2O$$
$$CH_3COCH_3 + 3I_2 + 4NaOH \rightarrow CHI_3 + CH_3COONa + 3NaI + 3H_2O$$

The sequence is:
propan-2-ol → propanone → iodoform + sodium *ethan*oate

Propan-1-ol does not undergo the iodoform reaction.

$$CH_3CH_2CH_2OH + [O] \rightarrow CH_3CH_2CHO + H_2O$$
$$CH_3CH_2CHO + I_2 + NaOH \rightarrow \text{no precipitate of iodoform}$$

Carboxylic acids

- Carboxylic acids as well as alcohols give steamy fumes of hydrogen chloride when phosphorus pentachloride is added:

$$RCOOH + PCl_5 \rightarrow RCOCl + HCl + POCl_3$$

An acid can be distinguished from an alcohol by the addition of a solution of either sodium hydrogencarbonate or sodium carbonate. With a carboxylic acid, bubbles of gas are seen:

$$RCOOH + NaHCO_3 \rightarrow RCOONa + H_2O + CO_2$$
$$2RCOOH + Na_2CO_3 \rightarrow 2RCOONa + H_2O + CO_2$$

The gas turns limewater cloudy.

$$CO_2 + Ca(OH)_2 \rightarrow CaCO_3 + H_2O$$

- If a carboxylic acid is warmed with ethanol in the presence of a few drops of concentrated sulfuric acid, an ester is produced. If the reaction mixture is poured into a dilute solution of sodium hydrogencarbonate, the characteristic smell of the ester can be detected.
- The infrared spectra of carboxylic acids have a broad band at approximately $3000\,cm^{-1}$ and a sharp band at approximately $1700\,cm^{-1}$.

Aromatic compounds

Aromatic compounds burn with a smoky flame.

Phenol and phenylamine rapidly turn bromine water from brown to colourless, with the formation of a white precipitate, for example:

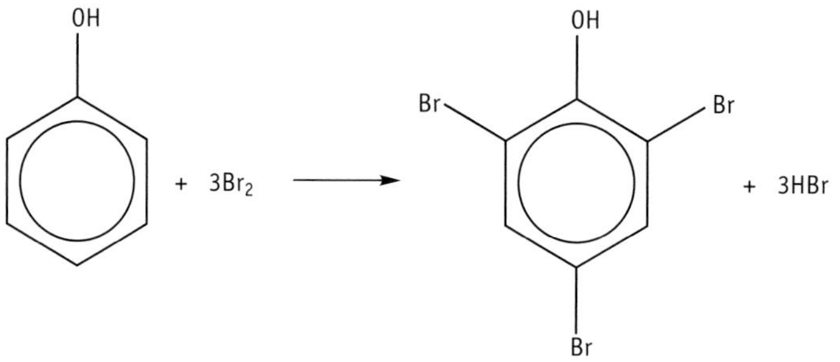

OH

+ $3Br_2$ ⟶

OH
Br Br
Br

+ 3HBr

Identification of the specific compound

Once the functional group in the molecule has been identified by chemical or spectral tests, the question remains as to what is the exact formula of the specific compound. There are a number of ways of determining this.

Infrared fingerprint

The region of an infrared spectrum below about $1300\,cm^{-1}$ is known as the fingerprint region because it is specific to a single compound. Thus, if the fingerprint region is checked against a database, the identity of the unknown can be found. This method requires the unknown sample to be pure, otherwise stray peaks caused by impurities will alter the spectral pattern.

R_f values in chromatography

R_f values can be used to identify the amino acids that result from the hydrolysis of proteins (see p. 276).

$$R_f = \frac{\text{distance moved by substance}}{\text{distance moved by eluent}}$$

NMR spectra

The number of peaks and their relative intensity can be used to distinguish between structural isomers. The identity can also be confirmed by examining the splitting pattern.

Worked example

A compound Y has the molecular formula C_4H_8O. It forms a precipitate with 2,4-dinitrophenylhydrazine. Its NMR spectrum is shown below. Identify compound Y.

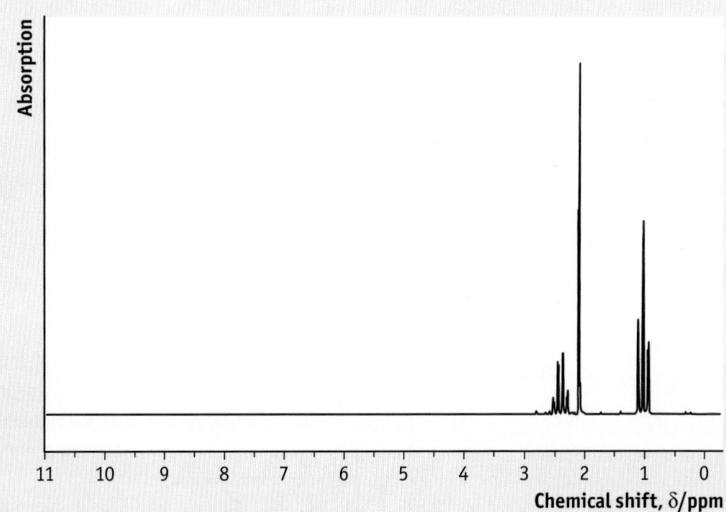

Answer

Compound Y is either an aldehyde or a ketone.

It could be:

- butanal, $CH_3CH_2CH_2CHO$
- 2-methylpropanal, $(CH_3)_2CHCHO$
- 2-butanone, $CH_3CH_2COCH_3$

Butanal, $CH_3CH_2CH_2CHO$, has hydrogen atoms in four different environments and so will have four peaks at different δ values. The NMR spectrum of Y has only three peaks at different δ values, so Y is not butanal.

2-methyl propanal, $(CH_3)_2CHCHO$, has hydrogen atoms in three different environments and so will have three NMR peaks. 2-butanone, $CH_3CH_2COCH_3$, also has hydrogen atoms in three different environments, so Y could be either 2-methylpropanal or 2-butanone.

The splitting pattern of the peaks can be used to distinguish between these two compounds.

2-methylpropanal, $(CH_3)_2CHCHO$, will have the peak due to the CHO hydrogen atom split into two by the neighbouring CH group.

The peak due to the (red) hydrogen atoms in the two CH_3 groups will also be split into two by the neighbouring CH group.

The peak due to the hydrogen atom in the CH group will be split into eight. This is caused by the six hydrogen atoms on the pair of neighbouring CH_3 groups and the hydrogen on the neighbouring CHO group.

Thus 2-methylpropanal will have a splitting pattern of 2,2,8 with one of the doublets as a very tall peak, since there are six hydrogen atoms causing this. The spectrum of Y has a splitting pattern of 4,3,1, so Y is not 2-methylpropanal.

The identity of Y is confirmed as follows.

2-butanone, $CH_3CH_2COCH_3$, will have a peak due to the CH_3 hydrogen atoms split into three by the two hydrogen atoms on the neighbouring CH_2 group.
The peak due to the hydrogen atoms in the CH_2 group will be split into four by the hydrogen atoms in the CH_3 group.

The hydrogen atoms in the other CH_3 group will not be split, as there are no hydrogen atoms on the neighbouring carbon atom. The splitting pattern of Y, 4,3,1, confirms that Y is 2-butanone.

e Remember the $(n+1)$ rule. If there are n hydrogen atoms attached to neighbouring carbon atoms, the peak will be split into $(n+1)$.

Boiling point determination

The boiling point of an organic liquid can be determined and the value compared with those in a database.

The apparatus for boiling point determination is shown in Figure 14.1.

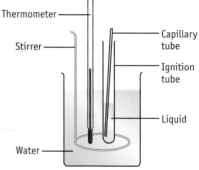

***Figure 14.1** Apparatus for the determination of boiling point*

The method is as follows:

- Place a small amount of the test liquid in the ignition tube and, using a rubber band, attach it to the thermometer.
- Place the capillary tube in the liquid, with its open end below the surface.
- Clamp the thermometer in the beaker of water.
- Slowly heat the water, stirring all the time. When the stream of bubbles coming out of the capillary tube is rapid and continuous, note the temperature and stop heating.
- Allow the beaker of water to cool, stirring continuously. Note the temperature when bubbles stop coming out of the capillary tube and the liquid begins to suck back into the capillary tube.

The average of these two temperatures is the boiling temperature of the liquid.

As the beaker of water is heated, the air in the capillary tube expands and bubbles of air slowly come out of the tube. These act as nuclei on which the bubbles of boiling liquid can form. This prevents superheating, which is when the temperature of the liquid rises above its boiling temperature.

The temperature measured on cooling is when the liquid stops boiling.

Both temperatures are slightly inaccurate because there is a time lag before the thermometer can register the boiling temperature. Averaging the two readings cancels out this error.

There are two problems with this method:

- The boiling points of similar substances are often quite close together and can differ by less than the experimental error.
- Impurities, and variation in atmospheric pressure, alter the boiling point.

Alcohol	Boiling temperature /°C	Alcohol	Boiling temperature /°C
Ethyl ethanoate	77.1	Ethanol	78.5
Methyl propanoate	78.7	Propan-2-ol	82.4
1-propyl methanoate	81.3	2-methylpropan-2-ol	82.5

***Table 14.1** Boiling temperatures of some esters and alcohols*

The terms 'boiling point' and 'boiling temperature' are interchangeable.

Melting point determination

The measurement of boiling point is not very reliable. Therefore, it is common practice for a liquid to be converted to a solid derivative. For example, a carbonyl compound can be converted to a 2,4-dinitrophenylhydrazine derivative. The derivative is then purified by recrystallisation and its melting point determined.

One method of determining melting point is:

- Insert some of the pure solid into a capillary tube and then attach the tube open end upwards to a thermometer with a rubber band.
- Place the thermometer into a bath of liquid. The liquid must boil at a higher temperature than the melting point of the solid being tested.
- Slowly heat the liquid bath, with constant stirring, and observe the solid in the capillary tube. Note the temperature when the solid melts.

Another method is:

- Place the solid in a boiling tube.
- Heat the boiling tube in a beaker of hot liquid (water if the solid melts below 100°C).
- When the solid melts, put a thermometer and stirrer in the molten solid and remove the boiling tube from the liquid bath.
- Allow the molten substance to cool, stirring the whole time, and read the temperature when the first crystals of solid appear.

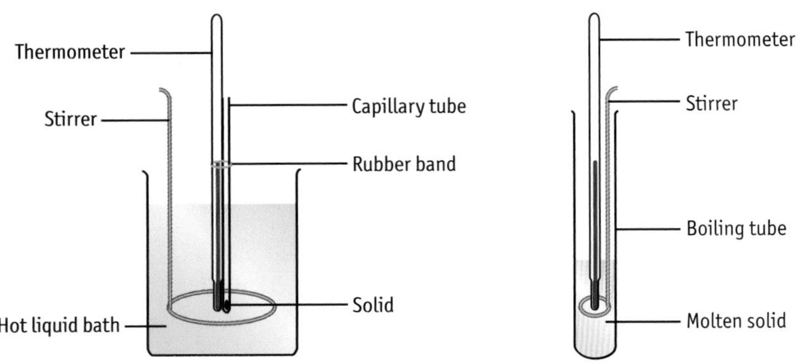

***Figure 14.2** Apparatus for the two methods of melting point determination*

Methods of separation and purification

For spectral analysis and for the measurement of melting and boiling temperatures, substances need to be very pure.

The result of any organic reaction will be a mixture of substances. The required product must be separated from the reaction mixture and then purified.

A stable liquid product

If the product does not decompose at the boiling temperature of the reaction mixture it may be possible to obtain it by distillation (see Figure 14.4).

If the product is insoluble in water, it is treated as follows:

- The distillate is washed with sodium carbonate solution in a separating funnel (Figure 14.3). Washing removes any acidic impurities that may be present. The pressure due to the carbon dioxide produced must be released from time to time. Washing is repeated until no more gas is produced.
- The aqueous layer is discarded and the organic layer is washed with water. This removes any unreacted sodium salts and any soluble organic substances, such as ethanol.
- This new aqueous layer is discarded and the organic layer dried, usually with anhydrous calcium chloride.
- The dried organic liquid is decanted off from the solid calcium chloride and distilled, collecting the liquid that boils off at ±2°C of the boiling temperature of the substance.

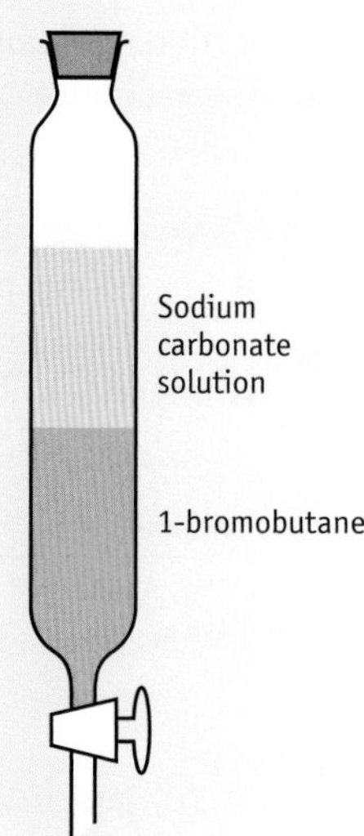

Figure 14.3 *Separation of 1-bromobutane from aqueous sodium carbonate*

Calcium chloride must not be used for drying amines or alcohols because it reacts with them. Potassium hydroxide or potassium carbonate are suitable alternatives.

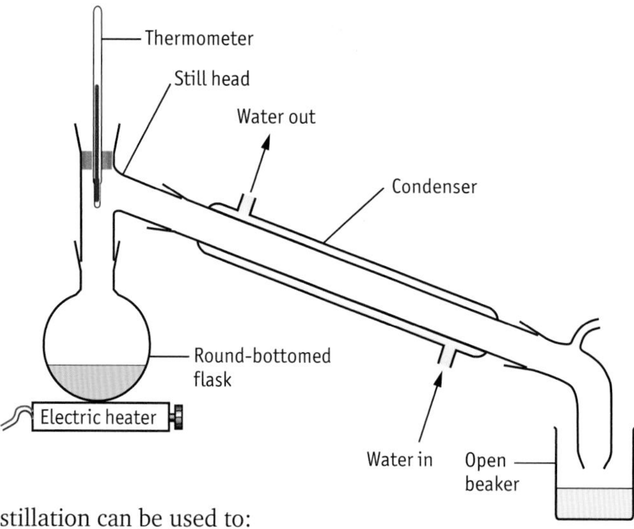

Figure 14.4 *Apparatus for simple distillation*

Simple distillation can be used to:

- remove a volatile liquid from non-volatile substances
- separate two volatile liquids that form a homogeneous solution

The latter works only if there is a small amount of one of the two liquids present as an impurity. The neck of the flask and the still head act as a mini-fractionating column. Larger amounts of an impurity require fractional distillation for separation.

Worked example

1-bromobutane can be prepared from butan-1-ol by reaction with a 50% solution of sulfuric acid and potassium bromide in a flask fitted with a reflux condenser. At first, steamy fumes of hydrogen bromide (HBr) are produced. Subsequently, bromine (brown) and sulfur dioxide (colourless) are produced. After reaction, 1-bromobutane and unreacted butan-1-ol are distilled out of the flask together with some acidic impurities.

The organic liquid is washed with sodium carbonate solution and then with water. Anhydrous calcium chloride is added and left for 1 hour. The liquid is decanted off and distilled.

	Boiling temperature/°C	Density/g cm^{-3}
Butan-1-ol	117	0.8
1-bromobutane	102	1.3

a What impurities are removed by washing with sodium carbonate solution?
b Why does the lower layer contain the 1-bromobutane?
c What impurities are removed by washing with water?
d Why is anhydrous calcium chloride added?
e Over what temperature range should the purified 1-bromobutane be collected?

Answer

a Sulfur dioxide, hydrogen bromide and bromine
b Because it is denser than water
c Some of the butan-1-ol and sodium salts such as sodium bromide from the reaction of acidic impurities with the sodium carbonate
d To remove any water dissolved in the organic layer
e Between 100°C and 104°C

An unstable liquid product

Steam distillation

Steam distillation is used to extract a volatile liquid that is immiscible with water, from a complex mixture. It is particularly useful for obtaining a substance that would decompose at its boiling point. It is used to extract perfumes from fruits and flowers. For example, lavender oil contains the compound linalool:

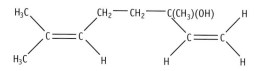

$$(H_3C)_2C{=}CH{-}CH_2{-}CH_2{-}C(CH_3)(OH){-}CH{=}CH_2$$

Linalool can be obtained by crushing lavender flowers and stalks with water and then subjecting the mixture to steam distillation. The liquid boils below the boiling temperature of both water and linalool, so no decomposition of the product takes place.

The procedure is as follows (Figure 14.5):

- Crush lavender flowers and stems in water and place the mixture in a flask.
- Steam distil and collect the distillate until pure water, with no oily droplets, distils over.
- Pour the mixture into a separating funnel and run off the aqueous layer.
- Add lumps of anhydrous potassium carbonate to the organic layer and leave until the product becomes clear.

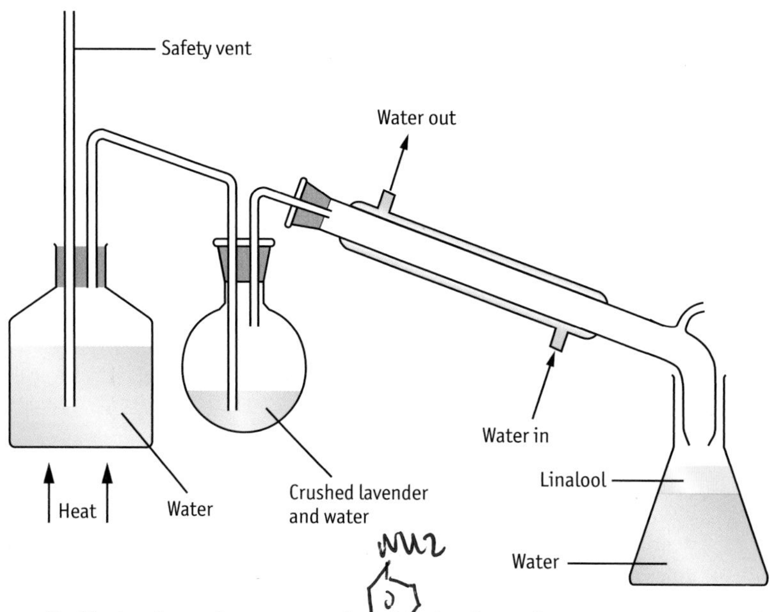

Figure 14.5 ***Apparatus for steam distillation***

Steam distillation is used to extract phenylamine from the reaction mixture after the reduction of nitrobenzene with tin and concentrated hydrochloric acid.

The nitrobenzene is warmed with a mixture of tin and concentrated hydrochloric acid. The tin is oxidised to tin(IV) chloride and the nitrobenzene is reduced to phenylamine. This reacts with the hydrochloric acid to form a solution of the salt $C_6H_5NH_3Cl$.

$$C_6H_5NO_2 + 6[H] + HCl \rightarrow C_6H_5NH_3Cl + 2H_2O$$

When alkali is added, phenylamine and a precipitate of tin hydroxide are produced.

$$C_6H_5NH_3Cl(aq) + NaOH(aq) \rightarrow C_6H_5NH_2(l) + H_2O(l) + NaCl(aq)$$
$$SnCl_4(aq) + 4NaOH(aq) \rightarrow Sn(OH)_4(s) + 4NaCl(aq)$$

The volatile phenylamine is removed from the suspension of tin hydroxide in alkali by steam distillation. Salt is added to the distillate to reduce the solubility of the phenylamine in water. The organic layer is separated from the aqueous layer using a separating funnel and then dried with solid potassium hydroxide.

A higher yield is obtained if the phenylamine is separated from the aqueous layer by solvent extraction.

Solvent extraction

It may be possible to separate an organic product from inorganic substances by solvent extraction. For example, 1-chloropentane can be prepared by adding excess phosphorus pentachloride to pentan-1-ol:

$$CH_3(CH_2)_3CH_2OH(l) + PCl_5(s) \rightarrow CH_3(CH_2)_3CH_2Cl(l) + HCl(g) + POCl_3(l)$$

Phosphorus oxychloride, $POCl_3$, has a boiling temperature of 105°C, which is very similar to that of 1-chloropentane (108°C). Therefore, it is not possible to separate the two by distillation of the reaction mixture.

The organic compound is separated by adding ethoxyethane (ether). This solvent dissolves 1-chloropentane but not phosphorus oxychloride. The solution of 1-chloropentane in ether is washed first with aqueous sodium carbonate and then with water. It is dried with anhydrous calcium chloride and the ether is distilled off by gentle warming.

Ether is highly flammable. It is distilled off in a fume cupboard by warming the flask with hot water, previously heated in the open laboratory.

Solid product

The solid product is removed from the reaction mixture by filtration. It is then purified by recrystallisation.

Recrystallisation

Recrystallisation is the method used to purify solids. A solvent has to be found in which the substance is soluble when the solvent is hot, but insoluble (or very much less soluble) when it is cold.

The method is:

- Dissolve the impure solid in the *minimum* of hot solvent.
- Remove any undissolved impurities by filtering the hot solution through a fluted filter paper, using a *warmed* stemless funnel, into a conical flask.
- Allow the solution to cool.
- Filter the mixture of the pure solid and the solvent under *reduced pressure*, using a Buchner funnel. Collect the solid on the filter paper and discard the liquid, which will contain the soluble impurities.
- Wash the solid on the filter paper with a little ice-cold solvent and leave the solid to dry.
- Carefully remove the pure solid from the filter paper.

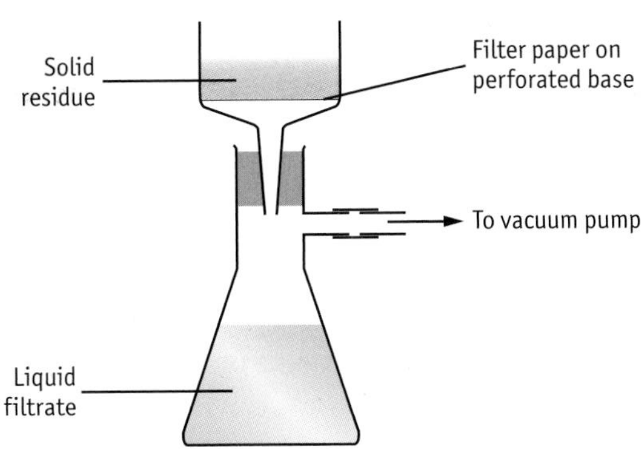

Figure 14.6 *Apparatus for reduced pressure filtration*

Benzoic acid is very soluble in hot water, but almost insoluble in cold water. Therefore, it can be purified by recrystallisation from hot water.

The compound formed by the reaction between an aldehyde or a ketone and 2,4-dinitrophenylhydrazine can be recrystallised from hot ethanol.

Organic synthesis

Organic synthesis is the construction of organic molecules using chemical processes. A particular synthesis may involve one or more intermediate compounds.

For a complex multi-step synthesis, it may be best to start by considering how the final product can be made and then work backwards to suitable original reactants.

An alternative technique is to determine whether the carbon-chain length has to be altered. If so, then the synthesis will involve one of the reactions below.

Increase in carbon-chain length

There are several ways in which the length of the carbon chain can be increased.

e Alcohols do not react with potassium cyanide. An alcohol must first be converted to a halogenoalkane and then reacted with KCN.

Nucleophilic substitution with KCN

Potassium cyanide reacts with primary, secondary and tertiary halogenoalkanes, RX. The reaction can be represented by:

$$RX + KCN \rightarrow RCN + KX$$

The conditions are heat under reflux in a solution of ethanol and water.

This reaction increases the carbon-chain length by one carbon atom. The product is a nitrile, which can be reduced to a primary amine or hydrolysed to a carboxylic acid:

$$RCN \xrightarrow[\text{2. Hydrolyse with } H_2SO_4(aq)]{\text{1. } LiAlH_4 \text{ in dry ether}} RCH_2NH_2$$

$$RCN \xrightarrow[\text{2. Add dilute acid}]{\text{1. Heat under reflux with } NaOH(aq)} RCOOH$$

Nucleophilic addition of HCN

Hydrogen cyanide reacts with aldehydes and ketones to form a compound with a –CN group and an –OH group on the same carbon atom:

$$C{=}O + HCN \rightarrow C(OH)CN$$

The conditions are that the reagents are mixed in the presence of a buffer at pH 8. The pH causes some of the hydrogen cyanide to be converted to cyanide ions, which are the catalyst for this reaction.

This reaction increases the carbon-chain length by one carbon atom. The product can be hydrolysed to a 2-hydroxycarboxylic acid:

$$>C(OH)CN \xrightarrow[\text{2. Add dilute acid}]{\text{1. Heat under reflux with } NaOH(aq)} >C(OH)COOH$$

For example, ethanal, CH_3CHO, can be converted to 2-hydroxypropanoic acid, $CH_3CH(OH)COOH$, by this method.

Friedel–Crafts reaction

A new carbon–carbon single bond can be formed when a halogenoalkane or an acid chloride reacts with an arene, such as methylbenzene. For example:

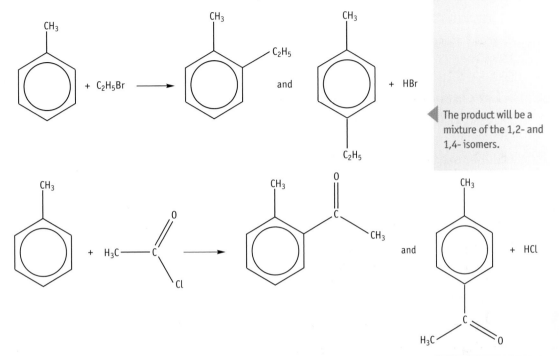

$$C_6H_5CH_3 + C_2H_5Br \longrightarrow \text{1-}CH_3\text{-2-}C_2H_5C_6H_4 \text{ and } \text{1-}CH_3\text{-4-}C_2H_5C_6H_4 + HBr$$

The product will be a mixture of the 1,2- and 1,4- isomers.

$$C_6H_5CH_3 + H_3C{-}C(=O){-}Cl \longrightarrow \text{1-}CH_3\text{-2-}(COCH_3)C_6H_4 \text{ and } \text{1-}CH_3\text{-4-}(COCH_3)C_6H_4 + HCl$$

A catalyst of anhydrous aluminium chloride is needed and the reagents must be dry.

Decrease in carbon-chain length

Iodoform reaction

Iodine and sodium hydroxide react with alcohols with a $CH_3CH(OH)$ group and carbonyl compounds with a $CH_3C{=}O$ group to form iodoform, CHI_3, and the salt of a carboxylic acid with one less carbon atom than the alcohol.

Propan-2-ol and propanone each give sodium ethanoate. The general reactions are:

$$\mathbf{RCH(OH)CH_3 \xrightarrow{I_2\ +\ NaOH(aq)} RCOONa}$$

$$\mathbf{RCOCH_3 \xrightarrow{I_2\ +\ NaOH(aq)} RCOONa}$$

The carboxylic acid can be formed from the salt by adding aqueous sulfuric or hydrochloric acid:

$$\mathbf{RCOONa + HCl \rightarrow RCOOH + NaCl}$$

Aromatic ketones, such as $C_6H_5COCH_3$, also undergo the iodoform reaction.

Other common transformations

Alkene to halogenoalkane

$$\mathbf{CH_2{=}CH_2 + HX \rightarrow CH_3CH_2X}$$

e Knowing these transformations will not only be useful in answering questions in synthesis. It is good revision of the earlier organic topics that may be tested in synoptic questions.

Reagent: hydrogen halide, for example hydrogen bromide
Conditions: mix gases at room temperature

Alkene to halogenoalcohol

$$CH_2{=}CH_2 + X_2 + H_2O \rightarrow CH_2XCH_2OH + HX$$

Reagent: aqueous halogen, for example bromine water
Conditions: pass gas through halogen in aqueous solution

Halogenoalkane to alcohol

$$RX + NaOH \rightarrow ROH + NaX$$

Reagent: aqueous sodium hydroxide
Conditions: heat under reflux in aqueous solution

Halogenoalkane to amine

$$RX + 2NH_3 \rightarrow RNH_2 + NH_4X$$

Reagent: ammonia
Conditions: concentrated aqueous solution of ammonia at room temperature or heat in a sealed tube.

Alcohol to halogenoalkane

$$ROH + PCl_5 \rightarrow RCl + HCl + POCl_3$$

Reagent: phosphorus pentachloride, PCl_5, for chlorination; KBr and 50% sulfuric acid for bromination; moist red phosphorus and iodine for iodination
Conditions: mix reagents at room temperature

Primary alcohol to aldehyde

$$RCH_2OH + [O] \rightarrow RCHO + H_2O$$

Reagent: acidified potassium dichromate(VI)
Conditions: add oxidising agent to hot alcohol and allow the aldehyde to distil off as it is formed

Primary alcohol to carboxylic acid

$$RCH_2OH + 2[O] \rightarrow RCOOH + H_2O$$

Reagent: acidified potassium dichromate(VI)
Conditions: heat under reflux

Secondary alcohol to ketone

$$RCH(OH)R' + [O] \rightarrow RCOR' + H_2O$$

Reagent: acidified potassium dichromate(VI)
Conditions: heat under reflux

Carboxylic acid to acid chloride

$$RCOOH + PCl_5 \rightarrow RCOCl + HCl + POCl_3$$

Reagent: phosphorus pentachloride
Conditions: mix dry reagents

Acid chloride to amide

$$RCOCl + 2NH_3 \rightarrow RCONH_2 + NH_4Cl$$

Reagent: aqueous ammonia
Conditions: mix at room temperature

Summary

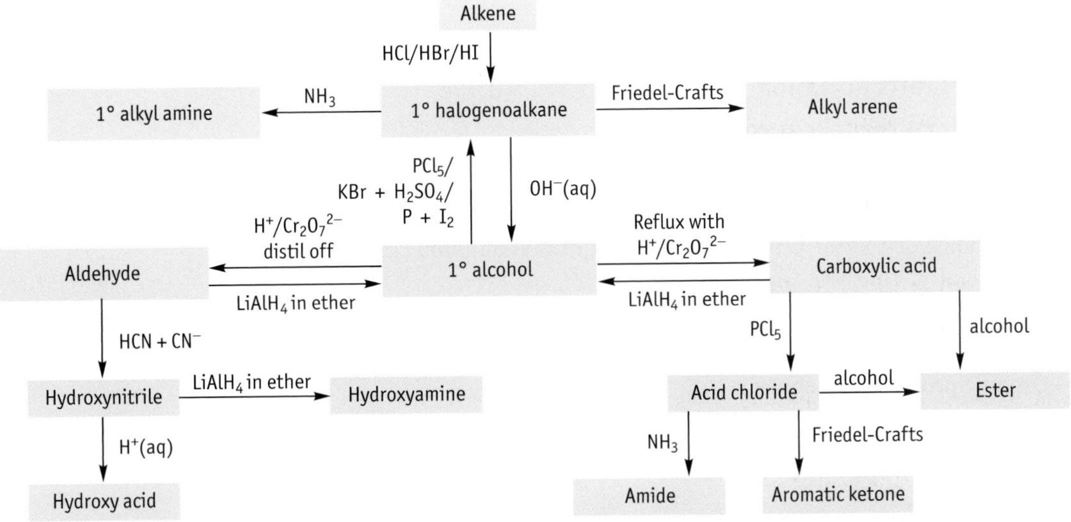

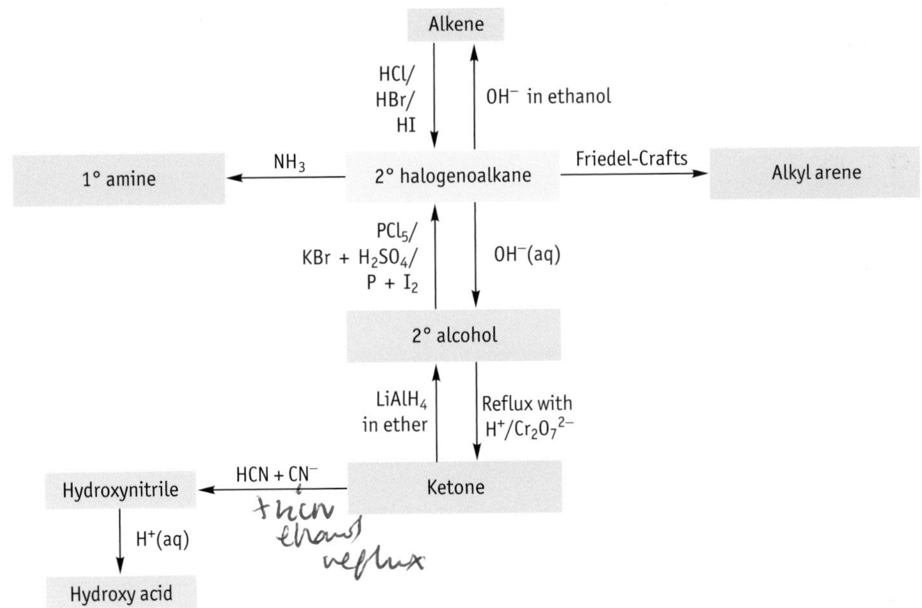

Reactions of arenes

Nitration

$$C_6H_6 + HNO_3 \rightarrow C_6H_5NO_2 + H_2O$$

Organic product: nitrobenzene
Reagent: concentrated nitric acid
Conditions: mix with concentrated sulfuric acid at 60°C

Friedel–Crafts alkylation

$$C_6H_6 + C_2H_5Br \rightarrow C_6H_5C_2H_5 + HBr$$

Organic product: ethylbenzene
Reagent: bromoethane
Conditions: dry, in the presence of an anhydrous aluminium chloride catalyst

Friedel–Crafts acylation

$$C_6H_6 + CH_3COCl \rightarrow C_6H_5COCH_3 + HCl$$

Organic product: phenylethanone
Reagent: ethanoyl chloride
Conditions: dry, in the presence of an anhydrous aluminium chloride catalyst

Halogenation

$$C_6H_6 + Br_2 \rightarrow C_6H_5Br + HBr$$

Organic product: bromobenzene
Reagent: liquid bromine
Conditions: dry with iron to form $FeBr_3$, which is the catalyst

Sulfonation

$$C_6H_6 + H_2SO_4 \rightarrow C_6H_5SO_3H + H_2O$$

Organic product: benzenesulfonic acid
Reagent: fuming sulfuric acid (concentrated plus some dissolved SO_3)
Conditions: heat under reflux

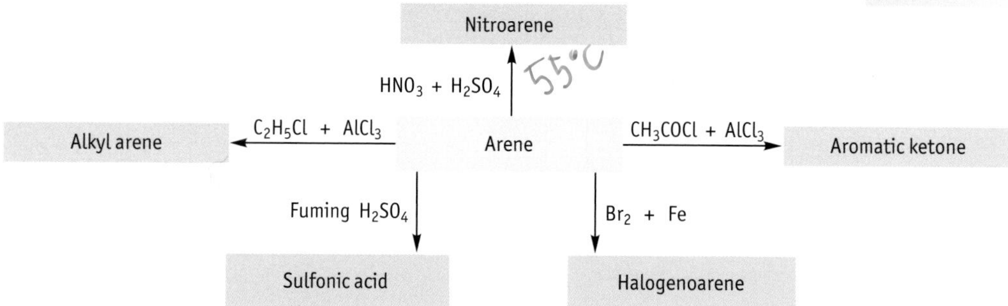

Reactions of compounds with two functional groups

At A-level, the assumption is made that functional groups behave independently. The exception to this is amino acids (see p. 274), where a zwitterion is formed, so there is no $-NH_2$ group to act as a nucleophile in the reaction with an acid chloride.

The assumption means that a compound such as 2-bromoethanol, CH_2BrCH_2OH, can be hydrolysed by aqueous alkali to a diol and can be oxidised to a halogenoacid:

$$CH_2OHCH_2OH \xleftarrow[\text{with } OH^-]{\text{heat under reflux}} CH_2BrCH_2OH \xrightarrow[\text{with } Cr_2O_7^{2-}/H^+]{\text{heat under reflux}} CH_2BrCOOH$$

Benzocaine:

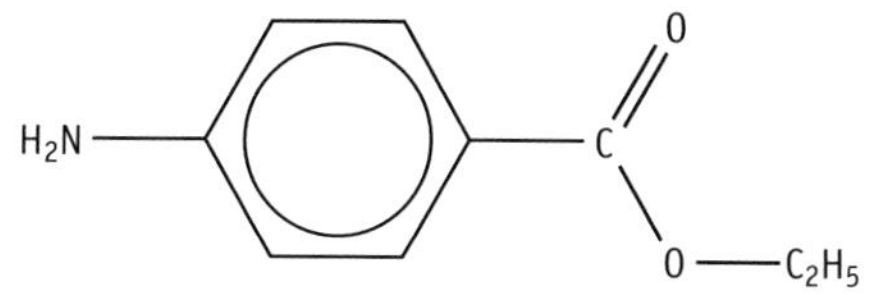

is used as a local anaesthetic. It contains an aromatic amine group and an ester group and so reacts in a similar way to both phenylamine and to ethyl benzoate.

Worked example

Write the formulae of the organic products of the reaction of benzocaine with:

a bromine water

b dilute hydrochloric acid at room temperature

c hot aqueous sodium hydroxide

Answer

a

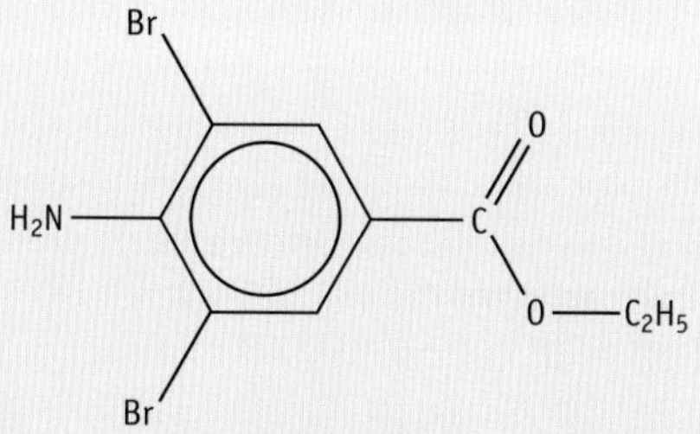

b

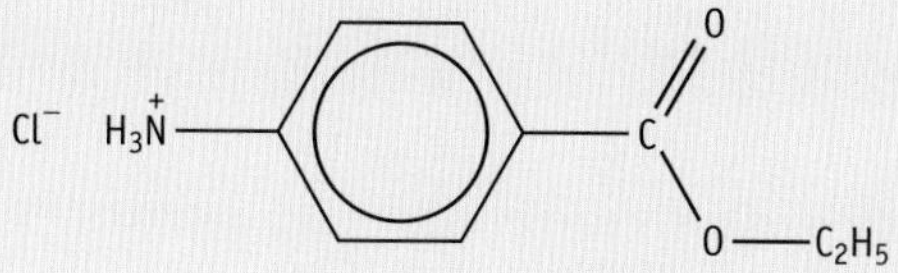

c

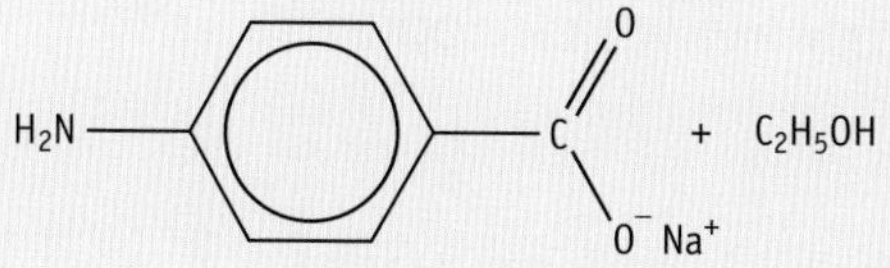

Multi-step syntheses

There are two distinct types of question on this topic.

The first type is when you are asked for an outline of a multi-step synthesis. In this case, what is required is a route that shows all the intermediate compounds and the reagents and conditions for each step. There are several ways of tackling this type of question:

- look to see if the carbon chain length has been increased or decreased
- start from the final product and work towards the starting substance
- start from the starting substance and think of the types of reaction that it can undergo

Worked example 1

Deduce the necessary reagents and conditions and the intermediates in the conversion of benzene into the following compound:

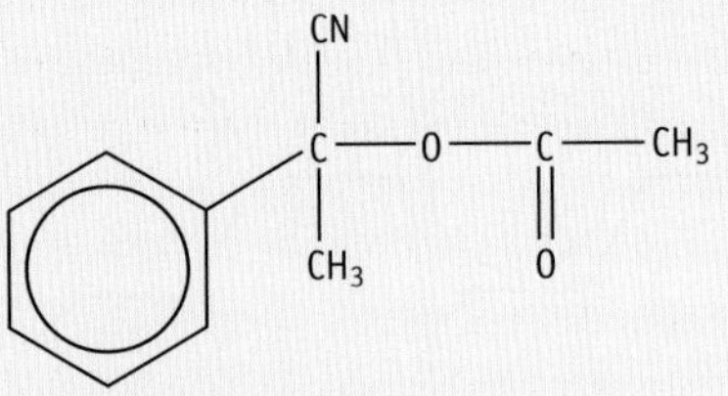

CN
C—O—C—CH$_3$
CH$_3$ O

Answer

Working backwards:

The final product is an ester of ethanoic acid and so can be made by the reaction between an alcohol and ethanoic acid or ethanoyl chloride.

The –CN and –OH groups on the same carbon atom come from the reaction of a ketone with HCN.

Working forwards:

A side chain has to be introduced into the benzene ring. This implies a Friedel–Crafts reaction. Since a ketone is required, the reaction is between benzene and ethanoyl chloride.

Step 1: reaction between benzene and ethanoyl chloride in the presence of anhydrous $AlCl_3$

$$C_6H_6 + CH_3COCl \rightarrow C_6H_5COCH_3 + HCl$$

Step 2: requires pH 8

$$C_6H_5COCH_3 + HCN \rightarrow C_6H_5C(OH)(CN)CH_3$$

Step 3:

$C_6H_5C(OH)(CN)CH_3 + CH_3COCl \rightarrow$

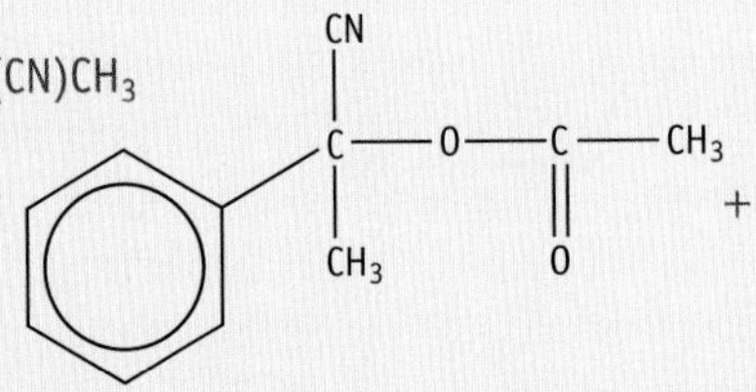

CN
C—O—C—CH$_3$ + HCl
CH$_3$ O

Worked example 2

Deduce the necessary reagents and conditions and the intermediates in the conversion of ethene into 2-aminoethanoic acid (glycine), NH_2CH_2COOH.

Answer

Working backwards:

An acid group can be made by oxidising a primary alcohol. An amine group can be made from a halogenoalkane, so a compound with a –Br and a –CH_2OH group is likely to be an intermediate.

Working forwards:

An alkene will react with a halogen to form a dihalogenoalkane or with a halogen dissolved in water to form a halogenoalcohol. Therefore, the first step is to bubble ethene into bromine water:

$$CH_2{=}CH_2 + H_2O + Br_2 \rightarrow BrCH_2CH_2OH + HBr$$

The second step is to react the product with excess concentrated ammonia:

$$BrCH_2CH_2OH + 2NH_3 \rightarrow NH_2CH_2CH_2OH + NH_4Br$$

Finally, the aminoalcohol is oxidised by heating under reflux with acidified potassium dichromate(VI) solution.

$$NH_2CH_2CH_2OH + 2[O] \rightarrow NH_2CH_2COOH + H_2O$$

The second type of question is where an intermediate has to be isolated before a subsequent step. The answer will include the reactants and conditions for each step, how the intermediate is isolated and the final product is purified.

Worked example

Describe how benzoic acid, C_6H_5COOH, can be converted in a two-step synthesis into methyl 3-nitrobenzoate.

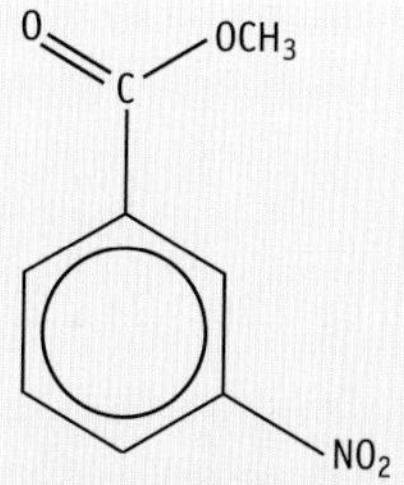

Show how the intermediate is isolated and a pure sample of the product obtained.

O, OCH_3, C, NO_2

Answer

Step 1 is the conversion of benzoic acid into the ester, methyl benzoate $C_6H_5COOCH_3$:

- Reflux together 10 g of benzoic acid, 20 cm^3 of methanol and 3 cm^3 of concentrated sulfuric acid for about 30 minutes.
- Allow the mixture to cool. Transfer it to a large separating funnel and add 40 cm^3 of water and 20 cm^3 of hexane.
- Shake the funnel, releasing the pressure from time to time.
- Allow the two layers to settle. Run off and discard the lower aqueous layer.
- Wash the organic layer with about 20 cm^3 of dilute aqueous sodium carbonate, releasing the pressure created by the production of carbon dioxide by the reaction of any remaining acid with the sodium carbonate.
- Discard the lower aqueous layer and wash the organic layer with about 20 cm^3 water. This removes salts and any excess methanol.
- Discard the lower aqueous layer and add some anhydrous calcium chloride to the organic layer.
- When the liquid clears, decant it into a flask and carefully distil off all the hexane (boiling temperature 69°C) leaving the methyl benzoate (boiling temperature 196°C) in the flask.

Step 2 is the nitration of methyl benzoate:

- Carefully add 10 cm^3 of concentrated sulfuric acid to the methyl benzoate and cool in an icebath to 5°C. Then add 10 cm^3 of a mixture of equal parts concentrated nitric and concentrated sulfuric acids dropwise, making sure that the temperature of the reaction mixture does not rise above 15°C.
- Leave the reaction mixture at room temperature for at least 20 minutes.
- Pour the reaction mixture onto some crushed ice and stir. Methyl 3-nitrobenzene is a solid and will crystallise out.
- Once the ice has melted, filter the mixture using a Buchner funnel.
- Collect the solid methyl 3-nitrobenzene and recrystallise it using the minimum amount of hot ethanol as the solvent.

e This would be a suitable experiment for assessment of practical skills. It requires four 1-hour periods.

Yields

Organic reactions rarely go to completion and there are usually competing side reactions. The yield is often as low as 60%. Purification processes also lower the yield. For instance, some solid will remain in solution during recrystallisation and a distillation apparatus will retain some vapour, so not all will be turned to liquid in the condenser.

If a synthesis requires three steps, each of which has a yield of 70%, the yield will be $0.7 \times 0.7 \times 0.7 = 0.34 = 34\%$. If the purification of the product is 80% efficient, the final yield will be $0.80 \times 34 = 27\%$.

Worked example

Benzene (molar mass $78.0\,g\,mol^{-1}$) can be converted into benzoic acid (molar mass $122.0\,g\,mol^{-1}$), which then has to be recrystallised. If 9.3 g of pure benzoic acid was made from 10.0 g of benzene, calculate the percentage yield.

Answer

$$\text{amount of benzene used} = \frac{\text{mass}}{\text{molar mass}} = \frac{10.0\,g}{78.0\,g\,mol^{-1}} = 0.128\,mol$$

theoretical yield of benzoic acid $= 0.128\,mol \times 122.0\,g\,mol^{-1} = 15.6\,g$

actual yield of pure benzoic acid $= 9.3\,g$

$$\text{percentage yield} = \frac{9.3 \times 100}{15.6} = 59.6 = 60\%\ \text{(to 2 s.f.)}$$

e A common error is to work out the yield as mass of product × 100 divided by the mass of reactant. Making this mistake in this example would give the incorrect answer of 93%.

Stereochemistry of reactions

Many drugs are stereospecific. This means that only one stereoisomer has a therapeutic effect. If possible, a reaction pathway that produces only the desired stereoisomer is used because this increases the atom economy of the manufacture.

Ibuprofen has a chiral carbon atom (marked with * on the formula below):

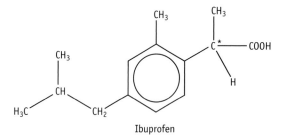

CH_3 CH_3 C* COOH CH_3 H CH H_3C CH_2

Ibuprofen

Only one isomer is effective as a pain-reliever and as an anti-inflammatory drug. The chemical reactions that produce it are non-stereospecific and so the drug is sold as the racemic mixture. In the body, an **isomerase** enzyme slowly converts the inert enantiomer into the active form. Therefore, there is no point in attempting to devise a stereospecific manufacturing process.

In the 1950s, thalidomide was used as a drug to relieve morning sickness in pregnant women. The molecule contains a chiral carbon atom (marked with * on the formula below):

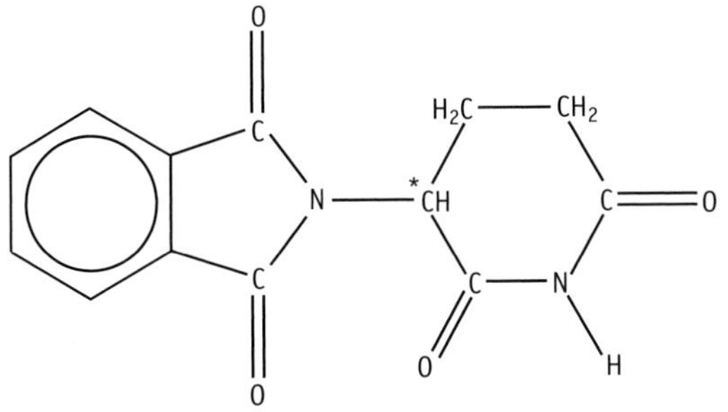

Thalidomide

It was later found that while one optical isomer relieved morning sickness the other isomer caused horrific birth defects. Unfortunately, the body spontaneously converts one optical isomer into the other. This means that even if a method were developed to produce the beneficial isomer only, the drug would still not be suitable for pregnant women. It is used today to relieve vomiting in patients undergoing chemotherapy.

Some reactions are stereospecific. Knowledge of the mechanism of a reaction allows chemists to predict whether the product will be a single isomer or a 50:50 mixture of the two stereoisomers.

Addition across a C=C bond

When bromine adds on across a double bond, the addition is *trans*. The reaction of bromine with cyclopentene gives the *trans* isomer only:

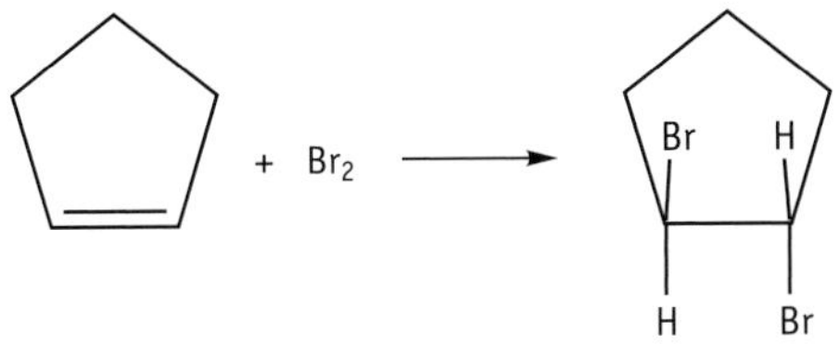

$$\text{cyclopentene} + Br_2 \longrightarrow \text{trans-1,2-dibromocyclopentane}$$

The first step is the addition of Br^+ to one of the carbon atoms. In the second step, the bromine atom in the carbocation intermediate sterically hinders the approach of the Br^- ion. It must therefore attack from the other side and so the *trans* geometric isomer is produced.

Hydrogen, in the presence of a platinum catalyst, adds on the same side. This is called *cis* addition. This is because the alkene is held onto the catalyst surface and hydrogen diffuses through the metal and adds across the C=C bond.

Reactions that take place on the surface of a heterogeneous catalyst are often stereospecific, as are most biochemical reactions.

If propene is polymerised using a Ziegler–Natta catalyst ($TiCl_4$ and $(C_2H_5)_3Al$), the addition is also *cis* and an isotactic polymer is produced in which all the $-CH_3$ groups are on the same side of the carbon chain.

Substitution on a carbon attached to a halogen

In a S_N2 mechanism, the nucleophile attacks from the opposite side to the halogen and so a single optical isomer with an *inverted* configuration is produced.

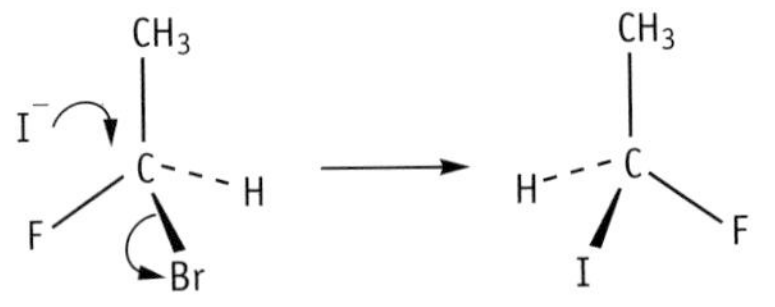

S-1,1-bromofluoroethane → R-1,1-fluoroiodoethane

Priority: $H < CH_3 < F < Br$ (or I)

The designation *R-* is given to the enantiomer that has the four groups arranged clockwise in order of priority, with the lowest priority (usually hydrogen) at the back. If the groups are arranged anticlockwise in order of priority, the enantiomer is designated *S-*. The priority is worked out in the same way as with *E-* and *Z-* isomers (see p. 123).

If the mechanism were S_N1, the product would be a racemic mixture. This is because the intermediate carbocation is planar at the reaction site and so the nucleophile can attack from either above or below.

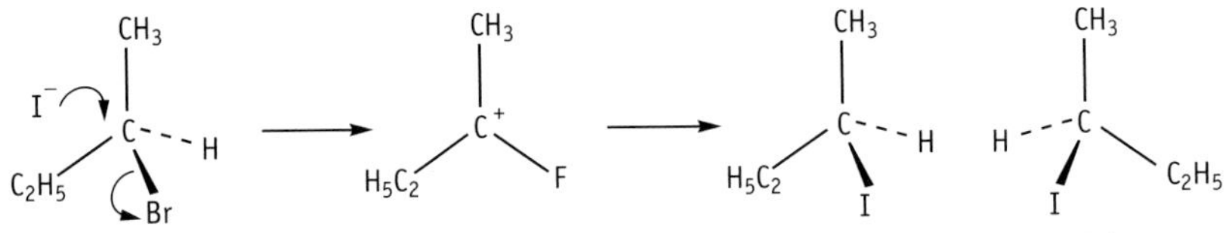

S-2-bromobutane → S-2-iodobutane + R-2-iodobutane

Priority: $H < CH_3 < C_2H_5 < Br$ (or I)

Stereoselective enzymes

Many enzymes catalyse a reaction of a non-chiral molecule to form one particular chiral product. A reductase enzyme catalyses the reduction of a ketone to just one enantiomer of the corresponding secondary alcohol, for example:

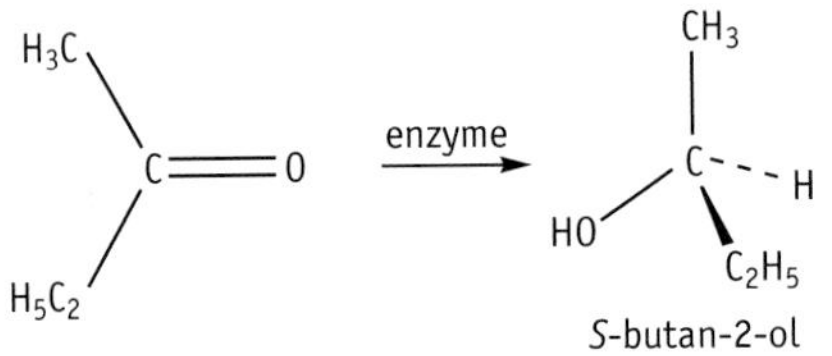

$H_3C(H_5C_2)C{=}O$ —enzyme→ S-butan-2-ol

Such stereoselective enzymes are particularly important in the pharmaceutical industry.

Safety issues

It is assumed in A-level questions that the following safety precautions are always taken:

- Safety glasses and laboratory coats are worn at all times in all practical classes.
- Hands are washed before leaving the laboratory, particularly if toxic substances have been used.
- Food and drink are never consumed in the laboratory.

e Since these safety precautions are assumed, they are never awarded marks.

Special care must be taken with certain chemicals:

- Almost all organic chemicals are flammable. Flasks and test tubes containing them should not be heated directly with a Bunsen flame. A water bath or an electric heater should be used.
- If a chemical is harmful and can be absorbed through the skin, gloves must be worn.
- If a chemical has a harmful, irritating or poisonous vapour, the experiment must be carried out in a fume cupboard.
- If a chemical is corrosive, harmful, irritating or poisonous (toxic), gloves must be worn and extra care must be taken.
- Any substance suspected of being a carcinogen must not be used in a school laboratory.

Hazard symbols

GARRY WATSON/SPL

ⓔ You are expected to be able to relate safety aspects to the specific hazards of the reaction or chemicals being handled.

Worked example

When methyl 3-nitrobenzoate is recrystallised, the solvent used is ethanol. This is a flammable liquid with a boiling temperature of 78°C. What specific safety precautions should you take when dissolving methyl 3-nitrobenzoate in the minimum amount of hot ethanol and filtering the solution?

Answer

The solid should be placed in a round-bottomed flask fitted with a reflux condenser. Some ethanol is added and the flask warmed until the ethanol is just boiling. Heating should be carried out using either an electric heater or a water bath. Extra ethanol is added down the condenser until all the solid dissolves. The heater is removed, followed by the condenser, and the hot solution is filtered using a warm stemless glass funnel.

Danger can be avoided by careful reading of hazard labels and appropriate precautions taken. The risk can be further reduced by using small quantities. If there is still doubt as to the safety of the experiment, it should be abandoned or replaced using safer materials.

Combinatorial chemistry and pharmaceuticals

The pharmaceutical industry maintains a huge library of compounds that it can test for activity as potential drugs to treat medical conditions. A modern technique is to use the principles of **combinatorial chemistry**.

The Nobel prizewinner Bruce Merrifield first used this technique to synthesise peptides. Peptides are chains of amino acids with a –CONH– group between every two amino acid residues. A dipeptide formed from glycine, (NH_2CH_2COOH) and alanine ($NH_2CH(CH_3)COOH$) is shown below:

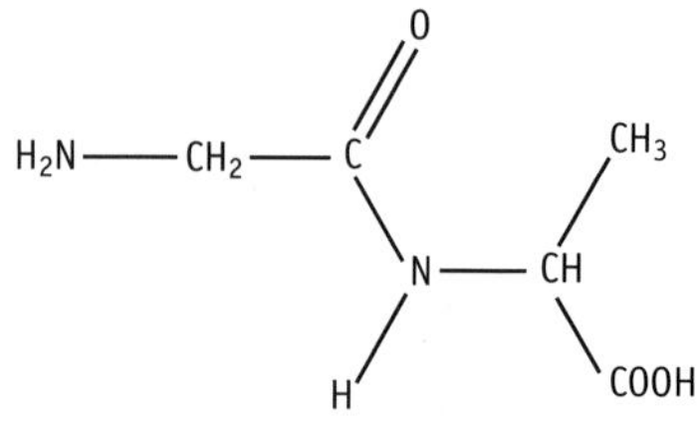

$$H_2N{-}CH_2{-}C(=O){-}N(H){-}CH(CH_3){-}COOH$$

A dipeptide

The synthesis is carried out on the surface of resin beads a few micrometers in diameter.

The principle of combinatorial chemistry can be illustrated by imagining forming a set of polypetides made up of just three amino acids, symbolised by the letters A, B and C. The resin bead is symbolised by the letter R.

Step 1: The amino acids are mixed with the resin and three products are obtained:

R–A R–B R–C

Step 2: The mix is then separated into three portions, each containing all three of the products from step 1. Amino acid A is added to one portion, amino acid B to the second and amino acid C to the third:

R–A R–B R–C	R–A R–B R–C	R–A R–B R–C
↓ + A	↓ + B	↓ + C
R–A–A R–B–A R–C–A	R–A–B R–B–B R–C–B	R–A–C R–B–C R–C–C

This gives nine dipeptides.

Step 3: These nine dipeptides are mixed and split into three portions, each containing all nine dipeptides. Amino acid A is added to one portion, amino acid B to the second and amino acid C to the third portion. The addition of amino acid A to the first portion results in the production of nine different tripeptides:

R–A–A–A	R–B–A–A	R–C–A–A
R–A–B–A	R–B–B–A	R–C–B–A
R–A–C–A	R–B–C–A	R–C–C–A.

Altogether, the three portions give rise to 27 tripeptides.

A further step will produce 81 tripeptides. The number of products is given by the formula N^x, where N is the number of amino acids and x is the number of steps.

At the end of the process, the polypeptides are removed from the resin beads and separated by chromatography. The process can be summarised as:

mix → split → react → mix → split → react and so on…

Because polypeptides are hydrolysed in the digestive system, they are unlikely to be useful as drugs. However, a similar combinatorial technique can be used to create a library of molecules that may be possible pharmaceuticals.

The first step is to identify:

- groups that may bind to either an enzyme or to a particular DNA, preventing its normal activity
- groups that will react with transmitter molecules, such as dopamine.

This is followed by the preparation of a variety of molecules containing these groups. The pharmaceutical activity of these compounds is then tested in vitro. Any substances that appear to work are then tested on animals.

Cocaine is an aromatic tertiary amino ester — the molecule contains a benzene ring, an ester group and a tertiary amine. It was used initially as a local anaesthetic, but it has harmful side effects. Research led to novocaine, which is also an aromatic tertiary amino ester, is a better anaesthetic and does not have side effects.

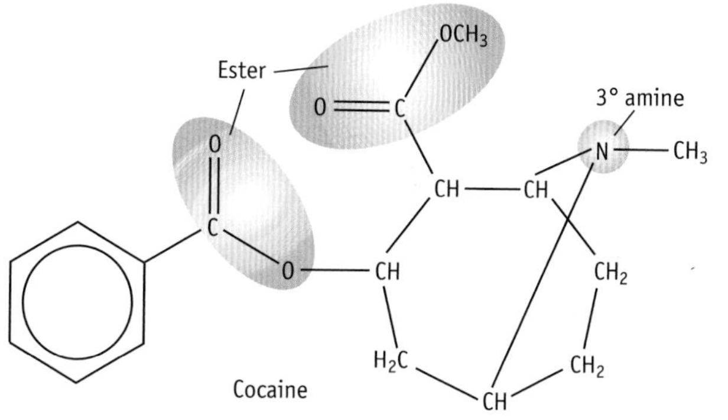

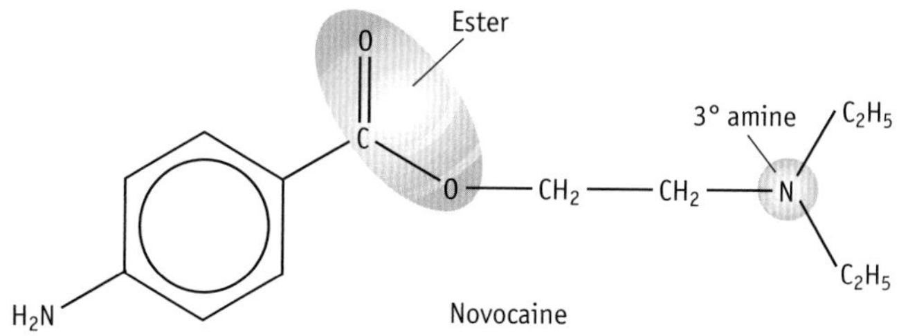

Cocaine

Novocaine

Cocaine remains in use as an anaesthetic for eye operations; novocaine is sometimes used as a local anaesthetic in dentistry, but it has been replaced mostly by lidocaine, which is an aromatic tertiary amino amide:

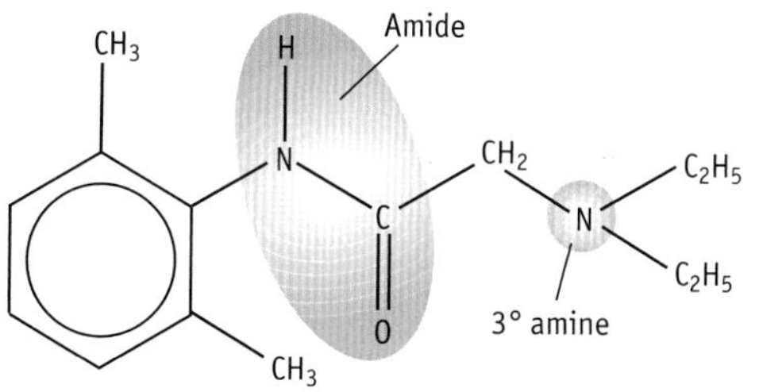

Amino esters and amino amides work by blocking specific nerve pathways and so prevent the transmission of pain signals to the brain.

Questions

1 An organic compound X is thought to contain chlorine and iodine bonded to carbon atoms. Describe the tests that you would carry out to prove this.

2 Linalool is the major component of lavender oil. Its formula is

$(CH_3)_2C{=}CHCH_2CH_2C(CH_3)(OH)CH{=}CH_2$

Describe the tests that would prove the presence of the functional groups in this molecule.

3 Spearmint oil contains the compound carvone. Use the results of the following tests that were carried out on carvone to deduce the functional groups present in the molecule. Justify your deductions from each test.

a It turned brown bromine water colourless.

b On warming with acidified potassium dichromate(VI), the solution remained orange.

c It gave a yellow precipitate with Brady's reagent.

4 Tests on compound Y, $C_3H_4O_3$, gave the following results:

a When phosphorus pentachloride was added, steamy fumes that turned damp litmus red were observed.

b When a solution of sodium carbonate was added, a gas was evolved that turned limewater cloudy.

c A yellow precipitate was obtained when a solution of 2,4-dinitrophenylhydrazine was added.

d Orange acidified potassium dichromate(VI) solution turned green when warmed with Y.

e Y gave a negative result in the iodoform test.

Deduce the structural formula of compound Y.

5 This question concerns citral, $C_{10}H_{16}O$ (found in lemon grass oil) and geraniol, $C_{10}H_{18}O$ (found in rose oil).

a Describe a test to show that citral is an aldehyde.

b Describe a test to show that geraniol is an alcohol.

c Describe a test to show that both compounds contain a C=C group.

d In what way would the infrared spectra of citral and geraniol differ?

e Outline how citral could be converted to geraniol.

6 Compound Z has the following composition by mass: carbon, 55.8%; hydrogen, 7.0%; oxygen, 37.2%.

a Calculate the empirical formula of compound Z.

b Use the information below to deduce the structural formula of compound Z:

- The largest m/e value in the mass spectrum of compound Z was 86.
- Bromine water remained brown on the addition of compound Z.
- No steamy fumes formed when phosphorus pentachloride was added to compound Z.
- Addition of 2,4-dinitrophenylhyd razine to compound Z gave a yellow precipitate.
- When Tollens' reagent was added to compound Z and the mixture warmed, a silver mirror formed.
- When warmed with a solution of iodine in sodium hydroxide, compound Z gave a yellow precipitate.

7 An ester E has the molecular formula $C_7H_{14}O_2$. When heated under reflux with aqueous sodium hydroxide, followed by the addition of excess acid, it was converted into two compounds F and G:

- Compound F has the following composition by mass: carbon, 54.5%; oxygen, 36.4%; hydrogen, 9.1%.
- Compound F gives off carbon dioxide when added to a solution of sodium hydrogencarbonate.
- Compound F has three peaks in its NMR spectrum.
- Compound G does not give a precipitate with iodine and sodium hydroxide.

Deduce the structural formula of the ester E and write the equation for its reaction with sodium hydroxide.

8 When phenylamine reacts with ethanoyl chloride, a solid is formed that is soluble in hot water but insoluble in cold water. Describe how you could purify this solid.

9 Cyclohexanol (boiling point 161 °C) can be converted into cyclohexanone (boiling point 156 °C).

a State the reagents and conditions for this reaction.

b State how you would obtain a pure sample of cyclohexanone from a mixture containing it and 2% cyclohexanol.

c State how you would obtain a pure sample of cyclohexanone from a mixture containing it and 25% cyclohexanol.

10 Before cyclohexanone and cyclohexanol (produced as in question 9) can be separated, the organic substances, which are insoluble in water, have to be extracted from the reaction mixture.

a Describe how this would be achieved.

b How are traces of water removed from the organic substances?

c How would you know when the water removal was complete?

11 An organic compound Z contains carbon, hydrogen and oxygen only. When 4.85 g of Z was burnt in excess oxygen 11.9 g of carbon dioxide and 4.85 g of water were produced.

The mass spectrum of Z had a peak due to its molecular ion at a mass/charge ratio of 72. Compound Z gave a yellow precipitate with Brady's solution but no precipitate with iodine and sodium hydroxide solution. Its NMR spectrum had four peaks in the ratio 3:2:2:1.

Showing all your working and using all the data provided, deduce the structural formula of Z.

12 a Give the reagents, conditions and intermediates in the conversion of but-2-ene to 3-aminobutan-2-one, $CH_3COCH(NH_2)CH_3$.

b Explain what effect the product would have on the plane of polarisation of plane-polarised light.

13 Write the structural formulae of the organic products of the reaction of

$$H_3C—\underset{\underset{O}{\|}}{C}—CH(OH)—COOH$$

with the following (you may assume that the functional groups behave independently):

a phosphorus pentachloride

b sodium hydrogencarbonate

c iodine dissolved in sodium hydroxide

d 2,4-dinitrophenylhydrazine

e lithium aluminium hydride, $LiAlH_4$, in ether followed by hydrolysis of the product.

14 Give the reagents, conditions and intermediates in the conversion of ethanol to ethanoyl chloride.

15 A compound $C_4H_8O_2$ was tested as follows:

- It did not give a precipitate with Brady's reagent.
- It turned bromine water from brown to colourless.
- When excess PCl_5 was added, 2 mol of HCl was given off per mol of the organic compound.
- The compound had three peaks in its NMR spectrum and a peak at $m/e = 57$ in its mass spectrum,

Suggest a formula for the compound. Justify your answer

16 Outline how 1-phenylethanol, $C_6H_5CH(OH)CH_3$, could be prepared in a two-step synthesis from benzene and one other organic reagent.

17 A student wished to convert propan-1-ol into 1-bromopropane. Two different methods were suggested:

(i) Heat the alcohol under reflux with a mixture of solid potassium bromide and concentrated sulfuric acid

(ii) Heat the alcohol under reflux with a mixture of solid potassium bromide and concentrated phosphoric acid.

Both methods initially produce gaseous hydrogen bromide. This then reacts with the alcohol to form the bromoalkane which is distilled off and condensed.

The hazards were identified as:

- concentrated sulfuric acid: corrosive and an oxidising agent
- concentrated phosphoric acid: corrosive
- hydrogen bromide vapour: irritant and a reducing agent
- propan-1-ol and 1-bromopropane: volatile and flammable
- bromine: volatile and irritant vapour

a Explain which method should be used to minimise the risk.

b For the method selected in a, suggest suitable special safety precautions that should be taken to minimise the risk

18 a Outline how you would convert $C_6H_5CH{=}CHCH_3$ into the following compound, M:

$$H_5C_6-\underset{\underset{O}{\|}}{C}-\underset{\underset{O}{\|}}{C}-CH_3$$

b Identify the product formed by reacting compound M with iodine and sodium hydroxide.

19 12.6 g of benzene was nitrated and the nitrobenzene produced was reduced with tin and concentrated hydrochloric acid. After addition of sodium hydroxide, the phenylamine was steam-distilled out of the reaction mixture, washed, dried and redistilled. The mass of pure phenylamine produced was 6.75 g. Calculate the percentage yield.

20 It was required to prepare the compound $(CH_3)_2C(OH)CONH_2$ from propanone. A student suggested the following outline synthesis:

$$(CH_3)_2C{=}O \xrightarrow{HCN} (CH_3)_2C(OH)CN$$
$$\downarrow H_2SO_4(aq)$$
$$(CH_3)_2C(OH)COOH$$
$$\downarrow PCl_5$$
$$(CH_3)_2C(OH)COCl$$
$$\downarrow NH_3$$
$$(CH_3)_2C(OH)CONH_2$$

Explain why this sequence would *not* give the required product.

21 Explain why the addition of hydrogen cyanide, HCN, to butanone, $CH_3COC_2H_5$, gives a racemic mixture, rather than a single optical isomer.

22 The rate equation for the nucleophilic substitution reaction of a single optical isomer of the halogenoalkane, 1-fluoro-1-iodoethane, CH_3CHFI, with cyanide ions is:

$$\text{rate} = k[CH_3CHFI][CN^-]$$

a Use this information to explain whether the product would be a single optical isomer or the racemic mixture.

b Suggest why the product is 2-fluoropropanenitrile, not 2-iodopropanenitrile.

23 This question is about the combinatorial reaction between two hydroxyacid chlorides, HORCOCl and HOR′COCl. Each was separately adsorbed onto resin beads through the –OH group. The results were mixed and divided into two portions. One sample was then reacted with HORCOCl and the other with HOR′COCl. The results of these reactions were mixed and divided again; one portion was reacted with HORCOCl and the other with HOR′COCl. This was repeated twice more. How many different polyester compounds were produced?

Index

A

absorption, colours from 218–20
absorption spectra 164–5
accuracy 204–5, 281
acetone *see* propanone
acid–base conjugate pairs 88–9, 104
acid–base equilibria 88–117
acid–base indicators 109–10, 114
acid buffer solutions 104
acid chlorides 144, 157–60
  in acylation reactions 246–7
  amides from 266, 298–9
  amine reactions with 260–1
  chain length increase with 297
  esters from 152, 159
  formation 150, 158–9, 298
  phenol reactions with 252, 254
  phenylamine reactions with 264
  polyamides from 267, 268–9
  polymers from 156
acid dissociation constant (K_a) 90–1, 98, 100, 102
  buffer solutions 105–8, 113–14
  carboxylic acids 146
acids 88, 90
  acid–base indicators 109–10
  amine reactions with 260
  amino acid reactions with 274
  benzoic acid 254–5
  in buffer solutions 104–9
  carbonate reactions with 6–7
  carboxylic acids as 148
  conjugate 102, 104
  K_a 90–1, 98, 100, 102
    titrations for 113–14
  neutralisation 115–16
  phenol 252–3
  phenylamine reaction with 263
  pH of 94–5, 98–102, 104
  salts 103
  and sodium thiosulfate decomposition 7–8
  structure of 116–17
  titrations 96–7, 111, 112–13, 114
    K_b from 114
activation energy 2–3, 24
  and rate constants 15, 16–17
  and rate of equilibrium 76
  and reaction feasibility 191, 192, 228
  and spontaneous change 36, 37
active sites 78, 82, 217
actuality of reaction 191–2
acylation 246–7, 300
acyl chlorides 157
  see also acid chlorides
addition
  alkenes, onto benzene rings 247–8
  free-radical 242
  nucleophilic 134–7, 277, 278
    increase chain length with 296
addition/elimination 134, 137–9, 144
addition polymerisation 270–1
adenine 271, 272
adrenaline 272
air, gases in, and pressure 61
alanine (2-aminopropanoic acid) 273
  optical isomers 274–5
alcoholic drinks and breathalysers 198–9
alcohols
  acid chloride reactions with 159
  aldehydes from 298
  aromatic, phenol 240, 249–54
  carboxylic acids from 298
  ester reactions with 152–3
  esters from 151, 152
  formation 137, 150, 298
  halogenoalkanes from 298
  halogenoalkanes in 298
  infrared spectra 285
  iodoform reaction with 141, 287, 297
  ketones from 298
  oxidation 133, 134, 148
  phenol 240, 249–54, 253–4
  tests for 285
aldehydes 130, 131, 134–41, 298
  distinguishing from ketones 140, 142, 286
    NMR spectra for 289–90
  hydrogen cyanide reacts with 296
  hydroxynitrile from 278
  infrared spectra 286
  laboratory preparation 133
  oxidation of 139–40, 148
  physical properties 132
  tests for 284, 285–7
aliphatic chemistry 236
alkali 4, 79
  in buffer solutions 104, 106–7
alkane side chains 248
alkenes 122, 137
  benzene reactions with 247–8
  halogenoalcohols from 298
  halogenoalkanes from 297–8
  hydrogenation, catalysis of 217
  synthesis with 302
  test for 284
alkylation 245–6, 300
alloys 201, 231, 232
alpha-helix 277
aluminium chloride catalysis 245–6, 281–2
amides 144, 266–71
  acid chloride reactions give 266, 298–9
    ammonia 159
    phenylamine reactions 264
  reduction, amines from 259–60
amine groups 271–2, 300
amines 258–62
  halogenoalkanes from 298
  hydrated hydroxides precipitated with 261
  as ligands 215, 225–6, 261
  polyamides from 267, 268–9
  secondary, adrenaline 272
amino acids 272–6, 300
  chromatography for 275–6, 288
  and peptide synthesis 307–8
  serine, in silk 269
amino amides 309, 310
4-aminoazobenzene 266
aminobenzene 240
amino esters 309, 310
aminoethane 258
  see also ethylamine
aminoethanoic acid *see* glycine
6-aminohexanoic acid, polymerisation of 270
1-aminopropane 258
2-aminopropanoic acid *see* alanine
ammine complexes, copper(I) 226
ammines, formation 223
ammonia 89, 90, 94
  acid chloride reactions with 159, 266
  in buffer solutions 104
  copper(I) chloride reaction with 226
  copper(I) iodide reaction with 230
  halogenoalkane reactions with 226, 259
  hydroxides precipitated with 221
  as a ligand 215, 220, 225–6
    copper(I) complexes 226, 230
    exchange reactions 223
    reduction potential 228
  oxidation, catalysis 25
  reaction forming 70, 80–2
    equilibrium constant 57, 68–9, 77
  titration with hydrochloric acid 113
ammonium chloride 36, 42–3
  in buffer solutions 104
  electrolyte, in batteries 195
ammonium hydrogensulfide 67
ammonium ions, as conjugate acids 102
ammonium nitrate, dissolving 37
ammonium salts, nitriles produce 148
amphoteric substances 91, 221–2
anaesthetics 155, 309

analysis 281, 282–91
angelic acid 123–4
angiograms 164
aniline 240
anions
  carboxylate anions 139
  as conjugate bases 102
  and dissolving 38, 44
  in electrochemical cells 182–3
  and lattice energy 39–40, 41
  size, and solubility 46
  in transition metal complexes 215
anodes 182, 195
  in fuel cells 196–7, 199
anti-clockwise rule 189
antiseptics 252
aqua ions 214, 221
arenes/aromatic compounds 236–56
  aromatic acids 147
  aromatic ketones 134, 139, 297
    iodoform reaction for 297
  with carbon side chains 248–9
  naming 240–1
  reaction summary 299–300
  tests for 139, 288
  see also benzene rings
Arrhenius equation 16, 17
aspartame 276
aspartic acid 273, 276
atom economy 79, 82, 83–4
  and stereochemistry 304
atomisation, enthalpy of 211
atoms, electron configuration 208
auto-ionisation of water 91–3

B
balances, errors in 204
barium, X-rays and 164
barium hydroxide 36, 70
barium sulfate, solubility 70
base dissociation constant (K_b) 91, 102, 103
  titrations for 114
bases 88–9, 90, 94
  amines as 260
  amino acid reactions with 274
  in buffer solutions 104
  carboxylic acid reactions with 149
  conjugate 102, 104, 108–10, 114
  in deprotonation 221
  dissociation constant 91, 102, 103, 114
  neutralisation 115–16
  phenylamine as 263–4
  pH of 94, 95–6, 104
  salts of 102, 103
  titrations 96–7, 111, 112–14
batch processes 80
batteries 195–6
Benedict's solution 140, 142, 286
  as copper(II) complexes 226
benzamide hydrolysis 267
benzene 241–8, 299–300
  ibuprofen from 281–2
benzenecarboxylic acid *see* benzoic acid
benzenediazonium compounds 264–6
benzene-1,4-dicarboxylic acid 147, 156
benzene rings 236–40
  in aromatic acids 147
  in aromatic ketones 134
  in phenol 249–50
benzenesulfonic acid 247, 300
benzocaine 300
benzoic acid 106, 147, 241, 254–5
  purification method 295
  in synthesis 303
benzoyl chloride 159, 254
beta-pleated sheet 269, 277
bidentate ligands 216
biodegradability, polymer 156, 157
biofuels 155, 198
  biodiesel 153, 154–5
biological molecules
  carboxylic acids 145
  and geometric isomerism 122, 123–4
    cisplatin, and DNA 123, 215
  and optical isomerism 125, 126, 127
  proteins 272, 273, 276–7
  see also natural products
Biopol 157
blood
  buffer solutions in 104
  haemoglobin 216
boiling, entropy and 31, 32, 48
boiling points 290–1
  amines 258–9
  carbonyl compounds 132
  carboxylic acids 146
  esters 151
bonding
  in aqua ions 214
  in benzene rings 236–7, 238–40
  and infrared absorption 165–6
  metallic bonding 211
  in ozone 238–9
  in peptides 277
  transition metal compounds 212–13
  see also hydrogen bonding; peptide links
Brady's reagent 139, 284, 286
  see also 2,4-dinitrophenylhydrazine
brass 201, 232
breathalysers 198–9
bromination 244–5
  phenol 251–2
  phenylamine 263
bromine 51
  addition across C=C 305
  benzene reaction with 242, 244–5, 300
  methanoic reaction with 5
  phenol reaction with 251–2
  phenylamine reaction with 263
  in tests 284, 285, 288
bromobenzene, formation 244–5, 300
1-bromobutane preparation 293
bromocresol green 110, 114
bromoethane 20, 277–8, 300
2-bromoethanol, reactions of 300
bromophenol blue 110, 114
1-bromopropane, hydroxide ion reaction with 16
bronze 232
buffer solutions 104–9, 114
  citric acid/citrate 145
burettes, accuracy of 204
burning *see* combustion
buses, zero-emission 196, 197
butanal, NMR spectra 289
butanenitrile 167
butanoic acid 145, 146, 147, 150
butan-2-ol 6, 148
butanone 148, 290
but-2-ene, geometric isomerism 122
butenedioic acid, isomers of 147
but-2-enoic acid, isomerism 122

C
calcium 67, 164
calcium carbonate 31, 36
calcium chloride, drying with 292, 293, 295
calcium hydroxide 67, 70
calomel electrodes 184, 185–6
cancer 123, 215, 242
caproic (octanoic) acid 145
carbocations 21, 128
carbon 35, 66, 195, 231
carbonates, acid reactions of 6–7, 149
carbon-chain isomerism 120
carbon-chain length 296–7
carbon dioxide 37, 165
  and blood pH 104
  carbonate reactions give 6–7, 31, 149
  in combustion analysis 282–3
  and fuel cells 197
carbonic acid 103, 116
carbon monoxide
  in iron oxide reduction 231
  methane from 63
  methanol from 198
  reactions forming 55–6, 57, 66, 84
carbon side chains 248–9
carbonyl compounds 139–42
  distinguishing between 140, 142, 286
  preparation 133–4
  reactions 134–41, 297
  see also aldehydes; ketones
carbonyl groups 130–1, 145
  test for 139, 141
carboxylate anions 139, 149
  see also ethanoate ions
carboxyl groups 145
carboxylic acids 144, 145–51
  acid chlorides from 158–9, 298
  derivatives 151–60
  ester reactions with 152
  esters from 152
  infrared spectra 286
  polyesters from 156
  reactions forming 148, 298
    aldehyde oxidation 134, 139–40
    amide hydrolysis 267
    iodoform reaction 297
  in tests 285, 287–8
cars 196, 197
cast iron 231–2
catalysts 24–5
  and collision theory 3

copper 233
and exothermic reactions 79
industrial processes 24–5, 79, 80
Contact process 83–4
and rate constants 15
and rate of equilibrium 78, 79, 82
titanium(IV) chloride 231
and transition metals 216–17
cathodes 183, 195, 196–7, 199
cations
as conjugate acids 102
and dissolving 38, 44–5
in electrochemical cells 182–3
and electrode potentials 183–4
electron configuration 208–9
hydrated metal ions, as acids 117
and lattice energy 39–40, 41
and solubility 46
transition metals, charges on 213
C=C group 284, 305
cell potentials 191–2
fuel cells 197
standard *see* standard cell potential
change of state, entropy and 31, 48
charge 40, 41
on transition metal cations 213
charge density, solubility and 46
chemical equations *see* equations
chemical shift 169, 170, 174
chemotherapy, platinum in 123, 215
chirality 124, 127–9
chloride complex, cobalt(II) 75
chloride ions
oxidation 189, 190–1
in transition metal complexes 215
chlorine
benzene reaction with 242, 244
carboxylic acid reactions with 150–1
electrode potential 187, 188
methylbenzene reaction with 248
phosphorus pentachloride decomposition gives 62–3
radicals from, blue light initiates 165
standard electrode potential 186
test for, in halogenoalkanes 285
chloroethane, benzene reaction with 245
1-chloro-2-methylpropane, ibuprofen from 281–2
2-chloro-2-methylpropane, hydrolysis 19, 21, 22
1-chloropentane, preparation 295
1-chloropropane 21, 167
2-chloropropanoic acid 151
chromate ions 79, 186–7, 229
chromatography 175–7, 275–6, 288
chromium
in redox reactions 227–30, 231
uses of 231
chromium(II) complexes 224–5
chromium(III) complexes 214, 216, 224
deprotonation of 221
chromium(III) hydroxide 229
reaction with acids 222
reaction with ammonia 223
cis addition 305
cisplatin, cancer treatment 123, 215
cis–trans isomerism 121–4
see also geometric isomers
citrate 145
citric acid 145
clock reactions 7–8, 16–17
coal, methane from 63
cobalt, ferromagnetic nature 220
cobalt(II) ions 75, 223, 261
cobalt(III) ions 228
cocaine 309
collision theory 2–3, 15, 18
colorimetry 5
colours 218–20
flame colours 164, 219
of hydroxide precipitates 223, 261
of light 162, 163, 164
of precipitates, halogenoalkane test 285
combinatorial chemistry 307–10
combustion
benzene 242
and entropy change 31, 34
hydrogen, extent of reaction 70
combustion analysis 282–3
complementary colours 218–19, 220
complex ions *see* transition metal complexes
compounds, oxidation numbers of 180
concentration
and electrode potential 186–7
and equilibrium constant K_c 52–60
and equilibrium position 78–9
and half-life measurements 10–12
of oxidising agent solutions 200–1
and reaction rate 2
of reducing agents 202–3
solutions of equal 94
concentration term (Q) 52, 73
condensation polymerisation 269–70
polypeptides from 276
condensation polymers 155–6, 267
condensation reactions, amino acid 276
conductivity, copper 232
conjugate acids 102, 104
conjugate bases 104, 108–10, 114
and salts 102
conjugate pairs, acid–base 88–9
Contact process 82–4
continuous flow production 80
coordination numbers 214
copper 201, 232–3
in cells 182–3
disproportionation 192, 226, 230
copper(I) chloride 226, 233
copper(I) complexes 226, 230
copper(I) iodide 230
copper(I) ions 226, 230
copper(I) oxide 230
copper(II) complexes 225–6, 261–2
copper(II) ions
redox reactions 182–3, 191–2
disproportionation 192, 230
copper(II) sulfate 182–3, 225–6
corrosion/rusting 188, 231
costs 79
coupling reactions, diazonium ion 265–6
covalent bonding 40, 213
dative bonds 214, 215
crops, fuel from 198
current, electrochemical cell 183
cyanide ions 135–6, 215
cyclohexane 236, 237, 240, 242
cyclohexatriene 237–8
cyclohexene 236, 237, 238
and isomerism 122
cysteine 273
cytosine 271, 272

D

dative bonds 214, 215, 216
d-block elements 208–33
reactions 221–31
amines 261–2
uses of 231–3
d–d transition 218
defrosting 168
d-electrons 210–11
and catalysis 217
paramagnetism 220
and transition metal colours 219–20
see also *d*-orbitals
delocalised electrons 211
benzene ring 237, 239–40, 243
phenol 249, 251
phenylamine 262
deprotonation 221–2
detergents 247–8
dextrorotatory 6
diacid chlorides, polymers from 156
1,2-diaminoethane 216, 224, 225–6
diazonium ions 252, 264, 265–6
dicarboxylic acids, polyesters from 156
dichlorobenzene, isomers of 240
dichromate(VI) ions 181–2, 227–8, 230
and breathalysers 198–9
cell potential 189
and electrode potential 186–7
equilibrium with chromate ions 79
oxidise hydrogen peroxide 229
diethylamine, as a base 260
diffraction 162
digital cameras 196
dilution of solutions 94
dimethylamine 258
dimethyl esters, polymers from 156
dinitrogen tetroxide 65, 77–8
2,4-dinitromethylbenzene 248
2,4-dinitrophenylhydrazine 138–9, 141
aldehydes and ketones react with 296
as Brady's reagent 284, 286
phenylethanone reaction with 249
2,4-dinitrophenylhydrazone 249
dipeptides 276
dipole moments, infrared absorption and 165
direction of change, predicting 73–4
disorder increase 30–1, 33
see also entropy
disposal of polyesters 156
disproportionation 192, 226, 230–1

dissolving, entropy and 37–8, 42–5
values for ionic solids 47, 70
see also solubility
distillation 292, 293–4, 295
and yield 304
diving 61, 188
DNA 123, 215, 271–2
d-orbitals 209
and catalysis 217
cation charges 213
and metal complex colours 165, 218–20
copper(I) 226
see also *d*-electrons
double bonds 121–2, 239
drugs 176, 272
ibuprofen 127, 167, 281–2, 304
isomerism 304–5
paracetamol 261
platinum, for chemotherapy 123, 215
see also pharmaceutical industry
dyes, diazonium ions in 265
dynamic equilibrium 51

E

EDTA 216, 225–6
chromium(III) ions react with 224
hydrated copper(II) ions react with 227
efficiency, buffer solution 109
electric energy in cars 196
electric fields 162, 174
electricity, reactions for 195–9
electrochemical cells 182–3, 195–8
electrode potentials 183–92
half-equations, altering 188
measurements of 185–6
non-standard conditions 186–8
standard *see* standard electrode potentials
electrodes 182–6
electrolysis, hydrogen from 197
electrolytes 183, 195
electromagnetic radiation 162, 169
see also infrared radiation; light; microwaves; radio waves; ultraviolet radiation; X-rays
electron configuration, *d*-block 208–9
electron releasing groups 135
electrons
in batteries 195
and covalent bonds 213
delocalised *see* delocalised electrons
in electrochemical cells 182–3
and emission spectra 163
and ionisation 174
and ionisation energies 210–11
light absorption 164
and complex ion colours 165, 218–20
lone pairs *see* lone pairs
in redox reactions 180–1, 182–3
oxidising agents 199
and reducing agents 202
electrophiles 134, 135
attack benzene 243, 244, 245, 247
attack phenol 249–50, 251–2
1-chloro-2-methylpropane gives 281–2
electrophilic substitution 243–8
phenol in 251–2
phenylamine in 263
elements
disproportionation 192
oxidation numbers of 180
eluent 175
emf *see* potential difference (emf)
emission spectra 163–4
empirical formulae 283
enantiomers 124–6, 127, 128
endothermic reactions 30, 33, 34
and dissolving 38–9
and electrode potential 188
and equilibrium 68–9, 74–6
spontaneous 36–7
energy
absorption of 164–5, 169
for change of state 48
and complex ion colours 218–20
conversions in cars 196
electrons
and complex ion colours 218–19
and covalent bonds 213
and emission spectra 163
orbitals 208
ionisation energies 209–11, 213
of light/radiation 162
Maxwell–Boltzmann distribution 2–3, 33
and microwave heating 168
and NMR spectra 169
nuclear spin and magnetic fields 168, 169
stabilisation energy 237, 238, 239, 243
temperature, and reaction rate 2–3
of ultraviolet radiation, and absorption 164
vibrational, and infrared 165
see also reaction profiles
enthalpy
and dissolving 38–9
hydration 38, 39, 41, 45–6
lattice energy 39–41, 45–6
and temperature 74–5
enthalpy of atomisation, *d*-block 211
enthalpy of formation 32
enthalpy of hydrogenation, benzene 237–8
enthalpy of neutralisation 115–16
enthalpy of solution 30, 38–9, 41–2, 45
and solubility 47–8
entropy 30–48
complex formation 226–7
and dissolving 37–8, 42–5
ionic solids 47, 70
and equilibrium constants 52–3, 68–9
and feasibility of reactions 34–6
and standard cell potential 190
of surroundings 33–4, 43
and solubility 47–8
of the system 31–3, 43–5, 46–7
and temperature 32–3, 68–9
see also total entropy change
enzymes 24, 78, 304, 306
equations
and equilibrium constants 53–4
half-equations 180–1
overall, half-equations give 181, 189
rate 8–15, 18
multi-step reactions 19, 22, 23
see also equilibria
equilibria
acid–base 88–117
redox 182, 186–92, 196, 199
see also reversible reactions
equilibrium 51–70
continuous flow production 80
and feasibility of reactions 35
position 74–85
predicting direction of change 73–4
equilibrium constants
acid dissociation *see* acid dissociation constant
base dissociation *see* base dissociation constant
and catalysts 79
and concentration K_c 52–60, 65–7, 73–8
Haber process 80
and direction of change 73–4
equilibrium position 74–8
heterogeneous reactions 66–7
ionic product of water (K_w) 91–2
and partial pressure K_p 60–7, 74–5
Contact process 82–3
Haber process 81
and standard cell potential 190
and temperature 68–9, 70, 84–5
and total entropy change 52, 68–70
and extent of reaction 69–70, 190–1
equivalence point 97, 109–10, 111
volume at 112
errors in measurements 204–5
essential amino acids 273
ester groups, benzocaine has 300
esterification 65–6, 149–50
of amino acids 274
esters 144, 151–7
of benzoic acid 303
hydrolysis 148, 152
natural 145, 153–5
reactions forming 152, 159
acid chloride/phenol 252, 254
synthesis with 302
in tests 285, 288
ethanal 130, 133, 138, 140
2-hydroxypropanoic acid from 296
iodoform reaction 134, 140, 287
nucleophilic addition 135, 136
polarity in 135
reduction to ethanol 137
testing for 142
ethanamide 144, 259–60, 267
ethanedioate ions, ligands from 216
ethanedioic acid (oxalic acid) 145
ethane-1,2-diol, esters from 156
ethanenitrile 259–60, 277
ethanoate ions 102, 140, 224–5
ethanoic acid 144, 146
in acid chloride preparation 158
in buffers 104, 105, 106–7, 108
esters from 151, 285
ethanoyl chloride from 150

reaction with ammonium carbonate 36
reaction with chlorine 151
reaction with ethanol 58, 65–6, 149–50
reactions forming
ethanamide hydrolysis 267
ethanoyl chloride/water 158–9
ethyl ethanoate/methanoic acid 152
solubility 147
synthesis with 302
titration with sodium hydroxide 112–13
as vinegar 147
as a weak acid 90, 94, 99
ethanol
and breathalysers 198–9
formation 137, 153
fuel cells using 198, 199
iodoform reaction with 141, 287
oxidation, ethanal from 133
reaction with ethanoic acid 58, 65–6, 149–50
reaction with phosphorus pentachloride 35
testing for 142
ethanoyl chloride 144, 150
in acylation 246–7, 300
alcohol reactions with 159
amine reactions with 260
paracetamol from 261
phenol reaction with 254
preparation 158
synthesis with 302
water reaction with 158–9, 159
ethene 35, 302
ethoxyethane (ether) 170–1, 295
ethylamine 258, 259–60, 261
as a ligand 215, 225
ethylammonium chloride 260
ethylbenzene 245–6, 300
ethyl ethanoate 144, 149–50
in nail varnish 155
reaction with methanoic acid 152
reaction with methanol 153
ethyl methanoate, formation 152
exchange energy 208
exchange reactions, ligand 222–7
exothermic reactions 30, 33, 34
and catalysts 79
and dissolving 38–9
and equilibrium 68–9, 74–6, 82
and gas solubility 37
and temperature 74–6, 81, 82
and electrode potential 188
experimental data
K_c from 54–8
K_p from 62–4
standard cell potential from 192–4
experimental methods
orders of reaction 9–15
reaction rates 4–8
safety issues 306–7
extent of reaction 69–70, 190–1
infrared spectroscopy for 167
eyes, retinal in, and vision 122
E/Z naming system 123

F

fats 145, 152, 153
in saponification 155
fatty acids 153–5
feasibility of reactions 34–6, 190–2
cell potential 190–4, 227–8
reduction potential data for 200
Fehling's solution 139, 140, 142, 286
as copper(II) complexes 226
fermentation 198
ferromagnetic substances 220
fertilisers, manufacture of 82
fibres, polymers make 156, 157
fingerprint region 166, 167, 288
first ionisation energies 209–10
first-order reactions 9, 10, 11–12
multi-step reactions 22
and rate–concentration graphs 14, 15
flame colours 164, 219
fluorine, oxidation number of 180
forces
and change of state 48
and hydration enthalpy 41
and lattice energy 39–40
see also intermolecular forces
formulae 130, 131, 282–4
fragmentation, molecular ion 174–5
free-radical addition 242
free-radical substitution 151, 165
frequency 162
in collision theory 3
and emission spectra 163
and infrared absorption 165–6
Friedel–Crafts reactions 134, 245–7, 252
acylation 246–7, 300
alkylation 245–6, 300
ibuprofen from 281
increase chain length 297
and phenylamine 263
synthesis with 302
fructose 6, 126
fuel cells 196–8, 199
functional-group isomerism 120–1
functional groups, tests for 284–8
fungicides 233

G

gases
in collision theory 3
disorder in 31
electrode potentials 183, 187–8
standard 186
and equilibrium 66, 76–8
Maxwell–Boltzmann distribution 2
partial pressure 60
solubility 37
volume evolved 6–7
gas–liquid chromatography 176–7
geometric isomers 121–4
benzene addition 242
carbonyl addition/elimination 138
carboxylic acids 147
glass electrodes 185
global warming 37
glucose 6, 125, 126, 216
glutamic acid 273, 277
glycerol esters, diesel from 154–5
glycerol (propane-1,2,3-triol) 145
esters from 153
glycine (aminoethanoic acid) 272, 273–4
lack of optical activity 275
synthesis with 302
in wool 269
group 1 metals
compound solubility 44–5
oxidation numbers of 180
group 2 metals
compound solubility 46, 47–8, 70
oxidation numbers of 180
groups, trends in 41, 45–8
guanine 271, 272

H

Haber process 25, 70, 80–2, 217
haemoglobin 216
half-equations 180–1
cell potentials from 188–90, 191–2
fuel cells 197, 198, 199
standard electrode potentials 185–6, 187
standard cell potentials from 188–90
see also redox equilibria
half-lives 10–12
halides 40, 47
halogenation, benzene 244–5, 300
halogenoacid, formation 150–1
halogenoalcohols, synthesis with 298, 302
halogenoalkanes
alcohols from 298
alkenes from 297–8
amine reactions with 261
amines from 298
ammonia 259
benzene reactions with 245
chain length increase with 297
hydrolysis 19–21, 126, 127–8
hydroxide ion reactions 18
nucleophilic substitutions 19–21, 127–8
infrared spectroscopy for 167
nitriles from 277–8
potassium cyanide reactions 167, 277–8, 296
synthesis with 302
test for 284–5
halogens
benzene reactions with 244–5
and isomerism 306
see also bromine; chlorine; iodine
hardness, *d*-block element 211
heat, conversion to light 163
heat exchangers 79
heat flow 30, 33
Henderson–Hasselbalch formula 107
Hess's law 39, 42
heterogeneous catalysts 24–5, 217
heterogeneous equilibria 66–7
heterogeneous reactions, rate of equilibrium 78
hexaaquacopper(II) ions 261–2

hexacyanoferrate(III) ions 228
hexahydroxy complexes 216
hexane, iodine solubility in 58–60
high-performance liquid chromatography 176
high-resolution NMR spectra 172–3
homogeneous catalysts 25, 217
homogeneous reactions 53–4, 78
homopolymers 157
hybrid cars 196
hydrated chromium(III) ions 214, 224
  see also chromium(III) complexes
hydrated copper(II) chloride 225
hydrated copper(II) hydroxide 223
hydrated copper(I) ions 226
hydrated copper(II) ions 214, 225
  reaction with ammonia 223
  reaction with 1,2-diaminoethane 226
hydrated hydroxides, precipitates of 221–2, 223
hydrated iron(II) ions 228
hydrated iron(III) ions 221
hydrated metal ions 214, 221–7, 228
  as acids 117
hydration enthalpy 38, 39, 41
  and group trends 45–6
hydrochloric acid
  and carbonate reaction 31
  copper(I) chloride dissolves in 226
  ethylamine reaction with 260
  in phenylamine preparation 262, 263, 294
  phenylamine reaction with 264
  as a strong acid 90, 94–5, 115
  in titrations 112, 113
  zinc in, half-equation of 180
hydrofluoric acid 100, 115, 116
hydrogen
  and acid structure 116–17
  in carbonyl reduction 137
  cars run on 197
  cis addition 305
  combustion, extent of reaction 70
  in fuel cells 196–8
  in the Haber process, catalysis 25
  in hydroxynitrile reduction 137
  and MRI spectra 174
  and NMR spectra 169, 170–3
  nuclear spin 168, 169
  oxidation number of 180
  reactions forming
    carbon monoxide/methanol 198
    carbon/steam 66
    magnesium/sulfuric acid 7
    methane/steam 25, 55–6, 57, 84–5, 197
  reactions with
    benzene 237, 242
    carbon 35
    carbon monoxide 63
    iodine 51–2, 73, 76–7
    nitrogen 57, 68–9, 70, 77, 80, 82, 217
    nitrogen(II) oxide 9–10
    oxygen 35, 191
  standard electrode potential 183
  in standard electrodes 184, 185
hydrogenation 153–4, 217, 237–8
hydrogen bonding
  amines 259, 266
  carbonyl compounds 132
  carboxylic acids 146, 147
  DNA 271–2
  peptides 277
  phenol 250
  phenylamine 262
  polyamide/nylon 268
hydrogen bromide 244–5, 251–2, 263
hydrogencarbonates, carboxylic acid reactions with 149
hydrogen chloride
  reactions forming 150
    acid chloride/amine 260, 267
    acylation 247
    chloroethane/benzene 245
    ethanoyl chloride/water 158–9
    phosphorus pentachloride/alcohol 285
    phosphorus pentachloride/carboxylic acid 287–8
hydrogen cyanide addition 135–6, 278
  increases chain length 296
  phenylethanone 249
  propanone 277
hydrogen iodide 51–2, 73, 76–7
hydrogen ions 88, 91–2
  and buffer solutions 108–9
  in carbonyl reduction 137
  and pH 92, 93
  reduction of 185
hydrogen peroxide 7, 229, 230
  as a reducing agent 202, 203
hydrolysis 6, 126
  amides 266–7
  esters 148, 152
  halogenoalkanes 19–21, 126, 127–8
  hydroxynitriles 136, 148
  nitriles 136, 148
hydronium ions, pH and 92
hydroxide ions 67, 91–2, 93
  and buffer solutions 108–9
  in halogenoalkane hydrolysis 19
  reaction with bromoethane 20
  reaction with 1-bromopropane 16
  reaction with halogenoalkanes 18, 128
  and strong bases 95
hydroxides 40, 46, 47, 70
  precipitates 221–2, 223, 261
hydroxyacids 157
  see also lactic acid
hydroxyamines 136
4-hydroxyazobenzene 265
3-hydroxybutanoate, polymers from 157
hydroxycarboxylic acids 136–7, 296
2-hydroxy-2-methylpropanenitrile 136
hydroxynitriles 136–7, 148, 278
N-4-hydroxyphenylethanamide (paracetamol) 261
2-hydroxypropanenitrile 135, 136, 148
2-hydroxypropanoic acid *see* lactic acid

I

ibuprofen 127, 167, 281–2, 304
immiscible solvents 58–60
indicators, acid–base 109–10, 114
industrial processes 79–84
  catalysis in 24–5, 79, 80, 217
  Haber process 25, 70, 80–2
inert gases, equilibrium and 78
infrared analyser breathalysers 199
infrared radiation 162, 165
infrared spectra 165–8, 288
  alcohols 285
  aldehydes 286
  aromatic compounds 238
  carboxylic acids 285, 288
  ketones 286
  propan-2-ol 166–7
infrared spectroscopy 5, 167, 199
initial rates 9–10
intermolecular forces 259
  amines 259
  benzene 241
  carbonyl compounds 132
  carboxylic acids 146
  phenol solubility 250–1
  phenylamine 262
  see also hydrogen bonding
invertase, sucrose hydrolysis with 6
iodate(V) ions 16–17, 200
iodide ion redox reactions 16–17, 25
  copper ions 191–2
  dichromate(VI) ions 181–2
  iodate(V) ions 200
  iron(III) 188–9
  persulfate ions 191, 217
iodination, propanone 22–3
iodine
  iodoform reaction 134, 140, 249
    decreases carbon-chain length 297
  propanone iodination forms 22–3
  reaction with hydrogen 51–2, 73, 76–7
  redox reactions 7, 181–2, 200–1
  solubility 58–60
  test for, in halogenoalkanes 285
  titrations, reactions rates from 4–5
iodine clock 7
2-iodobutane 6
iodoform reaction 134, 140–1
  carboxylic acids from 148
  decreases carbon-chain length 297
  phenylethanone in 249
  in tests 142, 287
ionic bonding 212, 277
ionic compounds
  dissolving 47, 70
  solubility of 38–48
ionic product of water (K_w) 91–2
ionic radii 40, 41, 46
  and metallic bonding 211
ionisation, mass spectroscopy and 174
ionisation energies 209–11, 213
ion pairs, standard electrode potential 186
ions
  complex *see* transition metal complexes

ethanedioate, ligands from 216
hydrated metal, as acids 117
and lattice energy 39–40, 41
oxidation numbers of 180, 220
in solution 2, 44
iron
and benzene/bromine reaction 24–5
ferromagnetic nature 220
Haber process catalysis 25, 80, 81, 217
in haemoglobin 216
in iodide ion oxidation 25, 217
in iron tablets 203
reaction with water 66
standard electrode potential of 185–6
as a transition metal 209
uses of 231–2
iron complex ions 221
iron(II) ions 67
redox reactions 193–4, 227–8
iron(III) ions 67
redox reactions 188–9, 193–4
iron(II) sulfate 56
in iron tablets, estimating percentage 203
iron oxide, reducing 231
isomerism 120–9
carbon chain 120
functional group 120–1
geometric 121–4
optical 124–9
positional 120, 240–1
stereoisomerism 121–9, 304–6
amino acids 274–5
structural 120–1, 289–90
see also geometric isomers

K
K_a *see* acid dissociation constant
K_b *see* base dissociation constant
K_c (equilibrium constant) 52–60, 65–7, 73–8
Haber process 80
ketones 131, 134–40, 141
aromatic, acylation reaction gives 247
and chain length increase 296
distinguishing from aldehydes 140, 142, 286, 289–90
enzyme catalysis 306
hydroxynitrile from 278
infrared spectra 286
laboratory preparation 134
methyl, iodoform reaction 134, 140, 148, 287
physical properties 132
secondary alcohols from 298
tests for 284, 285–7
ketone side chains 248–9
Kevlar® 268–9
kinetic energy 2–3, 168, 196
kinetics
actuality of reactions and 191
and reaction feasibility 228
see also rates of reaction
K_p (equilibrium constant) 60–7, 74–5
Contact process 82–3
Haber process 81
K_w (ionic product of water) 91–2

L
lactic acid (2-hydroxypropanoic acid) 136, 148
from ethanal 296
isomerism in 125, 126
in milk 145
polymers from 157
laevorotatory 6
lattice energy 39–41, 45–6
law of mass action 52–3
lead–acid batteries 195–6
lead(IV) oxide in batteries 195–6
leaving groups 144
Le Chatelier's principle 69, 74
leucine 273
Lewis acids 88
lidocaine 309
ligands 214–16
amines as 215, 225–6, 261
complex ion colours 218, 219–20
exchange reactions 222–7
and reduction potential 228
light 162
absorption, colours from 165, 218–20
in colorimetry 5
plane-polarised 125–7, 128
prisms split 162, 163
speed of 162
ultraviolet, and sunglasses 233
linalool 293
linear complex ions 215
linolenic acid 154
lipids 147
see also fats; oils
liquids
disorder in 31
mixing, and equilibria 66
molecular energy distribution 2
separation 177, 292–5
lithium aluminium hydride *see* lithium tetrahydridoaluminate(III)
lithium cells 196
lithium oxide in batteries 195
lithium tetrahydridoaluminate(III) 137
acid chloride reaction with 159
in amine preparation 259–60
in carbonyl reduction 137
in carboxylic acid reduction 150
in hydroxynitrile reduction 137
phenylethanone reaction with 249
lone pairs, electron 130
and acid/base definitions 88
and complex ions 214, 215, 216
on nitrogen 137–8, 159
in amines 258, 260, 261
in phenylamine 262, 263
nucleophilic substitution 135–6
on oxygen, in phenol 252
on water, acid chloride reactions 158
low-resolution NMR 170–2
lysine 273

M
magnesium, reactions, sulfuric acid 7
magnesium sulfate, solubility 70
magnetic fields 162, 168–9
high, MRI uses 173
in mass spectrometers 174
and NMR spectra 169
magnetic properties, transition metal 220
magnetic resonance imaging 173–4
malic acid 145
manganate(VII) ions
electrode potentials 184
redox reactions 186, 190–1, 202–3
see also potassium manganate(VII)
manganese(II) ions, electrode potentials 184
manganese(IV) oxide in batteries 195
margarine 152, 153–4
mass spectroscopy 174–5, 285
Maxwell–Boltzmann distribution 2–3, 33
measurements, uncertainty in 204–5
mechanisms, reaction 18–26
carboxylic acid reactions 144
and chirality 127–9
nucleophilic addition 135–6
and stereoisomerism 305–6
medical issues
antiseptics 252
cancer 123, 215, 242
magnetic resonance imaging 173–4
sickle cell anaemia 276–7
thalidomide 127, 305
use of X-rays 164
see also drugs
melting, entropy change and 31, 32, 48
melting points
acid chlorides 157
amides 266
and carbonyl tests 139
carboxylic acids 146
d-block elements 211, 212
phenol 250
substance identification with 291
titanium 231
mercury(II) oxide in batteries 195
metallic bonding 211
metals
catalysis by 25, 217
delocalised electrons in 211
group 1 180
compound solubility 44–5
group 2 180
compound solubility 46, 47–8, 70
oxidation numbers 180
standard electrode potentials 185–6, 187
transition *see* transition metals
methanal 130, 132, 135
methane 63, 165
hydrogen from 80, 84, 197
in hydrogen manufacture 25
reaction with steam 55–6, 57
methane hydrate deposits 85
methanoic acid 146, 147
in buffer solutions 105–6
reactions 5, 152
methanol

biodiesel from 153, 154–5
for fuel cells, and manufacture 198
in polymer manufacture 271
reactions 150, 153, 159, 303
methanoyl chloride 157
methylamine 258, 259
methylbenzene 240, 248, 297
2-methylnitrobenzene from 255–6
methyl benzoate 303
methyl butanoate, preparation 150
methyl esters 154–5, 156
N-methylethanamide 159
methyl ethanoate, formation 153, 159
methyl ketones, iodoform reaction 134, 140, 148, 287
2-methylnitrobenzene 255–6
methyl 3-nitrobenzoate, synthesis 303
methyl orange 110, 114
2-methylpropanal, NMR spectra 289–90
methylpropanoic acid, melting and boiling point 146
methyl red 110, 114
methyl secondary alcohols, iodoform reaction with 141, 287
microwaves 85, 162, 168
milk, esters in 145
molar mass 284
molecular formulae 283–4
molecular ions 174–5
molecules 2, 3, 33
mole fractions 60, 65
monodentate ligands 215
monomers 155–6, 157, 269–70
monounsaturated fatty acids 153
multi-step reactions 18–19, 20–4
myoglobin 277

N

naming substances
amines 258
arenes 240–1
carboxylic acids 146–7
esters 151
nitriles 277
2-naphthol, diazonium ion reactions with 265
natural gas, oceans hold 85
natural products 236
amine groups in 271–2
carboxylic acids in 145
esters 145, 153–5
polyamides 269
see also biological molecules
neutralisation, enthalpy of 115–16
neutrality 92
nickel 197, 220
in hydrogen manufacture 25, 84
nickel complex ions 223, 261
nitration 243–4, 252, 263, 299
of methyl benzoate 303
nitric acid 94, 116
benzene reaction with 243–4, 299
benzoic acid reactions with 255
methylbenzene reaction with 248
2-methylnitrobenzene from 255–6
phenol reaction with 252
phenylamine reaction with 263
sulfuric acid reaction with 89
nitric oxide *see* nitrogen(II) oxide
nitriles 277–8
chain length increases with 296
hydrolysis 136, 148
reduction 137, 259–60
in synthesis 136–7
nitrobenzene 243–4, 299
phenylamines from 262–3, 294
nitrogen
in air, and diving 61
lone pairs 137–8, 159
organic compounds containing 258–78
see also amines
reaction with hydrogen 80, 82, 217
catalysis 25
equilibrium constant 57, 68–9, 77
extent of reaction 70
reaction with oxygen 63–4
nitrogen dioxide 14–15, 65, 77–8
nitrogen(II) oxide/nitric oxide/nitrogen monoxide (NO) 165
lightning strike forms 63–4
reaction with hydrogen 9–10
reaction with oxygen 14–15
nitrophenol 252
nitrous acid, phenylamine reactions with 264
NMR *see* nuclear magnetic resonance
Nomex® 268
novocaine 309
nuclear magnetic resonance 168–74
spectra, in tests 285, 287, 289–90
nuclei, spin 168–9
nucleophiles 19
carbonyls attacked by 130, 131, 135
carboxylic acids attacked by 144, 150
nitrogen 137–8, 260
oxygen in phenol 252, 254
phenylamine 264
in pi-bond reduction 260
water 158
nucleophilic addition 134–7, 277, 278
increase chain length with 296
nucleophilic substitutions 296
halogenoalkanes 19–21, 127–8
infrared spectroscopy for 167
nitriles from 277–8
nylon 267–8, 269, 277

O

oceans, methane hydrate in 85
octahedral ions 214
octanoic (caproic) acid 145
oils 145, 152, 153–4, 155
oleum 84
omega-3 acid 145
omega-3 oils 154
omega-6 oils 154
optical activity 5–6, 126–7
optical isomerism 124–9
amino acids 274–5
orbitals 208
see also d-orbitals, *p*-orbitals
order of reaction 8–15, 25–6
and enzyme catalysis 24
and mechanism 19, 22, 23, 127–8
single-step reactions 18
orientation, collisions and 1, 15
oxalic acid (ethanedioic acid) 145
oxidation 180, 181
of alcohols 133, 134, 148
of aldehydes 148
ammonia, catalysis 25
of carbonyl compounds 134, 139–40
in electrochemical cells 182–3
of iodide ions 16–17, 25
of iodine 7
by manganate(VII) ions, chloride ions 190–1
of propan-2-ol 5
of zinc in copper sulphate 182
see also redox reactions
oxidation numbers (states) 180–2
and disproportionation 192
transition metals 212, 220, 227
vanadium 192–3
oxidation potentials 229
see also electrode potentials
oxides, lattice energies of 40
oxidising agents 199–201
for aldehyde oxidation 139–40
and cell potential for reactions 189–90
for chromium 229
dichromate(VI) ions 227–8, 230
electrode potentials 186, 187–8
oxo-acids 116
oxygen
electrode potential 187–8
in fuel cells 196–8, 199
as a nucleophile 252, 254
oxidation number 180
reactions with
benzene 242
hydrogen 35, 191
nitrogen 63–4
nitrogen monoxide 14–15
phosphorus 31, 32, 34
sulfur dioxide 53
in sulfuric acid manufacture 82, 83–4
from sulfur trioxide 55, 78
ozone 164, 238–9

P

paints, titanium(IV) oxide in 231
paracetamol, formation of 261
paramagnetism 220
partial order of reaction 8, 9
partial pressure 60–1
and equilibrium constants 60–5, 81
partition coefficients 58–60
peak height (NMR) 170–2
peptide links 269, 276
peptides, synthesis 307–9
percentage composition 282–3
perfumes 132, 155, 293
period 3 elements, first ionisation energies

209–10
permafrost, methane hydrate in 85
persulfate ions 25, 191, 217
PET (polyethylene terephthalate) 156
pH 94–104
of acids 94–5, 98–102
of bases 94, 95–6, 104
and buffer solutions 104–9
and dilution of solutions 94
at equivalence point 109
and indicators 109, 110, 114
measurements, reaction rates from 8
and reduction potential 228–9
rule of two 104
and salts 102–3
and titrations 96–7, 111–14
pharmaceutical industry 79, 80, 85
analysis in 176, 281
combinatorial chemistry 307–10
see also drugs
phases, equilibria and 66
phenate ion 253, 265
phenol 240, 249–54
reactions 251–4, 265
tests for 253, 288
phenolphthalein 110, 114
phenylalanine 276
phenylamine 240, 262–4, 266
preparation 262–3, 294
test for 288
phenylammonium salts 263
phenyl benzoate 254
phenylethanoate 254
1-phenylethanol 249
phenylethanone 134, 240, 248–9
acylation gives 247, 300
phenyl group 240
pH meters 185
phosphorus, oxygen reaction with 31, 32, 34
phosphorus pentachloride 62–3
reactions with
alcohols 35, 285, 295
carboxylic acids 150, 158, 287–8
phosphorus trichloride 158
photochromic sunglasses 233
pH scale 92
physical properties
acid chlorides 157–8
amides 266
amines 258–9
amino acids 273–4
benzene 241–2
carbonyl compounds 132
carboxylic acids 146, 147
d-block metals 211
esters 151–2
phenol 250–1
phenylamine 262
polyamides 269
pi-bonds 239, 137
and aromatic substitution 255
benzene 237
and carbonyl compounds 130, 134
in geometric isomerism 121–2
reduction of 260
pipettes, accuracy of 204
pi-system
benzene 237, 239–40, 243
phenol 249–50, 251
phenylamine 262
pK_a 93, 99, 100
pK_w 93
plane-polarised light 125–7, 128
plastics 156, 157
platinum 25, 137
electrodes from 184, 197
platinum complexes, chemotherapy with 123, 215
pOH 93
polar bonds
carbonyl compounds 130, 131, 132
C=O group 144
polarimetry 5–6, 126
polarisation of light 125–7, 128
pollution, atom economy and 83–4
polyamides 267–9
polyatomic ions, oxidation numbers of 180
polydentate ligands 216
polyesters 155–6, 270, 271
poly(ethenol) (polyvinylalcohol) 271
polylactic acid 157
polymerisation
addition 270–1
condensation 269–70
polypeptides from 276
stereoisomerism in 300
polymers 270–1
condensation 155–6, 267
homopolymers 157
nylon 267–8, 269
polyamides 267–9
polyesters 155–6, 270
polypeptides 276, 308–9
poly(propenamide) 270–1
poly(propene) 270
poly(propenoic acid) 270–1
polyunsaturated fatty acids 153, 154
polyvinylacetate 271
polyvinylalcohol 271
p-orbitals 239, 249–50
positional isomerism 120, 240, 240–1
positive ions *see* cations
potassium cyanide 167, 277–8
chain length increases with 296
potassium dichromate(VI)
and alcohol oxidation 133, 134, 148
and breathalysers 198–9
propan-2-ol 5
tests for alcohols 285
aldehydes oxidised by 140, 284, 287
tests with 285, 287
potassium manganate(VII) 139, 202
in alkene test 284
see also manganate(VII) ions
potassium methanoate, buffers with 105–6
potential difference (emf) 183, 185
precipitation
in halogenoalkane test 284–5
hydroxides 221–2, 223
with amines 261
silver 67
silver oxide 222
preparation
acid chlorides 158
amides 266
amines 259–60
benzenediazonium compounds 264
1-bromobutane 293
carbonyl compounds 133–4
carboxylic acids 148
1-chloropentane 295
esters 152
2-methylnitrobenzene 255
nitriles 277–8
phenylamine 262–3
pressure
and collision theory 3
and electrode potential 187–8
and equilibria 76–8, 85
in industrial processes 79, 81–2, 83
partial pressure 60–1
primary alcohols 137, 150
oxidation 133, 148, 298
tests for 285
primary amines 258, 261
primary halogenoalkanes, hydrolysis 19–20, 21
primary structure, protein 277
prisms, light splitting by 162, 163
progress of reaction, spectroscopy for 167
propanal, NMR spectrum of 171
propanenitrile 277–8
propane-1,2,3-triol (glycerol) 145
esters from 153
propanoic acid 100, 146, 151
mass spectrum of 175
preparation 148
propan-1-ol 148, 172–3
propan-2-ol
iodoform reaction 287, 297
in redox reactions 5, 134, 137
spectra 166–7, 171–2
propanone (acetone) 131, 132, 134, 135
2,4-dinitrophenylhydrazine reactions with 138–9
hydrogen cyanide reactions with 136, 277
iodination of 22–3
iodine titrations 4–5
iodoform reaction 287, 297
in redox reactions 5, 137
propenamide, polymer from 270–1
propene, polymerisation 270, 300
propenoic acid, polymer from 270–1
1-propylamine 258
proteins 272, 273, 276–7
pseudo-zero-partial order reactions 24
purification 292–6, 303, 304
purity, testing for 168, 176

R

racemic mixtures 126, 127, 128, 136
radicals, chlorine 165, 242
radio waves 162, 168
 MRI with 173–4
 NMR with 168–74
rainbows 162, 163
rate constants 8, 9–10, 15–17, 23
rate-determining steps 19, 22–3, 24
rate equations 8–15, 18
 multi-step reactions 19, 22, 23
rates of reaction 2–26, 167
 and equilibrium position 76, 78, 81, 81–2
 and temperature 2–3, 76, 81
reaction mechanisms 18–26
reaction profiles 18, 20, 24
reaction quotient (*Q*) 52, 73
reactions
 actuality of 191–2
 electricity from 195–9
 feasibility of 34–6, 190–4, 227–8
  reduction potential data for 200
 standard cell potential 188–90
recrystallisation 295–6, 304
redox equilibria 182, 186–92
 for fuel cells 197, 199
 in rechargeable cells 196
redox reactions 180–3
 in batteries 195–6
 cell potentials for 188–90
 chromium compounds 229–30
 feasibility of 190–1, 200
 for fuel cells 197, 199
 transition metals in 227–30
  vanadium 192–4
 see also oxidation; reduction
reducing agents 202–3
 aldehydes as 134
 and cell potentials 189–90, 193–4
 in hydroxynitrile reduction 137
 and standard electrode potentials
  186, 187
 transition metal ions as 228
reduction 180, 182–4, 193–4
 of carbonyl compounds 137
 of carboxylic acids 150
 of hydrogen ions 185
 of hydroxynitriles 136–7
 by lithium tetrahydridoaluminate(III) 159
 see also redox reactions
reduction potentials *see* electrode potentials
reference electrodes 185
relaxation time 169
residue, amino acid 276
resonance structures 238–9, 251, 253
retention time 176
retinal 122
reversible reactions 51, 88, 196
 see also equilibria
rhodium in fuel cells 197
RNA 271, 272
rotational energy 168
rule of two 104, 112

S

safety issues 306–7
salicylic acid 241
salt bridge 182, 183, 185
salts 102–3, 104
 amine/halogenoalkane reactions produce
  261
 in buffer solutions 104–9
 of carboxylic acids 149, 267, 297
 phenylammonium 263
 pH of 94
saponification 155
saturated carboxylic acids 145
saturated fatty acids 153
scandium 208, 209, 212
secondary alcohols 133, 137, 298
 iodoform reaction with 141, 148, 287
 tests for 285
secondary amides 264
secondary amines 258, 260, 261, 272
secondary halogenoalkanes 21
secondary structure, protein 277
second-order reactions 9, 10, 12, 18
 multi-step reactions 22
 and rate–concentration graphs 14–15
separation methods 292–6, 303
serine in silk 269
sickle cell anaemia 276–7
silk 269, 277
silver, precipitation of 67
silver chloride 233
silver halides, solubility 47
silver(I) ions 67, 223, 261
silver nitrate 56, 284
silver oxide 195, 222
single-step reactions 18
smell of organic compounds 132, 147
S_N1 reactions, halogenoalkane 19, 20–1
 and chirality 127–8
S_N2 reactions, halogenoalkane 19–20, 21
 isomerism 127–8, 306
soaps, preparation 152, 155
sodium
 in hydroxynitrile reduction 137
 phenol reaction with 253
sodium benzoate 106, 267
sodium carbonate 253
 purification with 292, 293, 295
 in tests 285, 288
sodium ethanoate 102–3, 297
 in buffers 104, 105, 107, 108
sodium hydrogencarbonate 253
 in tests 285, 288
sodium hydroxide 90, 94, 112–13, 115
 in batteries 195
 and benzoyl chloride reactions 254
 in buffer solutions 106–7
 chromium complexes from 216
 detergents/soaps 155, 247–8
 in halogenoalkane test 284
 hydrated chromium(III) ions react with
  224
 in hydrolysis 152, 267
 hydroxides dissolve in 221–2
 iodoform reaction 140, 141, 142, 249
  carbon-chain decrease 297
 in phenol/diazonium ion reactions
  265
 phenol reaction with 253
sodium thiosulfate 4, 7–8, 201
 see also thiosulfate ion
solids 3, 66
 and entropy 31, 37–8, 42–5
 separating and purifying 295–6
 solubility of 37–48
solubility 30, 37–48
 amides 266
 amines 259
 benzoic acid 254
 carbonyl compounds 132
 carboxylic acids 147
 esters 152
 group 2 hydroxides 46, 47, 70
 group 2 sulfates 47–8, 70
 phenol 250–1
 phenylamine 262
solutes 58–60, 66
solution, enthalpies of 30
solutions 2, 66
 dilution and pH 94
 oxidising agent concentration 200–1
 standard, errors in 204
solutions of equal concentration 94
solvent extraction 294–5
solvents 54
 distribution of solute between 58–60
 esters as 155
 phenol solubility in 250–1
 phenylamine solubility in 262
 propanone 132
 for recrystallisation 295–6
spectra 163–5
 infrared *see* infrared spectra
 NMR 169, 170–3
spectrophotometers 5
spectroscopy
 infrared 5, 167, 199
 mass 174–5, 285
speed of light/radiation 162
speed, particle 2, 3
spin–spin splitting 172–3
spontaneous change 30, 33, 34, 42
 endothermic reactions 36–7
 redox reactions 190–1
stabilisation energy 237, 238, 239, 243
stability constants 222–3, 226–7
stainless steel 231
standard cell potentials 188–90
 experimental results for 192–4
 and feasibility 190–4, 227–8
 fuel cells 197
standard electrode potentials 183–4,
  186, 187
 cell potentials from 188–90
 and disproportionation 192
 of metals 185–6, 187
  transition metals 193, 228
 oxidising agents 199, 227

reducing agents 202, 227
standard enthalpies of formation 32
standard entropy 32
standard hydrogen electrodes 184, 185
standard reduction potentials *see* standard electrode potentials
standard solutions, errors in 204
stationary phase 175–6
steam
in hydrogen manufacture 25, 84, 197
reaction with carbon 66
reaction with methane 55–6, 57, 80, 84
steam distillation 293–4
stereoisomerism 121–9, 274–5, 304–6
see also geometric isomers; racemic mixtures
steric acid 145, 153
steric hindrance 21
strength, metal 211, 231
strong acids 90, 103, 115
pH of 94, 94–5
titrations 96–7, 111, 112, 113, 114
strong alkali in buffer solutions 106–7
strong bases 90, 103
neutralisation 115–16
pH of 94, 95–6
titrations 96–7, 111, 112–14
structural formulae 130, 131
structural isomerism 120–1, 289–90
structure of proteins 277
substitution
acid chlorides in 158
alkenes, onto benzene rings 247–8
aromatic, rules for 255
free-radical 151, 165
halogens, and isomerism 306
see also electrophilic substitution; nucleophilic substitution
successive ionisation energies 210–11
sucrose, hydrolysis 6, 126
sulfate solubility 46, 47–8, 70
sulfides, lattice energies 40
sulfonation 247, 300
sulfur clock 7–8
sulfur dioxide 53, 55, 78
sulfuric acid from 82, 83–4, 217
sulfuric acid 95, 116
in batteries 195–6
benzene reaction with 247, 300
in esterification 149–50
magnesium reactions with 7
manufacture 82–4, 217
in 2-methylnitrobenzene preparation 255–6
in nitric acid/benzene reaction 243–4
in nitric acid/benzoic acid reaction 255
in nitric acid/methylbenzene reaction 248
nitric acid reaction with 89
phenylamine reaction with 263
sulfur trioxide
oxygen/sulfur dioxide reaction 53, 55, 78, 217
sulfuric acid manufacture 82, 83–4
in sulfuric acid/benzene reaction 247, 300
sunglasses 233
superconductors 173
surface area, collision theory and 3
surroundings 31
entropy of 33–7, 43, 47–8
synthesis 136–7, 296–303
system 31
entropy change 31–7, 43–5, 46–7

T

tangents, drawing 13
temperature
and electrode potential 188
and endothermic reactions 84–5
and energy distribution 2, 3, 33
and entropy 32–3, 68–9
and reaction feasibility 34, 35–6
and equilibria 68–9, 85
position 74–6, 82–3
and gas solubility 37
in the Haber process 70, 81
and kinetic experiments 11
and rate constants 15–17
and reaction rate 2–3, 76, 81
terephthalic acid 147, 156
tertiary amines 258, 261
tertiary halogenoalkanes 19, 20–1
Terylene 156, 270
tests
for alcohols 285
for aldehydes 139, 284, 285–7
distinguishing from ketones 140, 142
for benzene 242
for carbonyl groups 139, 141
for carboxylic acids 287–8
functional groups 284–8
iodoform test 142, 287
for ketones 139, 284, 285–7
distinguishing from aldehydes 140, 142
for phenol 253, 288
for phenylamine 288
tetrahedral complex ions 215
tetramethylsilane (TMS), NMR with 169
thalidomide 127, 305
thermodynamics, laws of 30–1, 32
thermodynamic stability 191
thin-layer chromatography 176, 275–6
thionyl chloride 150, 158
thiosulfate ion oxidation 200–1
see also sodium thiosulfate
third-order reactions 23
thymine 271, 272
thymol blue 110, 114
tiglic acid 123–4
time, reaction 2, 7
tin 232, 262, 263, 294
tin hydroxide 294
tin(II) ions, reduction by 194
titanium 209, 231
titanium(IV) chloride 231
titanium(IV) oxide 231
titanium(IV) sulfide in batteries 196
titrations 96–7, 111–14
equivalence point 97, 109–10
reactions rates from 4–5
Tollens' reagent 139, 140, 142, 285, 286
toluene 240
total entropy change 34–7
and cell potential 190
and dissolving 42–3, 47, 70
and equilibrium constants 52, 68–70
and equilibrium position 74, 75, 76
trans addition 305
transesterification 152–3, 154, 271
trans fatty acids 153, 154
transition metal complexes 214–16, 221–31
amine reactions with 261–2
cancer treatments with 123, 215
colours of 165, 218–20
geometric isomerism in 123
transition metal ions
chromium 226, 227–30, 231
copper(I) 226, 230
copper(II) 182–3, 191–2, 230
iron 188–9, 193–4, 227–8
oxidation states 220
physical properties 218–20
standard electrode potentials 228
transition metals 208–33
and catalysis 216–17
chemical properties 212–17
definition 209
reactions 221–31
uses 231–3
vanadium 192–4
transition states 18, 19–20, 24
2,4,6-tribromophenol 251–2
2,4,6-tribromophenylamine 263
triglycerides 153
triiodomethane (iodoform) 140
see also iodoform reaction
trimethylamine 258
2,4,6-trinitrotoluene (TNT) 248

U

ultraviolet radiation 162, 164, 233
uncertainty in measurements 204–5
units
equilibrium constants 57–8, 61–2
rate constants 15
unknown substances, identifying 284–91
unsaturated carboxylic acids 145
unsaturated fatty acids 153
uracil 272

V

valence-shell electron-pair repulsion theory 214
valine 273
vanadium 192–4
vanadium(V) oxide catalysis 83, 217
vibrational energy, infrared and 165
vinegar 147

visible light 162–3, 164–5
volume, gas, and equilibrium 76–8

W

water
acid chloride reactions with 158–9
amide solubility 266
amine reactions with 260
amine solubility 259
ammonium nitrate reaction with 37
and aqua ions 214
auto-ionisation of 91–3
benzoic acid solubility 254
carbonyl compound solubility 132
carboxylic acid reactions with 148–9
in combustion analysis 282–3
deprotonation by 221
in equilibrium expressions 54, 90
formation from elements 35
as a greenhouse gas, and dipole moments 165
heating with microwaves 168
iodine solubility in 58–60
ionic product of (K_w) 91–2
iron reaction with 66
as a ligand 214, 220, 222–3, 225
phenol reaction with 253
phenol solubility in 250–1
phenylamine reaction with 263
phenylamine solubility in 262
special polymers absorb 270–1
see also steam
wavelength 162
wavenumber 102–3, 104–10, 162
weak acids 90
carboxylic acids as 148
dissociation constant 90–1, 98, 100, 102
neutralisation 115–16
pH of 94, 98–102, 104
titrations 111, 112–14
weak bases 90, 101–2
in buffer solutions 104
dissociation constant 91, 102, 103
pH of 94, 104
salts of 102, 104
titrations 111, 113, 114
wool 269

X

X-rays 162, 164

Y

yield 79, 81, 83, 304

Z

zero-order reactions 9, 12, 22, 24, 25
Ziegler–Natta process 231
zinc 195, 201, 209, 232
oxidation state 212, 213
in redox reactions 180–1, 182–3, 230
zinc ions 223, 261
zinc sulfate, cells with 182–3
zwitterions 273–4, 300

The periodic table

Key: Relative atomic mass / Atomic symbol / name / Atomic (proton) number

Group

Period	1	2											3	4	5	6	7	0
1	1.0 H hydrogen 1																	4.0 He helium 2
2	6.9 Li lithium 3	9.0 Be beryllium 4											10.8 B boron 5	12.0 C carbon 6	14.0 N nitrogen 7	16.0 O oxygen 8	19.0 F fluorine 9	20.2 Ne neon 10
3	23.0 Na sodium 11	24.3 Mg magnesium 12											27.0 Al aluminium 13	28.1 Si silicon 14	31.0 P phosphorus 15	32.1 S sulfur 16	35.5 Cl chlorine 17	39.9 Ar argon 18
4	39.1 K potassium 19	40.1 Ca calcium 20	45.0 Sc scandium 21	47.9 Ti titanium 22	50.9 V vanadium 23	52.0 Cr chromium 24	54.9 Mn manganese 25	55.8 Fe iron 26	58.9 Co cobalt 27	58.7 Ni nickel 28	63.5 Cu copper 29	65.4 Zn zinc 30	69.7 Ga gallium 31	72.6 Ge germanium 32	74.9 As arsenic 33	79.0 Se selenium 34	79.9 Br bromine 35	83.8 Kr krypton 36
5	85.5 Rb rubidium 37	87.6 Sr strontium 38	88.9 Y yttrium 39	91.2 Zr zirconium 40	92.9 Nb niobium 41	95.9 Mo molybdenum 42	[98] Tc technetium 43	101.1 Ru ruthenium 44	102.9 Rh rhodium 45	106.4 Pd palladium 46	107.9 Ag silver 47	112.4 Cd cadmium 48	114.8 In indium 49	118.7 Sn tin 50	121.8 Sb antimony 51	127.6 Te tellurium 52	126.9 I iodine 53	131.3 Xe xenon 54
6	132.9 Cs caesium 55	137.3 Ba barium 56	138.9 La lathanum 57	178.5 Hf hafnium 72	180.9 Ta tantalum 73	183.8 W tungsten 74	186.2 Re rhenium 75	190.2 Os osmium 76	192.2 Ir iridium 77	195.1 Pt platinum 78	197.0 Au gold 79	200.6 Hg mercury 80	204.4 Tl thallium 81	207.2 Pb lead 82	209.0 Bi bismuth 83	[209] Po polonium 84	[210] At astatine 85	[222] Rn radon 86
7	[223] Fr francium 87	[226] Ra radium 88	[227] Ac actinium 89	[261] Rf rutherfordium 104	[262] Db dubnium 105	[266] Sg seaborgium 106	[264] Bh bohrium 107	[277] Hs hassium 108	[268] Mt meitnerium 109	[271] Ds darmstadtium 110	[272] Rg roentgenium 111	Elements with atomic numbers 112–116 have been reported but not fully authenticated						

140.1 Ce cerium 58	140.9 Pr praseodymium 59	144.2 Nd neodymium 60	144.9 Pm promethium 61	150.4 Sm samarium 62	152.0 Eu europium 63	157.2 Gd gadolinium 64	158.9 Tb terbium 65	162.5 Dy dysprosium 66	164.9 Ho holmium 67	167.3 Er erbium 68	168.9 Tm thulium 69	173.0 Yb ytterbium 70	175.0 Lu lutetium 71
232 Th thorium 90	[231] Pa protactinium 91	238.1 U uranium 92	[237] Np neptunium 93	[242] Pu plutonium 94	[243] Am americium 95	[247] Cm curium 96	[245] Bk berkelium 97	[251] Cf californium 98	[254] Es einsteinium 99	[253] Fm fermium 100	[256] Md mendelevium 101	[254] No nodelium 102	[257] Lr lawrencium 103